Lecture Notes in Physics

317

C. Signorini S. Skorka
P. Spolaore A. Vitturi (Eds.)

Heavy Ion Interactions Around the Coulomb Barrier

Proceedings of a Symposium,
Held in Legnaro, Italy, June 1–4, 1988

Springer-Verlag
Berlin Heidelberg GmbH

Editors

C. Signorini
P. Spolaore
INFN Padova, Laboratori Nazionali di Legnaro
Via Romea 4, I-35020 Legnaro, Italy

S. Skorka
Sektion Physik, University of Munich
Am Coulombwall 1, D-8046 Garching, FRG

A. Vitturi
University and INFN Trento
I-38100 Trento, Italy

ISBN 978-3-662-13692-8 ISBN 978-3-540-46058-9 (eBook)
DOI 10.1007/978-3-540-46058-9

Originally published by Springer-Verlag Berlin Heidelberg New York in 1988
Softcover reprint of the hardcover 1st edition 1988

2158/3140-543210 – Printed on acid-free paper

Organizing Committee

C. Signorini (Padova - Salerno)
S. Skorka (München)
P. Spolaore (Legnaro)
A. Vitturi (Trento)

Conference Secretariat

A. Spalla (Padova)
M. Stefani (Legnaro)

Sponsors

The Symposium was organized under the sponsorship of:

- INFN, Laboratori Nazionali di Legnaro and Sezione di Padova;
- Sektion Physik Universität München;
- Dipartimento di Fisica, Università di Padova
- Dipartimento di Fisica, Università di Salerno

The Symposium received support from:

- Balzers S.p.A, Milano, Italy
- Bruker Analytische Messtechnik GmbH, Karlsruhe, F.R.G.
- Danfysik A/S, Jyllinge, Denmark
- De Pretto-Escher Wyss S.p.A., Schio, Vicenza, Italy
- Leybold S.p.A., Milano, Italy
- Rial S.p.A., Parma, Italy
- Varian S.p.A, Segrate, Milano, Italy

PREFACE

Extensive investigations in recent years have revealed the great complexity and diversity of heavy ion interactions at energies around the Coulomb barrier. With the growing interest in this field of research many new data have been obtained and more sophisticated theoretical approaches and improved experimental techniques have been developed.

In a new effort to review the recent results, a Symposium was held at the Laboratori Nazionali di Legnaro, in Italy, from June 1-4, 1988.

It was a third international meeting on this subject, following the one at MIT in 1984 and the one at ANL in 1986. About hundred participants attended.

This volume contains the invited talks and contributed papers selected for oral presentation. The last day was devoted to a discussion of recoil mass spectrometers.

We wish to thank all participants for their stimulating contributions. We are very grateful to the sponsors whose support helped to make the Symposium possible. It is also a great pleasure to acknowledge the assistance of the L.N.L staff in preparing and organizing the conference.

C. Signorini
S. Skorka
P. Spolaore
A. Vitturi

TABLE OF CONTENTS

FUSION AND QUASI-ELASTIC REACTIONS 91

SPIN DISTRIBUTION 155

SUB- AND NEAR-BARRIER FUSION

DEVELOPMENTS IN SUB-BARRIER REACTIONS

S. Landowne
Physics Division, Argonne National Laboratory
Argonne, IL 60439-4843 USA

1. INTRODUCTION

The study of sub-barrier reactions has developed in recent years to include a broad range of interconnected phenomena. The initial discoveries of enhanced sub-barrier fusion cross sections and the attempts to understand them in terms of couplings to other reaction channels have stimulated investigations into all aspects of heavy-ion collisions at low bombarding energies. As a result, new effects have been observed and new ideas are being put forward. The present talk gives an overview of some current themes in this field.

Perhaps the most encompassing development is that one can no longer think of the low energy fusion reaction as an isolated phenomenon, since the fusion rates depend crucially on the presence of other reaction channels. Thus, one wants to know what are the properties of these channels and then how to use this knowledge to explain the fusion rates quantitatively. Generally speaking, quite a number of sub-barrier fusion reactions have been measured [1] but the other reactions which occur have been isolated in a relatively few cases. It is also generally true that theoretical sub-barrier fusion calculations have been successful for a limited range of fairly light mass systems. Thus, the field continues to be a challenging area of research.

Going beyond the problem of understanding the fusion mechanism, there have been interesting new developments in all of the reaction classes that have been studied at sub-barrier energies, namely, elastic scattering, inelastic excitation, transfer reactions and deep inelastic collisions. A brief discussion of each of these subjects and how they relate to fusion will be given below. In addition, the important subject of compound nuclear spin distributions produced in fusion reactions will be noted.

2. POLARIZATION POTENTIAL FOR ELASTIC SCATTERING

The observation of a marked energy dependence in low-energy heavy-ion elastic scattering has been an important development. It has been found that the real part of the nuclear optical potential increases in strength and almost doubles as the bombarding energy is lowered toward the barrier, while the corresponding imaginary potential decreases abruptly [2]. The correlation between the energy dependence of the real "polarization" potential and the absorption can be understood in general terms by a dispersion relation [3]. A recent case of Si+Ni is shown in Figure 1.

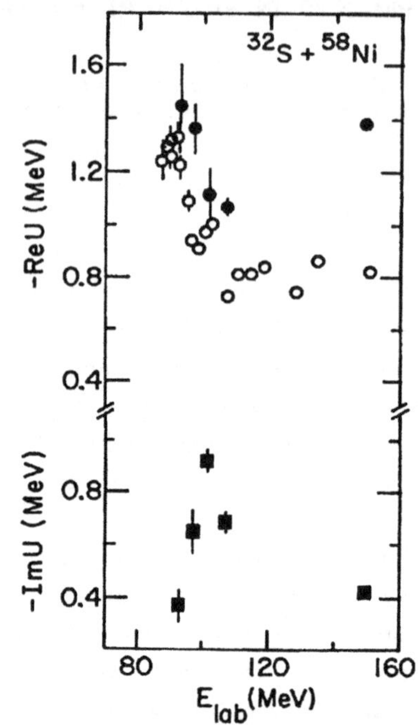

Figure 1. Effective potential strengths extracted from elastic scattering data (full points) and from fusion measurements (open points). Ref. [4]

The polarization potential is caused by the couplings of the elastic channel to other reaction channels. An expression for the effective potential can be found by treating the couplings in second-order time-dependent perturbation theory [5]. The result is

$$\Delta U_{00}(r) = -i \sum_{\gamma} \frac{|V_{\gamma 0}(r)|^2}{\hbar\tau} \sqrt{2/\pi} \int_{-\infty}^{\infty} dt \exp\left(-\frac{t^2}{\tau^2} - i\omega_\gamma t\right) \int_{-\infty} dt' \exp\left(-\frac{t'^2}{\tau^2} + i\omega_\gamma t'\right) \qquad (1)$$

This expresses the effective potential in terms of the coupling interactions $V_{\gamma 0}$, the channel excitation energies $\hbar\omega_\gamma$ and the collision time τ. The real part of ΔU_{00} is the polarization potential and the imaginary part is the absorption.

There are two limiting cases for ΔU_{00}. When the excitation frequencies are small compared to the inverse of the collision time one has

$$\Delta U_{00}(r) \rightarrow -i \sum_{\gamma} \frac{|V_{\gamma 0}(r)|^2}{\hbar/\tau} \sqrt{\pi/2}\,, \qquad \omega_\gamma \ll 1/\tau \qquad (2)$$

which is purely absorptive. The other extreme is for high frequencies or long collision times

$$\Delta U_{00}(r) \rightarrow -\sum_{\gamma} \frac{|V_{\gamma 0}(r)|^2}{\hbar\omega_{\gamma}}, \qquad \omega_{\gamma} >> 1/\tau \tag{3}$$

in which case, ΔU_{00} is purely real.

3. EFFECTIVE COUPLING FOR INELASTIC EXCITATION

The expression for the effective potential can be generalized to give an effective coupling interaction for transitions out of the elastic channel [6],

$$\Delta U_{\beta 0} = -i \sum_{\gamma} \frac{V_{\beta\gamma}(r)\ V_{\gamma 0}(r)}{\hbar\tau} \sqrt{2/\pi} \int_{-\infty}^{\infty} dt \exp\left[-\frac{t^2}{\tau^2} + i\omega_{\beta\gamma}t\right] \int_{-\infty}^{t} dt' \exp\left[-\frac{t'^2}{\tau^2} + i\omega_{\gamma 0}t'\right] \tag{4}$$

Thus one expects to find similar energy dependent effects in transitions to excited states. Indeed, low energy inelastic scattering offers an important way to study such phenomena because it is sensitive to Coulomb-nuclear interferences [7].

An analysis of low-energy O+Pb inelastic scattering is shown in Figure 2. The different curves show how the interaction must be increased to fit the data at the low energies. Additional data for this case has recently been analyzed and has confirmed that there is a large energy dependent effect similar to what has been observed for the elastic scattering [8].

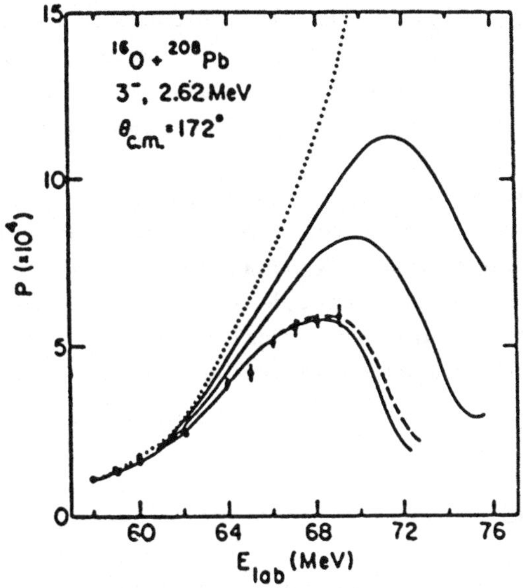

Figure 2. Low energy inelastic excitation probability [7].

4. CONNECTION TO SUB-BARRIER FUSION

Inasmuch as the couplings to other channels cause the energy dependence in the elastic optical potential and in the inelastic effective interaction, these effects are correlated with the enhancements in sub-barrier fusion reactions. Recently there has been an attempt to make a direct connection between these phenomena. Namely, it has been suggested that one can take the energy-dependent real potential which results from optical model analyses of elastic scattering data and use it in a one-dimensional barrier penetration calculation to obtain the sub-barrier fusion [9]. Some apparent support for this idea is indicated by the analysis shown in Figure 1. Here energy dependent potentials used to fit sub-barrier fusion cross sections are compared to the ones used to fit elastic scattering data. There is, however, a counter-example which will be presented below.

Independently of examples for or against describing fusion in terms of an energy-dependent barrier, it should be noted that the fusion process involves many channels of the interacting nuclei at short distances, while an optical model analysis of elastic scattering only determines the wave function in one channel at large distances. Thus, one should not expect to find a strong correlation at the phenomenological level.

Theoretically, the only limit in which the full solution of the many-channel scattering problem can be reduced to an effective one-channel calculation is the adiabatic one where the excitation frequencies of the various channels are all large compared to the collision frequency. As indicated above by Eq. (3), the real potential in the elastic channel is renormalized by the couplings and no flux goes into the excited states. However, in this limit the polarization potential is independent of the bombarding energy (collision time). Thus, an energy dependent real potential in the elastic channel and the appearance of an absorptive potential are signals that the full problem cannot be mapped into an effective one-channel calculation.

It is also important to consider the sudden limit where the excitation frequencies are small compared to the collision frequency. In this case the coupled-channels barrier penetration problem reduces to taking averages over a distribution of one-dimensional barriers [10]. A familiar and realistic example is the case involving a deformed nucleus where one averages over the intrinsic orientations [11]. This averaging over a distribution of barrier penetration factors cannot be reconciled with penetrating a single, energy-dependent barrier.

5. SPIN-DISTRIBUTIONS

The counter-example of the energy-dependent barrier idea mentioned above is provided by the fusion and spin-distribution measurements reported recently in Ref. [12]. These data and their analyses are reproduced in Figure 3. They also present a particular challenge since they are not well reproduced by the types of coupled-channels calculations which have been done so far. Of course, the fusion cross sections can be fit

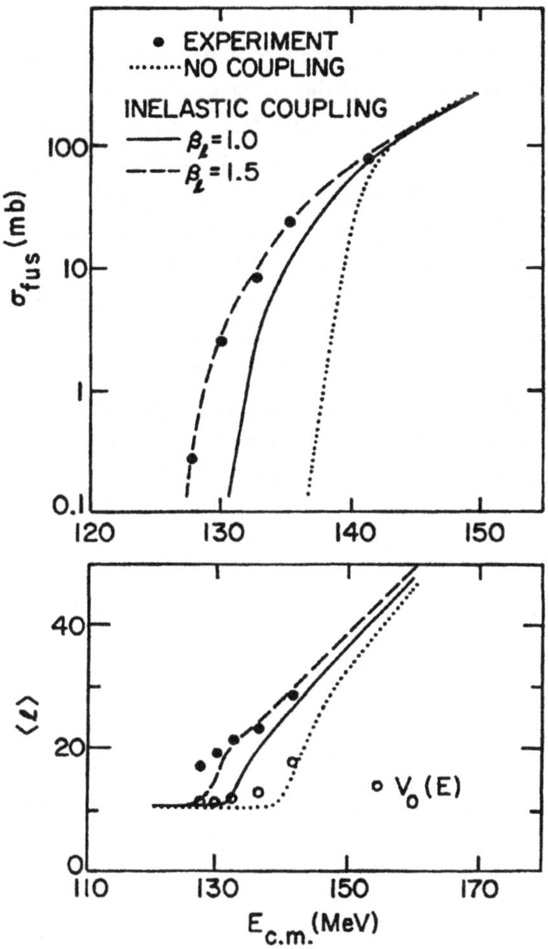

Figure 3. Fusion cross sections and average compound spins for $^{64}Ni+^{100}Mo$. Ref. [12].

by using an energy-dependent potential. However, as the open circles show, the corresponding average spins do not describe the measurements.

The dashed curves in Figure 3 were obtained with a coupled-channels calculation using abnormally large coupling interaction strengths. The resulting spin distributions describe the data rather well, except at the lowest two energy points. This discrepancy is significant since one expects that the energy dependence of the cross section determines the spin-distribution.

In particular, the mean-square-value of the compound nucleus spin-distribution at the energy E should be related to the fusion cross section $\sigma_f(E)$ by [13]

$$\langle \ell^2 \rangle_E = \frac{2\mu r_b^2}{\hbar^2} \frac{1}{E\sigma_f(E)} \int_{-\infty}^{E} dE' E' \sigma_f(E') \ . \qquad (5)$$

This relation is based on the following assumptions: (i) the fusion cross section corresponds to the flux transmitted by a distribution of barriers which are independent of E; (ii) the role of the partial wave angular momentum ℓ is essentially to increase the barrier heights by an amount $\hbar^2\ell^2/2\mu r_b^2$, where the moment of inertia at the barrier, μr_b^2, is assumed to be independent of ℓ and E; (iii) usually μ and r_b are taken to be the asymptotic reduced mass and unperturbed barrier radius. These assumptions are incorporated in the calculations shown by the curves in Figure 3.

The disagreement with the average spin at the lowest energies does not necessarily mean that these assumptions are breaking down. An alternative conclusion would be that the fusion cross section at lower energies are actually larger than the continuation of the model fit shown by the dashed curve. That is to say, the lower energy contributions to the integral in Eq. (5) have not been measured.

It should also be emphasized that the present understanding of the compound nuclear spin distributions is not clear. The case of $^{16}O+^{232}Th$ fusion reported in Ref. [14] is an important example. An analysis of this data is shown in Figure 4.

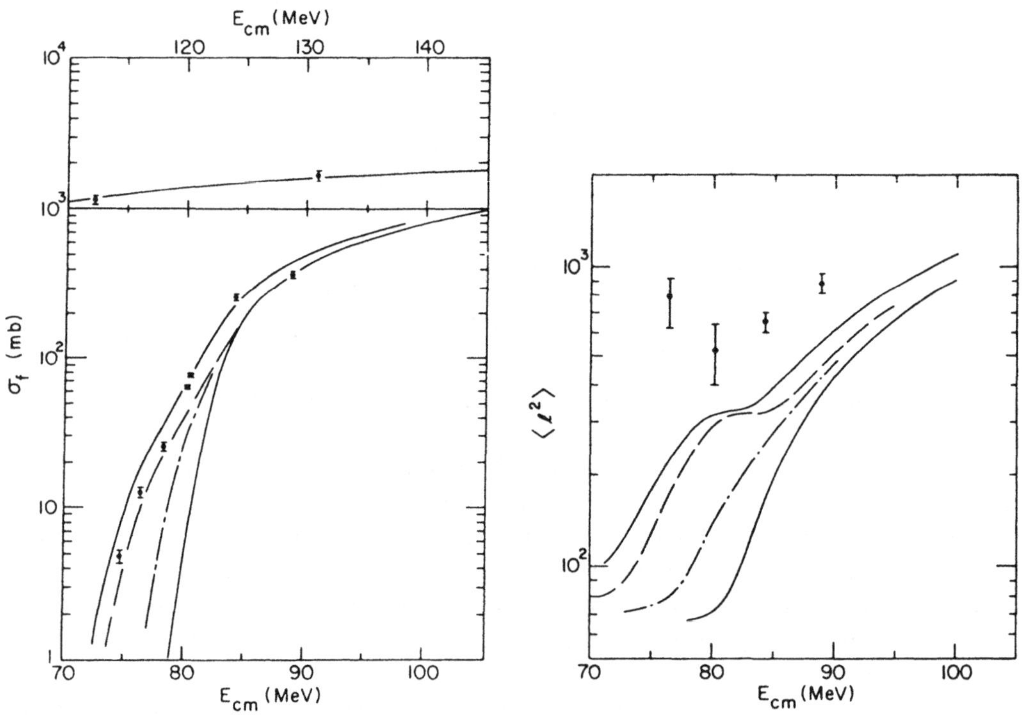

Figure 4. Fusion cross sections and average compound nuclear spins for $^{16}O+^{232}Th$. Ref. [15]

The calculations allow for the static quadrupole and hexadecpole moments of ^{232}Th as well as an octupole vibration [15]. They are able to account for the total fusion cross sections rather well, but the corresponding average spin values are not reproduced. It is interesting to note that the disagreement is pronounced even for the highest energy data point. This corresponds to a large total fusion cross section that is not very sensitive to the coupling effects.

6. FUSION AND ROTATIONAL/VIBRATIONAL EXCITATION

The previous case of ^{16}O+^{232}Th shown in Figure 4 is a good example of how the fusion cross section is influenced by rotational excitation. Because of the low excitation energies associated with the rotation, the coupled-channels calculation can be well approximated by averaging over the cross sections calculated for fixed orientations of the deformed surface [15]. This procedure accounts for all possible rotational states.

In other cases, like Ni+Ni, the vibrational model has been used to describe the couplings to the lowest states of a given angular momentum and parity. Here the potential radius is allowed to vibrate and a first-order expansion is made to obtain the transition form factors. Usually the expansion is not carried out further, with the result that only independent one-phonon states of the projectile and target are allowed to couple with the ground-state entrance channel.

A recent calculation of the Ni+Ni case has studied the effect of taking the vibrational model to the next higher order [16]. Thus, in the expansion of the nuclear potential there appear quadratic terms in the surface amplitudes in addition to the usual linear ones. Their effect on the sub-barrier fusion was found to be substantial, as can be seen in Figure 5.

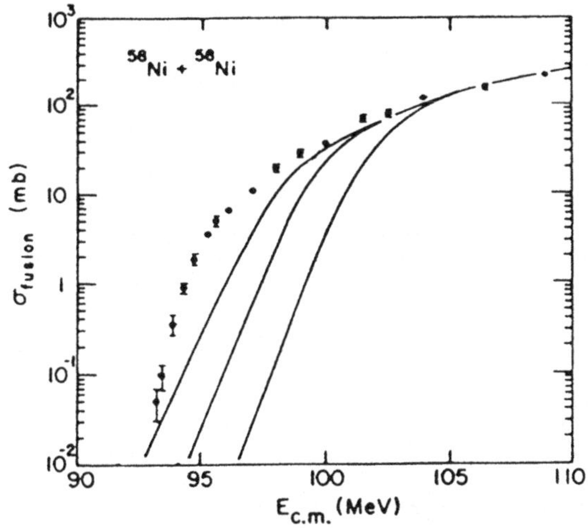

Figure 5. Fusion calculations for ^{58}Ni+^{58}Ni allowing for first- and second-order vibrational couplings [16].

These second-order calculations bring in the mutual excitation process where the projectile and target are simultaneously excited. This is found to account for about half of the difference with respect to the first-order calculation shown in Figure 5. Most of the remainder comes from the fact that the second-order terms have diagonal matrix elements which lower the potential barriers in the excited states. The ability to directly populate two-phonon states in the projectile or target does not play a strong role in these calculations. The relatively high energy of the phonons in this case is the reason why, and it also causes the third-order effects to be relatively small [16].

7. FUSION AND TRANSFER REACTIONS

Transfer reaction channels are not included in the calculations shown in Figure 5. It is generally believed that they should also play a role, particularly in explaining the differences between similar systems, such as $^{58}Ni+^{64}Ni$ as compared to $^{58}Ni+^{58}Ni$. Measurements for $^{58}Ni+^{64}Ni$ transfer cross sections have been carried out [17]. It is a challenge for the theory to incorporate these channels consistently with the higher-order vibrations.

A recent study has been made in this direction [18]. It incorporates an idea, based on a macroscopic point of view, that the transfer coupling interaction should depend on the distance between the nuclear surfaces which are allowed to vibrate in the same way as for the nuclear potential. Transfer form factors were parameterized in this way and were adjusted to simultaneously fit the quasi-elastic and one-particle transfer data for $^{58}Ni+^{64}Ni$, as shown in Figure 6.

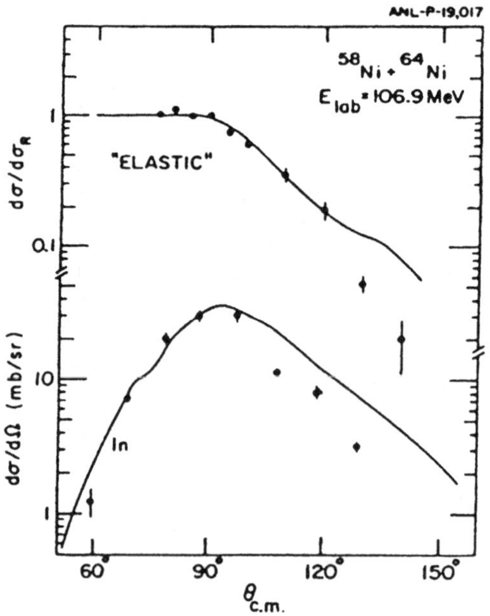

Figure 6. Quasi-elastic and one-particle transfer cross sections for $^{58}Ni+^{64}Ni$ compared to coupled-channels calculations [18].

These coupled-channels calculations have an ingoing-wave boundary condition to describe the fusion and also allow for vibrational excitations to second-order. No imaginary potential is used. Some residual absorption could be introduced to improve the agreement with the data at the more backward angles. However, it is interesting to see how well the calculations reproduce the data without an imaginary potential.

The corresponding $^{58}Ni+^{64}Ni$ fusion calculations are shown in Figure 7. The uppermost solid curve shows that although the relatively large one-particle cross sections are reproduced, their effect on top of the second-order vibrational couplings is not so strong and they do not change the low-energy slope. Allowing for a direct two-neutron transfer seems to be required, as indicated by the dashed curve.

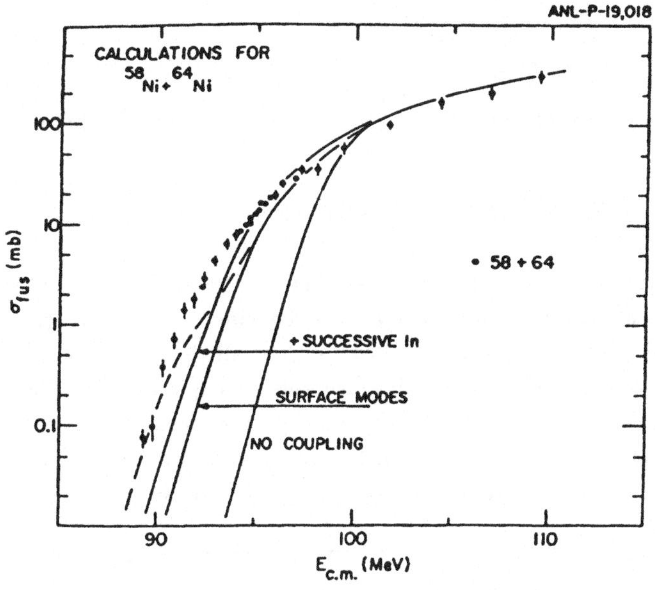

Figure 7. Fusion calculations for $^{58}Ni+^{64}Ni$. Ref. [18]

8. SUB-BARRIER TRANSFER REACTIONS WITH DEFORMED NUCLEI

The idea that transfer couplings depend on the distance between vibrating surfaces has a natural analogue in the case when permanently deformed nuclei collide. Here one would expect that the transfer should depend on the relative orientation of the nuclear surfaces. Consequently, the low energy transfer cross section for two well deformed prolate nuclei should be larger than for an equivalent spherical case. The curves in Figure 8 show some recent estimates for two-neutron transfer between different Sm isotopes [19]. The ^{144}Sm is treated as a sphere and ^{154}Sm is taken to be deformed. A significant part of the differences shown in Figure 8 are due to nuclear sizes and reaction Q-values. These have to be taken into account in order to isolate the effect of deformation.

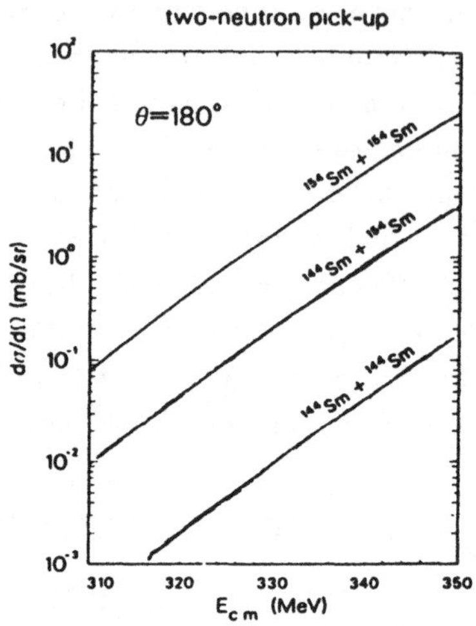

Figure 8. Estimates of low-energy two-neutron transfer reaction Sm+Sm collisions [19].

9. LOW ENERGY DEEP INELASTIC REACTIONS

Deep inelastic reactions constitute a large part of the total reaction cross section in heavy-ion collisions at energies well above the barrier. Recently, the deep inelastic process was followed down to lower energies in the collisions of $^{58}Ni+^{124}Sn$ [20]. The percentage of the total cross section, divided according to quasi-elastic, deep inelastic and fusion is shown in Figure 9. At the lowest energy, the deep inelastic and fusion cross sections are about the same. Thus, any attempt to account for the total cross section for such collisions must also incorporate the deep inelastic processes.

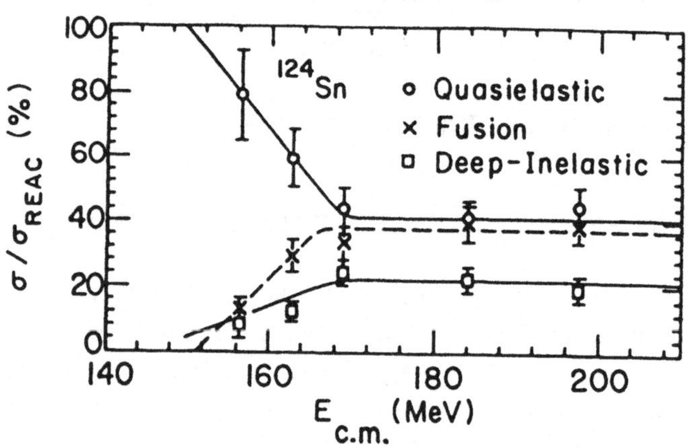

Figure 9. Total cross section percentages in the $^{58}Ni+^{124}Sn$ collision [20].

ACKNOWLEDGEMENTS

This work was supported by the U. S. Department of Energy, Nuclear Physics Division, under contract W-31-109-ENG-38.

REFERENCES

[1] M. Beckerman, Phys. Rep. 129 (1985) 145.

[2] J. S. Lilley, B. R. Fulton, M. A. Nagarajan, I. J. Thompson and D. W. Banes, Phys. Lett. B151 (1985) 181.

[3] M. A. Nagarajan, C. Mahaux and G. R. Satchler, Phys. Rev. Lett. 54 (1985) 1136.

[4] A. M. Stefanini, D. Bonamini, A. Tivelli, G. Montagnoli, G. Fortuna, Y. Nagashima, S. Beghini, S. Signorini, A. DeRosa, G. Inglima, M. Sandoli, G. Cardella and F. Rizzo, Phys. Rev. Lett. 59 (1987) 2852.

[5] C. H. Dasso, S. Landowne, G. Pollarolo and A. Winther, Nucl. Phys. A459 (1986) 134.

[6] C. H. Dasso, G. Pollarolo and S. Landowne, Nucl. Phys. A443 (1985) 365.

[7] S. Landowne, C. H. Dasso and G. Pollarolo, Phys. Lett. B178 (1986) 336.

[8] C. S. Lim, M. D. Rebgetz, R. H. Spear, W. J. Vermeer, M. P. Fewell, B. A. Robson and R. Smith, Phys. Lett. B200 (1988) 263.

[9] M. A. Nagarajan and G. R. Satchler, Phys. Lett. B173 (1986) 29.

[10] C. H. Dasso, S. Landowne and A. Winther, Nucl. Phys. A405 (1983) 381; A407 (1983) 221.

[11] C. Y. Wong, Phys. Rev. Lett. 31 (1973) 766.

[12] M. L. Halbert, J. R. Beene, D. C. Hensley, K. Honkanen, T. M. Semkow, V. Abenante, D. G. Sarantites and Z. Li, Sixth Adriatic International Conference on Nuclear Physics, Dubrovnik, Yugoslavia (1987).

[13] C. H. Dasso, H. Esbensen and S. Landowne, Phys. Rev. Lett. 57 (1986) 1498.

[14] R. Vandenbosch, T. Murakami, C. C. Sahm, D. D. Leach, A. Ray and M. J. Murphy, Phys. Rev. Lett. 56 (1986) 1234.

[15] H. Esbensen and S. Landowne, Nucl. Phys. A467 (1987) 136.

[16] H. Esbensen and S. Landowne, Phys. Rev. C 35 (1987) 2090.

[17] K. E. Rehm, F. L. H. Wolfs, A. M. van den Berg and W. Henning, Phys. Rev. Lett. 55 (1985) 280.

[18] H. Esbensen, Winter Workshop on Nuclear Dynamics V, Sun Valley, Idaho (1988).

[19] S. Landowne and C. H. Dasso, Phys. Lett. B202 (1988) 31.

[20] F. L. H. Wolfs, W. Henning, K. E. Rehm and J. P. Schiffer, Phys. Lett. B196 (1987) 113.

MULTIDIMENSIONAL TUNNELING DESCRIPTION OF SUBBARRIER FUSION

J. Schneider, H.H. Wolter

Sektion Physik, Universität München
D–8046 Garching, Germany

Abstract

We introduce an alternative method for the exact quantum mechanical treatment of subbarrier fusion, namely the time–dependent wave packet method where a wave packet in the radial and in additional internal collective degrees of freedom of the fusing system is propagated on the multidimensional potential surface. An advantage of the method is to yield a very intuitive picture of the fusion process. It also solves directly a given model for the energy surface without the need to expand in a basis set. In this way we can clearly ascertain the effects of the spreading in the additional degrees of freedom (zero–point–motion and excitation), the dynamics of the motion on the surface, and the tunneling through the barrier. In applying the approach we have considered as a new degree of freedom in subbarrier fusion the formation of a neck and treated it in the adiabatic approximation. Neck formation is found to be very important for the enhancement of fusion giving as much or more enhancement as the deformation effect alone. For the fusion of ^{58}Ni + ^{58}Ni we are thus able to account very well for the experimenal data.

Introduction

Subbarrier fusion of nuclei – and the inverse process of fission – have been studied with renewed interest recently as a manifestation of degrees of freedom in addition to the radial separation. The investigation and identification of the relevant degrees of freedom for fusion is a major goal of this conference. In this contribution we want ot treat this problem from a novel viewpoint, namely the time–dependent wave packet method. It is developed not only to give us an intuitive picture of the fusion process but also to explore the importance of a neck degree of freedom, which is more difficult to deal with in the usual approaches.

The simplest description includes only the radial separation of the two ions without any internal excitation but it is clear that this is not sufficient for the fusion of heavier systems [1]. It was realized early that deformation of the ions is important to lower the barrier [2,3]. Furthermore the role of transfer is presently discussed intensively in order to explain isotopic differences in the fusion of neighbouring systems, which may be related to positive Q–values of specific transfer channels [4,5].

It should be noted that the degrees of freedom discussed above are those of the separated ions and are in fact the same ones used in direct nuclear reactions above the barrier. One should consider also the possibility that fusion can be much better represented by degrees of freedom which are appropriate for nuclei in close contact, i.e. in a fusion configuration. A particular and interesting case is the formation of a neck between the fusing ions which could be an important initiator of fusion [6]. In the language of direct reaction degrees of freedom a neck might describe more general shapes than e.g. small quadrupole deformation. It may also describe the flow of matter between the ions and thus provide a kind of macroscopic description of transfer reactions. For asymmetric systems there should be in addition a mass asymmetry coordinate [6].

One should also mention the possibility of incoherent single particle degrees of freedom which would provide a consistent description of dissipative effects in this framework. These have been treated macroscopically so far [7], but could possibly be investigated in the socalled Multistep Direct Reaction (MSDR) formalism [8].

The different degrees of freedom clearly make the subbarrier fusion a multidimensional tunneling problem. The important questions are then which are the relevant degrees for fusion and what are the dynamical properties of this process, i.e. which is the path of the system on the multidimensional potential surface, especially when it moves through the barrier. Also it is an interesting problem in itself how to treat multidimensional tunneling problems.

Because of the complexity of this problem different approximation schemes have been used to describe subbarrier fusion. The one–dimensional barrier penetration corresponds to a straight trajectory on the surface, while in the adiabatic picture the system is constrained to move along the minimum path. The dynamic path results if the classical equations of motion are solved for the system and some prescription is supplied where to tunnel through the barrier [9]. Quantum effects in the additional coordinate are added if the spreading of the system in this degree of freedom is considered. If applied to the straight trajectory one obtains the zero–point–motion (ZPM) picture [10]. Two methods have been proposed to treat the problem in full generality. One is the stationary Coupled Channels (CC) method where the additional degrees of freedom are expanded in a basis system of modes of the separated systems, i.e. at infinity. While such a method is in principle exact it might not be the most efficient one and indeed, calculations using it have in some cases difficulty to obtain enough enhancement inspite of large coupling schemes [3].

The method proposed here is the time dependent wave packet method. A wave packet (WP) in the radial and additional coordinates is propagated according to the Hamiltonian on the energy surface coming from infinity. It will thus choose its own path and shape on the surface. The WP will split at the barrier and the norm of the transmitted part will give the fusion probability. Since there is no expansion into modes this method may also treat more complicated degrees of freedom, as e.g. neck formation, provided the corresponding energy surface and mass parameters are given. This approach has the added bonus of providing an intuitive picture of the quantum mechanical process of multidimensional scattering and fusion.

In this contribution we present for the first time a model study for the latter approach. It will be rather realistic in constructing the energy surfaces, including for the first time also a neck degree of freedom. However it will make similar crude assumptions about the mass parameters as in CC approaches. Since in our method the number of coordinates treated explicitely is the limiting factor we apply it to the symmetric system ^{58}Ni + ^{58}Ni, where the number of degrees of freedom can be reasonably reduced.

II. Wave packet scattering

In this chapter we will discuss the time–dependent wave packet method. Scattering of large WPs is treated in standard text books [11], however, the application to finite WPs on multidimensional surfaces presents some new features. Greater detail will be given elsewhere [12].

Let r denote the radial coordinate and ξ the collection of additional degrees of freedom. Then at time t = 0 a normalized initial WP is constructed as

$$\Psi_o(r,\xi) = \psi(z)\phi_o(\xi) \tag{1}$$

where ϕ_o represents the ground state in the ξ–coordinates and is $\psi(z)$ a wave packet in beam direction (taken parallel to the z–axis) with infinite extent perpendicular to it and with mean wave number k_o. The initial WP is propagated in time according to ($\hbar = 1$)

$$\Psi(r,\xi,t) = e^{-iH(r,\xi)t}\,\Psi_o \tag{2}$$

The differential scattering cross section is then obtained by finding the flux into a cone $d\Omega$ for large times

$$\frac{d\sigma}{d\Omega} = \lim_{t\to\infty} \int r^2dr \int d\xi\, |\Psi(r,\xi,t)|^2 \tag{3}$$

Since the initial momenta are in one direction only a partial wave expansion can be employed

$$\Psi(r,\xi,t) = \sum_{\ell}(2\ell+1)\, i^{\ell}\, \psi_{\ell}(r,\xi,t)\, P_{\ell}(\cos\Theta) \tag{4}$$

and the WP algorithm applied with one spatial dimension. The propagation of eq.(2) will usually couple different partial waves, corresponding to angular momentum transfer to the other coordinates. However, it has been found that this is not an important feature in subbarrier fusion [3]. Also the potential models we use later on will only depend on the modulus of r and we will thus neglect angular momentum transfer. However the coupling between the radius r and ξ will be fully considered.

Absorption into the fusion degree of freedom will be considered by a sort of ingoing wave boundary condition by extending the potential smoothly to $r \to -\infty$. For large times the partial WP $\psi_{\ell}(r,\xi,t)$ will have split into a reflected and a transmitted part ψ_{ℓ}^{r} and ψ_{ℓ}^{t} at positive and negative r, respectively. The fusion cross section is then obtained as

$$\sigma_{fus}^{WP} = \sum_{\ell}(2\ell+1) \lim_{t\to\infty} \int r^2dr \int d\xi\, |\psi_{\ell}^{t}(r,\xi,t)|^2 \tag{5}$$

The connection to the stationary scattering theory is obtained by a Fourier representation of the incident WP

$$\psi(z) = \int dk\, a(k)\, e^{ikz}/\sqrt{2\pi} \tag{6}$$

A transmission coefficient $t_\ell(k)$ is obtained from the asymptotic stationary wave for each k–component after integration over ξ. Then the fusion cross section has the form

$$\sigma_{fus}^{WP} = \frac{\pi}{k_o^2}\sum_\ell(2\ell+1)\int dk(k_o^2/k^2)|a(k)|^2|t_\ell(k)|^2 = \int dk\,|(k_o/k)a(k)|^2\,\sigma_{fus}^{stat}(k) \tag{7}$$

In the last equality the fusion cross section of the stationary theory has been introduced. It is seen that the WP cross section appears as a folding of the stationary cross section over the momentum spectrum of the initial wave packet. This is, of course, due to the energy spread introduced by the finite extension (in z) of the wave packet which is an artificial effect and which in our calculations is much larger than the energy definition. Therefore the WP cross section should be compared to a folded experimental cross section $<\sigma_{fus}^{exp}>$, which is formed as in eq.(7) with the known a(k) and the experimental fusion cross section. Alternatively $\sigma^{(WP)}$ could be unfolded. This artificial energy smearing is not essential to the time–dependent method but could be avoided by using other boundary conditions.

The time evolution of eq.(2) has been performed following a method of ref.[13]. The Hamiltonian of the system is split into two parts

$$H(r,\zeta) = H_r + H_\xi = T_r + V(r) + \frac{1}{2}W(r,\xi) + T_\xi + V(\xi) + \frac{1}{2}W(r,\xi) \tag{8}$$

where the interaction $W(r,\xi)$ has been distributed equally into the two parts. The time evolution from time t to $t+\tau$ is then

$$\Psi(t+\tau) = e^{-iH\tau}\Psi(t) = \left[e^{-iH_r\tau}e^{-iH_\xi\tau}(1+\frac{\tau^2}{2}[T_r - T_\xi, W]) + O(\tau^3)\right]\Psi(t) \tag{9}$$

If the order of the r and ξ evolution in (9) is symmetrized the term in τ^2 can also be eliminated. In any case the time evolution will factorize giving

$$\Psi(t+\tau) = e^{-iH_\xi\tau}\hat{\Psi}\ ;\ \hat{\Psi} = e^{-iH_r\tau}\Psi(t)\,. \tag{10}$$

In order to preserve unitarity each propagator is treated in the Cayley approximation

$$\exp(-iH_r\tau) = \frac{1-iH_r\tau/2}{1+iH_r\tau/2} + O(\tau^3) \tag{11}$$

and similarly for ξ. Eq.(11) then leads to solving two sets of inhomogeneous Schrödinger equations, where for each set the other coordinate is a parameter. To complete the procedure the system is enclosed in a box in (r,ξ) space and the equations are integrated with the condition that the wave function vanish on the boundaries of the box. The box will be suitably shifted during the evolution of the system. The numerical implementation of the above procedure is quite extensive. The step size in the discretization of time and space have to be chosen carefully because they will determine the loss of the high frequency and short wave length components which sensitively influence the fusion cross section. However, a careful consideration can keep the errors under control [12].

It may be instructive to demonstrate the wave packet procedure in the simplest case of one–dimensional barrier penetration with only the relative coordinate r. We take the case of ^{58}Ni + ^{58}Ni using the nuclear potential specified in the next section. Fig.1a shows the wave packet evolution for a mean energy close to the barrier height. The width of the Gaussian WP is $\sigma = 7$ fm which corresponds to an energy spread of 2.4%. Thus a substantial part of the WP is actually above the barrier. The transmission coefficient is obtained by the norm of the penetrated WP (where the reflection at the edge is insubstantial as long as the WP does not cross the barrier again). The fusion cross sections for this model is shown in Fig.1b. It is compared to the stationary and the folded stationary cross section, eq.(7). The practical identity of the latter with the WP cross section demonstrates the relation eq.(7) and shows that the WP method is feasible. Shown is also the folded experimental cross section of eq.[1]. Similar checks on the numerical accuracy have been made for two–dimensional applications [6,12].

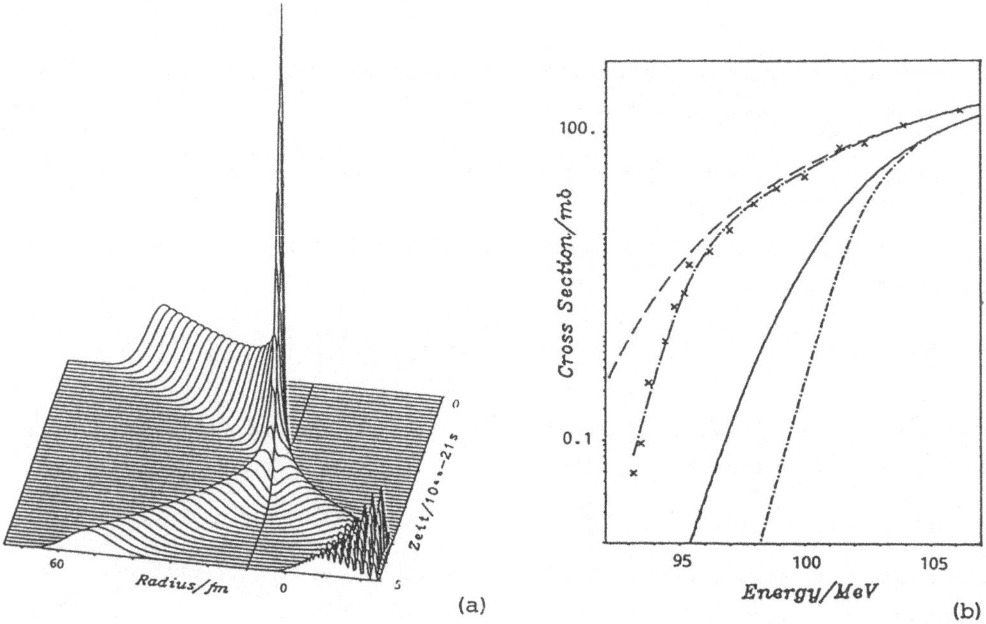

Fig.1:
Barrier penetration in the wave packet picture for ^{58}Ni + ^{58}Ni at $E_{cm} = 102$ MeV and for $\ell = 0$. Part (a) shows the modulus $|\Psi|^2$ of the WP at different times increasing from back to front. The position of the barrier ($r_b \approx 10$ fm, $V_b \approx 102$ MeV) is shown by a line. Initially a Gaussian WP is incident from the left. Later one sees a reflected and a fused WP (the latter being eventually reflected at the boundary). (b) Fusion cross sections. Shown are the cross sections for the stationary barrier penetration (dash–dot), the stationary penetration folded by the momentum distribution (dashed), the WP penetration of part (a) (solid), the experiment[1] (dash–dot and crosses) and the experiment folded with the momentum distribution (long dash)

III. Application to symmetric systems

III.1. Specification of model

As discussed previously we apply the model to the fusion of symmetric systems, in particular Ni + Ni, in order to keep the number of coordinates as small as possible. In such systems it is reasonable to assume that the radius, the equal deformations and the neck are the most important coordinates. However, these are already too many for our approach and we will have to make further approximations below.

We specify the potential energy surface by the model of Krappe, Nix and Sierk [14]. For a suitably family of shapes the nuclear potential V_{KNS} is given by a double folding integral with a Yukawa plus exponential effective interaction and equivalently for the Coulomb potential. The constants of the effective interaction–model have been determined from systematics of masses, fission data and radii in refs. [14] and [15]. For the family of fusion shapes we use the three quadratic surface parametrization of ref.[16]. Here two cut–off ellipsoids are smoothly fit to a hyperboloid or another spheroid to form necks or noses. Typical shapes are shown in fig.2. The condition of continuity and

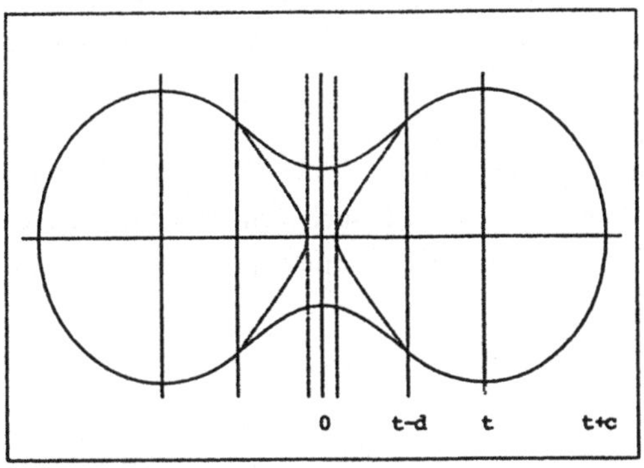

Fig.2:
Typical three quadratic surface shapes for fusion configurations showing shapes with a neck between nuclei and with noses on the nuclei. Parameters are the centers of the nuclei t, the symmetry half axis c and the position where the neck or nose is smoothly connected $\pm(t - d)$. (In this figure the nose is not smoothly connected)

volume conservation leaves three independent symmtry half axis c and the neck position d. However it is not reasonable to use shape parameters directly as physical coordinates since their mass parameters are badly defined. Instead we take the first and second moments of the mass distribution

$$r = \frac{2}{M} \int_{z>0} z\,dm \quad ; \quad s = \left[\frac{5}{M} \int_{z>0} (z - \tfrac{r}{2})^2 dm\right]^{1/2} - R \tag{12}$$

Here M is the mass of one nucleus and R the radius of the equivalent sphere. Then r is the c.m. separation and s is a measure of the deformation of one nucleus. For non–touching spheroids (without neck) $s = c - R$. Thus we have the coordinates r,s and the neck parameter d. Due to computational limitations we could only treat two coordinates fully dynamically, which we take as r and s. For the treatment of the neck we have considered two limiting situations:

(a) only spheroidal deformation, i.e. no neck. For touching configurations this means two connected cut–off spheroids (with volume conservation). This is the diabatic limit where the neck is so stiff that it is not excited.

(b) $d = d_{min}(r,s)$, i.e. the potential energy $V_{KNS}(r,s,d)$ is minimized for fixed (r,s) with respect to d. This is the adiabatic limit, where the neck degree of freedom is so fast (mass parameter so small) that it will instantaneously adjust to its most favourable value.

Physically the adiabatic limit seems more justified, since the mass involved in forming a neck is relatively small. However it should overestimate the fusion probability.

The Hamiltonian of the system is given as

$$H_\ell = -\frac{\hbar^2}{2\mu}\frac{1}{r}\partial_r\, r\, \partial_r + \frac{\hbar^2}{2\mu r}\ell(\ell+1) + \frac{\hbar^2}{2B}\frac{\partial^2}{\partial s^2} + V_{KNS+Coul}(r,s,d) \tag{13}$$

with d fixed as above. For the radial motion we use a constant mass parameter $\mu = M/2$. Certainly at least an r–dependent $\mu(r)$ would be more appropriate, but little is known about its value and the same approximation is used in CC–methods [3]. We have also assumed that the rotational motion is strictly coupled to the radial motion in such a way that the nuclei always maintain a tip–to–tip (belly–to–belly) configuration. Also the internal rotational motion is neglected thus conserving the orbital angular momentum ℓ. A more complete treatment of the rotation would require additional variables. The vibrational inertial parameter B is also taken to be constant and fixed by the vibrational properties of the isolated nuclei. At large distances the curvature of the potential surface in s direction is found. B is then fixed to yield the correct energy of the first 2^+ state in the vibrational model

We believe that the above model adds an important new aspect to the discussion of fusion, namely the influence of neck formation. With respect to other quantities, such as inertial parameters, rotational motion etc. it is of similar quality as CC–models. It could certainly be improved at the expense of enlarging the numerical calculations.

III.2. Results for $^{58}Ni + {}^{58}Ni$.

We have applied the above model to the fusion of $^{58}Ni + {}^{58}Ni$, since this system has been widely investigated both experimentally [1] and theoretically [3,9]. It should be noted that by the above specifications the model is completely defined and there are no further parameters to be adjusted. Thus we can expect a meaningful comparison with experiment.

In fig.3 we show the potential surface for the adiabatic model with the shape of the system superimposed on the contour diagram. The adiabatic path, i.e. the path of minimum potential energy over the saddle, is shown. It is also seen how the potential is smoothly extended to small separations in order to obtain an ingoing fusion flux (the potential of eq.(13) would have a strong Coulomb repulsive core). A typical feature is that the adiabatic path for large distances is shifted slightly to oblate shapes because of the long range Coulomb force. Close to the barrier it bends sharply to prolate shapes with a snapping–on of the neck. The saddle point configuration is rather elongated. A crucial question will be how much the system will be able to make use of the saddle point. The picture for the diabatic (spheroid) model is qualitatively similar but the saddle is higher (1–dim. barrier $r_b = 10.1$ fm , $V_b = 102.1$ MeV; spheroid saddle $r_b = 12.1$ fm, $s_b = 1.3$ fm, $V_b = 96.0$ MeV; adiabatic neck saddle $r_b = 11.9$ fm, $s_b = 1.2$ fm, $d_{min} = 2.6$ fm, $V_b = 94.4$ MeV).

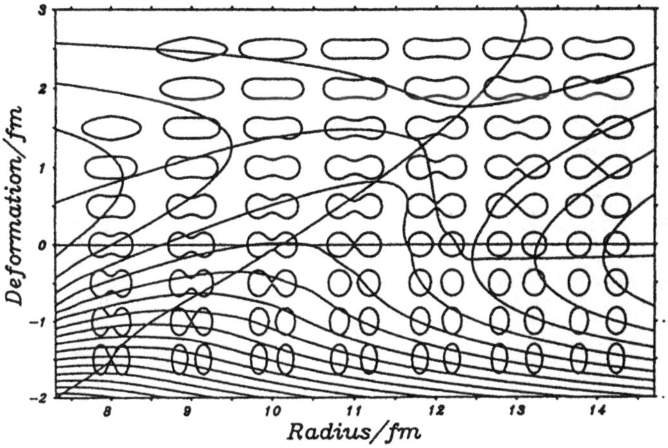

Fig.3:
Contour plot of potential surface for the adiabatic neck model for ^{8}Ni + ^{58}Ni fusion (ℓ=0). The straight trajectory, the adiabatic path and the potential ridge are indicated. Superimposed are the shapes corresponding to the respective position on the surface.

A WP is then propagated on the potential surface of fig.3. The initial WP has a wave function in s–direction corresponding to the ground state of the vibrational motion of the isolated nucleus. The WP is Gaussian in r–direction with a width of $\sigma = 7$ fm. A picture of the fusing system at some instances is shown in fig.4. The mean energy is 97.0 MeV which is above the saddle point but below the spherical barrier. It is seen how the bulk of the incoming WP is deflected to oblate deformations due to the Coulomb force and even remains there until the barrier. However at reflection the WP deformes strongly in the direction of the saddle. The tail of this deformed WP creeps over the saddle and streams down the fusion valley while the main part of the WP is reflected.

This intuitive and instructive picture demonstrates several important aspects of the fusion process. First of all the spread of the WP in deformation direction, i.e. the ZDM, is very important for fusion [10]. It is further enhanced by the strong deformation of the WP during the reflection at the barrier corresponding in the CC language to strong inelastic vibrational excitation of the nuclei. This is counteracted by the Coulomb repulsion acting over large distances deflecting the system to oblate shapes toward increasing barriers. This competition was already found in a semiclassical analysis by Landowne and Nix [9] and is seen to be important for a realistic description of the fusion process.

Results for the fusion cross section for the above model are shown in fig.5. It is seen that there is the well known enhancement due to spheroidal vibrational deformation which, however, does not explain the data for this system [3]. There is a further enhancement of about the same amount due to the neck formation. In this calculation we obtain very good agreement with the experimental data.

This encouraging result should not be taken without reservation, however. We expect, that the adiabatic neck model should overpredict the fusion particularly at the higher energies and, indeed, the calculation is slightly above the data there. In the calculation of fig.5 the force parameter set of ref.[14] was used. The set of ref.[15] yields qualitatively similar results but underpredicts the data by a factor of three at lower energies.

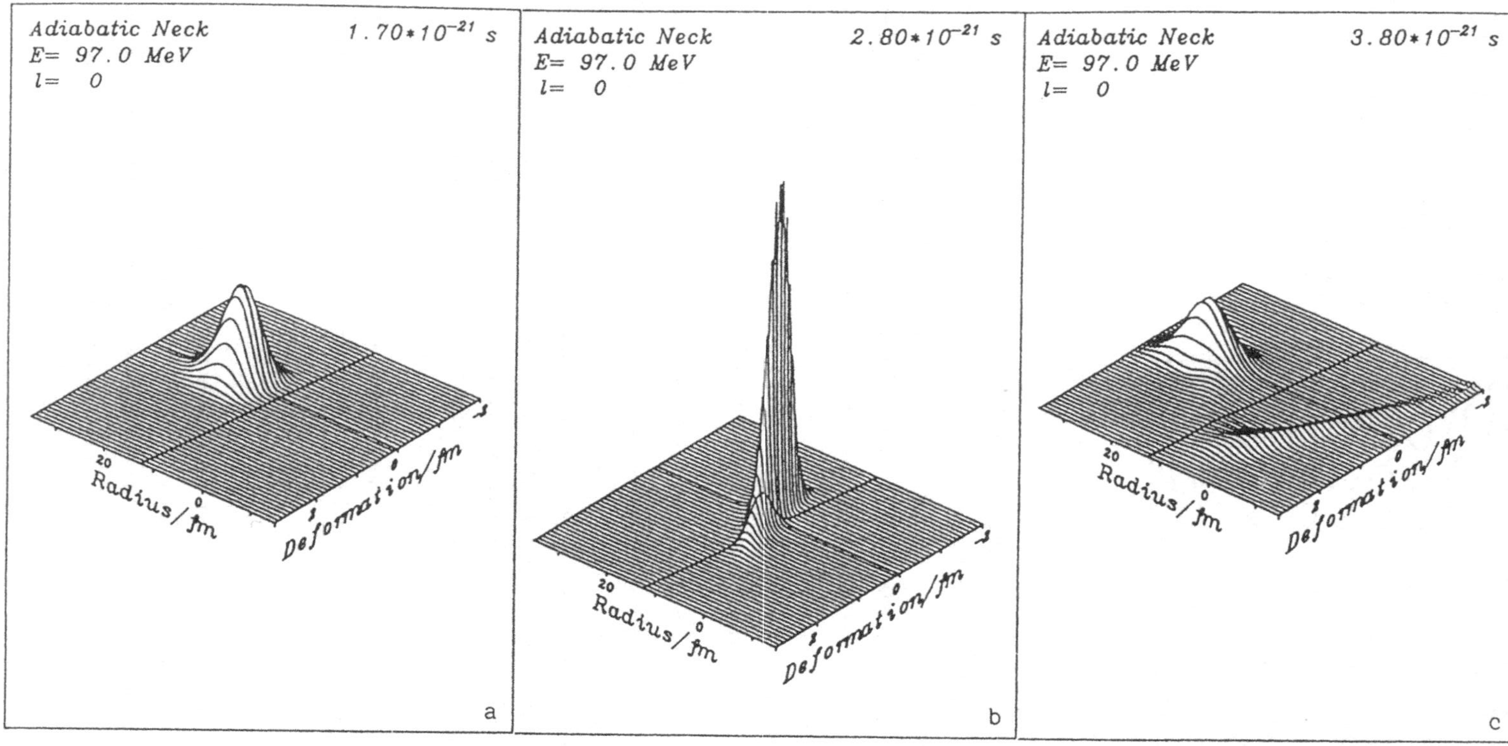

Fig.4:
Wave packet picture of Ni+Ni fusion corresponding to the energy surface of fig.3 for $\ell=0$. The position of the potential ridge is given by a line. The WP is shown at three instances (a) before reaching the barrier, (b) at the barrier, and (c) after the scattering

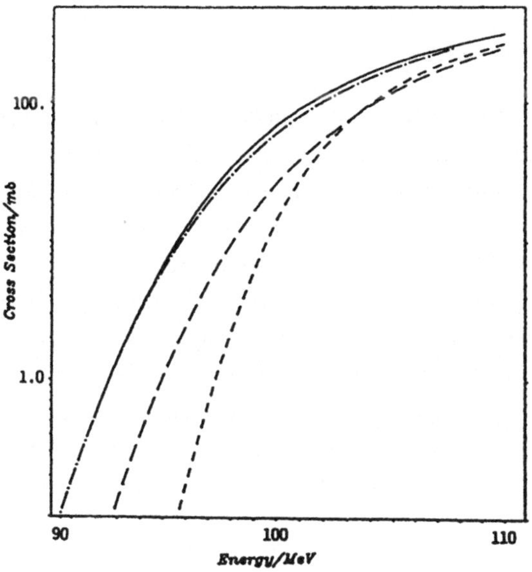

Fig.5:
Wave packet fusion cross section for ^{58}Ni for various models. Shown are the one–dimensional barrier penetration (i.e.fusion of spheres, short dash), the spheroid model (i.e. fusion of ellipsoids, long dash), and the adiabatic neck model (solid). Also shown are the experimental data [1] folded with the momentum distribution of the WP (dash–dot).

The partial wave fusion cross sections are another interesting observable which appear as the spin distributions of the compound nuclei and have recently been extracted from experiment [17] and investigated theoretically [18]. The results for the different cases of our model are shown in fig.6. It is seen that inclusion of deformation and neck degrees of freedom not only increases the cross section but also introduces higher spins as has been seen in other calculations [3]. However the detailed picture becomes quite involved at or above the barrier. One should note that our distributions do not show any structure as may be introduced by specific coupled states in the CC–method [3].

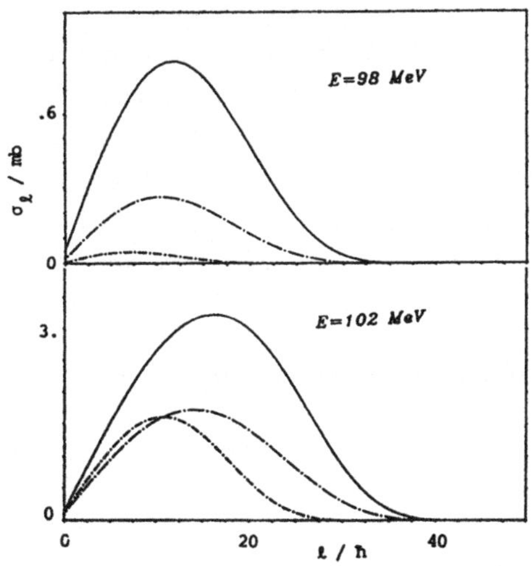

Fig.6:
Partial cross sections for ^{58}Ni + ^{58}Ni fusion for two mean energies below and at the one–dimensional barrier. Shown are the results for the barrier penetration (short dash–dot), the spheroid model (long dash–dot), and the adiabatic neck model (solid).

IV. Discussion and Conclusion

The purpose of this contribution is twofold: (1) to introduce an alternative method for the exact quantum mechanical treatment of scattering and subbarrier fusion with additional degrees of freedom, and (2) to investigate with this method quantitativly the importance of a new degree of freedom in subbarrier fusion, namely the formation of a neck.

The time–dependent wave packet method starts with a multidimensional potential energy surface spanned by the coordinates considered in the process. In principle this should be supplemented by similar maps of the inertial parameters, which, however, were taken constant in our calculations since little is known about them. A WP is then propagated on this surface by the Hamiltonian of the system. In this way the dynamics of the system and its internal structure are treated exactly within the model space. The method is particularly suited to investigate generalized ccordinates specific to fusion configurations because it does not rely on expansions in a given (mostly asymptotic) basis set. The procedure used here for implementing the WP method is numerically rather extensive limiting us to considering only two coordinates explicitely. The WP method in the form given here relies on a macroscopic description of the degrees of freedom. It could, however, be combined with a consideration of WPs in different channels [11]. This might be used to treat the coupling to specific transfer channels.

The method has the attractive feature of providing a very intuitive picture of the fusion process since it allows a direct classical interpretation.We can thus see the importance of the spreading of the wave packet in the direction of the additional coordinates (ZPM) and of its deformation due to the asymmetric shape of the potential surface at the barrier. We also observe the dynamical path of the bulk of the WP under the influence of nuclear and Coulomb fields. It is seen that these two effects oppose each other in subbarrier fusion.

We applied this method to a new degree of freedom in subbarrier fusion, the formation of a neck. In order to reduce the number of dynamical coordinates we have treated the neck in two limits: either by neglecting it or in the adiabatic limit. This last limit is an overestimate but probably a good approximation since the mass in the neck is small. It is found that the neck formation gives an important contribution to the subbarrier fusion enhancement of about the same amount as the deformation effect alone. In the Ni+Ni case studied here we obtain a very good description of the experimental data. In terms of direct reaction degrees of freedom, i.e. in the CC language, a neck corresponds to inclusion of more complicated shapes, in particular octopole deformations. It could also be interpreted as a flow of matter between the nuclei, i.e. to transfer processes treated macroscopically.

In conclusion it is hoped that this new approach will aid in the understanding of the mechanism of subbarrier fusion.

Acknowledgements

We would like to thank in particular J.R.Nix for the suggestion and the help in considering the neck formation. We further want to thank C.H.Dasso, S.Landowne, H.J.Krappe ,and K.Möhring for many very helpful discussions and the latter for providing programs to check our calculations. We also thank for illuminating discussions and encouragement from S.Skorka. The work was partly supported by a grant from the BMFT of the Fed. Rep. of Germany.

References

1. e.g.M. Beckerman et al., Phys. Rev. **C29** (1984) 1938
2. C.H. Dasso et al., Nucl. Phys. **A405** (1983) 381; Nucl. Phys. **A407** (1983) 221
3. S. Landowne, S.C. Pieper, Phys. Rev. **C29** (1984) 1352
4. A.M. Stefanini et al., Nucl. Phys. A456 (1986) 509; Phys. Lett. **B185** (1987) 15
5. S. Landowne, S.C. Pieper, F. Videdaek, Phys. Rev. **C35** (1987) 597
6. W. Swiatecki, Nucl. Phys. **A376** (1982) 275; H.J. Krappe, et al., Z. Phys. **A314** (1983) 23
7. P. Fröbrich, Phys. Rep. **116** (1984) 337
8. T. Tamura, et al., Phys. Rev. **C26** (1982) 379
9. S. Landowne, J.R. Nix, Nucl. Phys. **A368** (1981) 364
10. H. Esbenson et al., Nucl. Phys. **A411** (1983) 275;
11. L. Rodberg, S. Thaler, *Introduction to the Quantum Theory of Scattering*, Academic Press 1960
12. H.J. Schneider, H.H. Wolter, to be published
13. A. Goldberg, et al., Amer. J. Phys. **35** (1967) 177; K. Hahn, Thesis, Munich 1978
14. H.J. Krappe, J.R. Nix, A.J. Sierk, Phys. Rev. **C20** (1979) 992
15. P. Möller, J.R. Nix, Nucl. Phys. **A361** (1981) 117
16. J.R. Nix, Nucl. Phys. **A130** (1969) 241
17. e.g. R. Vandenbosch et al., Phys. Rev. Lett. **56** (1986) 1234
18. H. Esbensen, L. Landowne, Nucl. Phys. **A467** (1987) 136
19. J. Schneider, Thesis, Munich 1982

INVESTIGATION OF MECHANISMS LIMITING THE NEAR-BARRIER FUSION OF MASSIVE NUCLEI

W. Reisdorf

GSI, Postfach 110 552, D-6100 Darmstadt, Fed. Rep. of Germany

Our subject is the experimental observation that an extra kinetic energy above the expected fusion barrier is needed if massive nuclei are to undergo complete fusion. The phenomenon sets in for products of nuclear charges $Z_p \times Z_t$ roughly around 1600 and has been predicted by theory. The calculations of Blocki, Feldmeier and Swiatecki [1] are displayed as a solid line in fig. 1. The energy in excess of the expected [2] Coulomb barrier, E_{xx}, the so-called extra-extra-push energy, is plotted versus a mean fissility x_m which is a weighted average of the entrance channel fissility x_e, characteristic of the binary system prior to fusion and the fused system's conventional fissility x. This quantity was found numerically to be a good scaling variable for E_{xx} in the macroscopic dynamical approach [1] The figure also shows the experimental situation as of 1985 [3-6]. In order to avoid any ambiguities in the definition of fusion the data are limited to evaporation residue data. Although the calculation seems to follow the data globally, three observations can be made:

1) evidence for finite E_{xx} values can be found already below the theoretical threshold, $x_m = 0.723$,

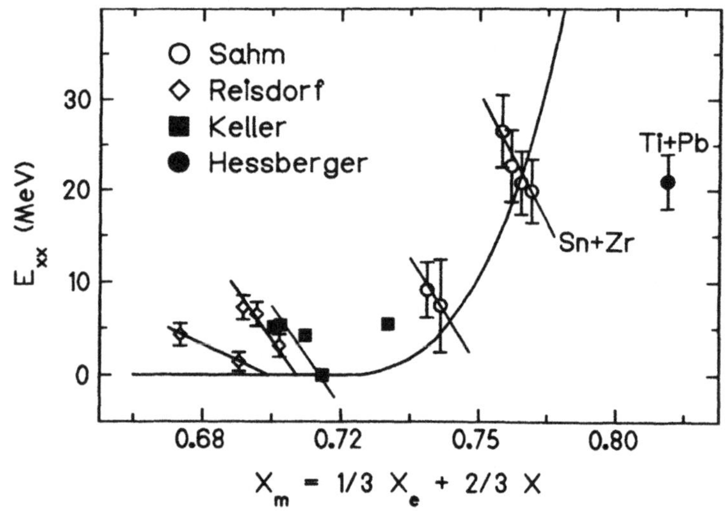

Fig. 1. Extra-extra push energies for massive systems. The data are from ref. [3] (open circles), ref. [4] (full squares), ref. [5] (diamonds) and ref. [6]. Isotopic series are joined by straight lines to guide the eye. The solid line is a calculation from Blocki et al. [1]

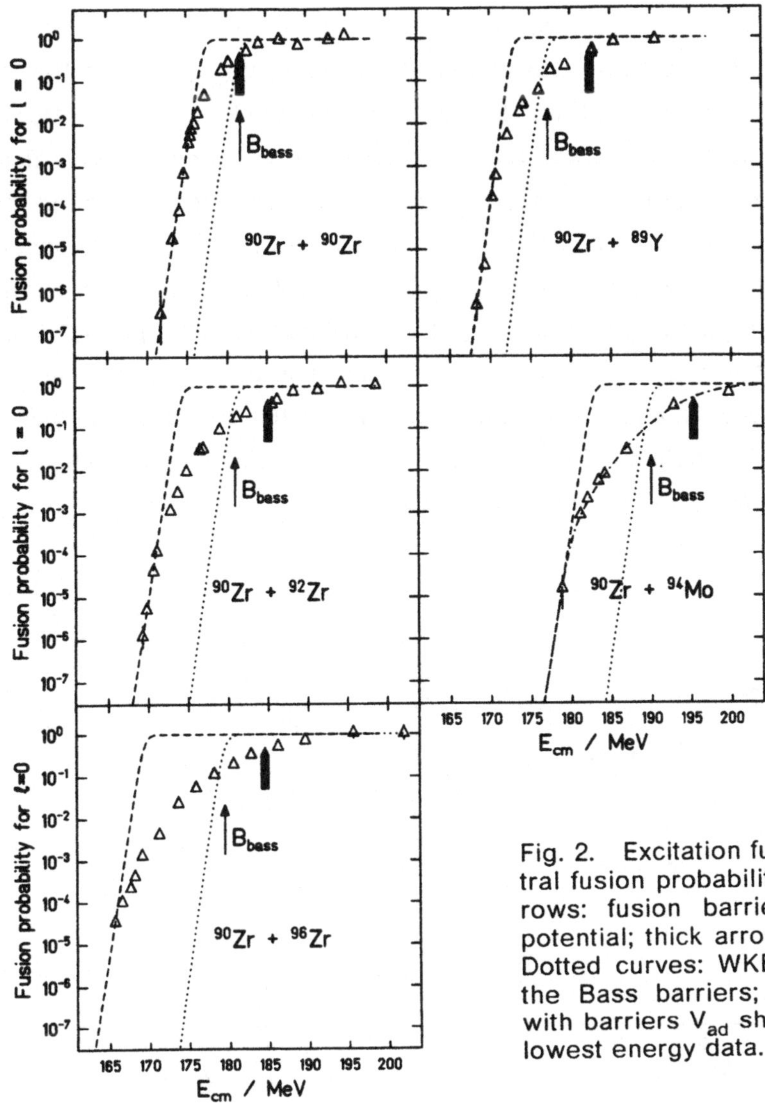

Fig. 2. Excitation functions for the central fusion probability P_{fus} [4]. Thin arrows: fusion barriers from the Bass potential; thick arrows: $P_{fus} = 1/2$ points. Dotted curves: WKB calculations using the Bass barriers; dashed: same, but with barriers V_{ad} shifted to describe the lowest energy data.

2) while most of the data points relate to systems close to symmetry, an E_{xx} value determined for a more asymmetric system, $^{50}Ti + ^{208}Pb$ [6], lies well outside the theoretical prediction (full circle in fig. 1); and

3) there is a conspicuous isotopic trend in the data opposite to the tendency predicted by the calculation.

Strong evidence that these isotopic trends are connected with the shell structure of the colliding nuclei came from the work of Keller at al. [4] for systems around Zr + Zr, in a region where the extrapush phenomenon is just beginning to be observable. The data are summarized in fig. 2. Plotted are excitation functions for the 'central fusion probabilities', P_{fus}, a term which we shall define more quantitatively later. For reference, we indicate by thin arrows the location V_{Bass} of the barrier obtained from the global potential of Bass [2], the location of the $P_{fus} = 1/2$ points, i.e. the experimental barriers (thick arrows), and the result of a no-coupling WKB calculation with the Bass potential (dotted lines). The dashed curves that run through the lowest energy data determine the 'adiabatic' barriers V_{ad}. Running down the panels on the left side, we notice a rapid increase of $V_{ad} - V_{Bass}$ with the number of added neutrons. Changing Z (right-hand side) does not change this quantity, provided the neutron number is untouched. One can conclude that it is the neutrons that are causing the shift of the adiabatic barrier. Furthermore, for the closed-shell system $^{90}Zr + ^{90}Zr$ (upper left panel), the Bass barrier is equal to the experimental barrier. In contrast, any deviation from the shell symmetry causes an 'inelasticity', i.e. the $P_{fus} = 1/2$ points move up in energy.

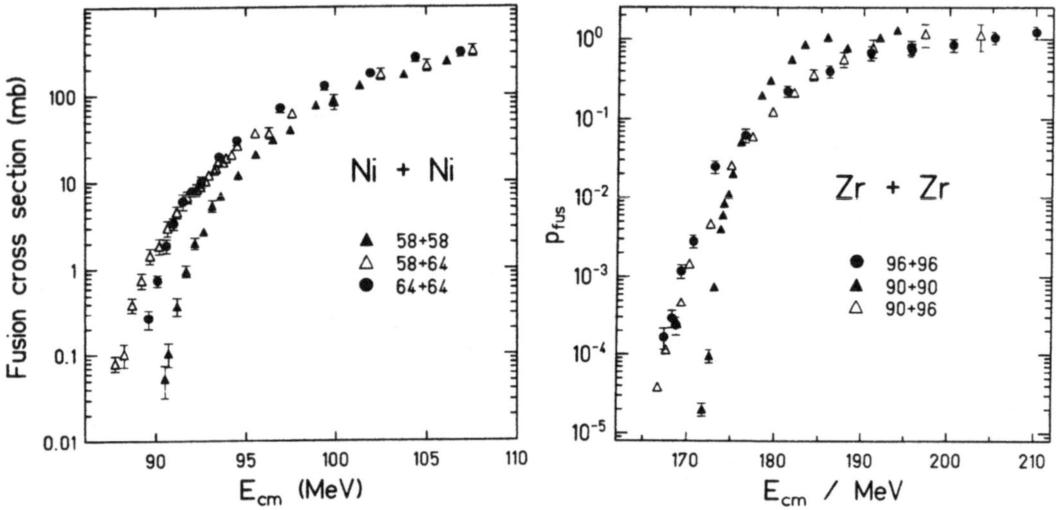

Fig. 3. Left panel: reduced fusion excitation functions for the system Ni + Ni [7]. Right panel: reduced excitation functions of the central fusion probability for the system Zr + Zr [4] [8].

Other evidence for a qualitative change above the extrapush threshold is illustrated in fig. 3 where we compare the famous Ni + Ni 'triade' [7] with a similar Zr + Zr triade [4] [8]. In both panels 'reduced' excitation functions have been plotted, i.e. the global influence of nuclear radii on the nucleus-nucleus potential has been removed (see ref. [9]). In the Ni + Ni system the excitation functions for the two symmetric systems run sort of parallel, the 'mixed' system $^{64}Ni + ^{58}Ni$ shows a remarkably different behaviour. In the mixed system we expect a strong trend towards neutron/proton equilibration via neuton pickup by ^{58}Ni or proton pickup by ^{64}Ni. Qualitatively, at least, this could explain the different slope of the fusion excitation function. For the Zr + Zr system the comparison looks different. The relevant parameter seems to be the presence or absence of valence neutrons: the excitation function for the closed-shell system $^{90}Zr + ^{90}Zr$ has a strikingly steeper slope, whereas the barrier shift shows up for the other two systems only. Investigations of the transfer reactions in these systems are under way.

Before presenting new data and discussing possible extrapush mechanisms, let us reconsider the fusion of lighter 'unhindered' systems. Most theoretical approaches to subbarrier fusion are actually just barrier-passing theories, in other words one does not consider in any detail the evolution of the di-nuclear system *after* the barrier is passed. This simplification is only justified if fusion is irreversibly initiated each time the barrier is passed. For systems like O + Pb, this may indeed be a reasonable assumption, since the dominant binary reactions in the vicinity of the fusion barrier are relatively simple direct reaction channels, involving moderate excitation of the collision partners and/or single nucleon transfer (see for example ref. [10]). Is the equation *fusion = barrier passing* correct for all systems that do not exhibit the extrapush phenomenon?

We have studied the binary reactions in the system $^{40}Ar + ^{122}Sn$ at an energy of about 28 MeV above the barrier [11]. Fig. 4 shows the angular distribution of the mass and energy integrated binary products for various indicated nuclear charges.

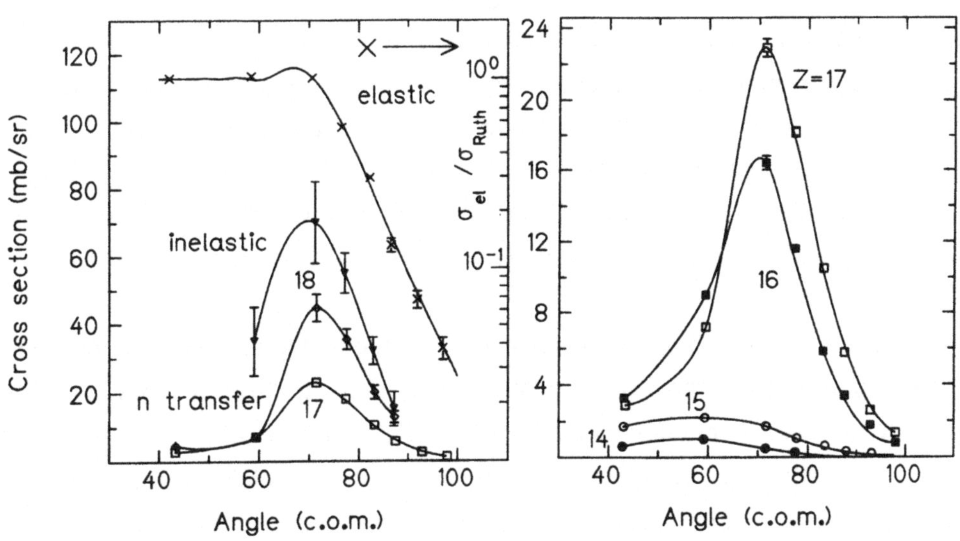

Fig. 4. Angular distributions of direct-reaction products in the system $^{40}Ar + ^{122}Sn$ at an incident energy of 180 MeV [11].

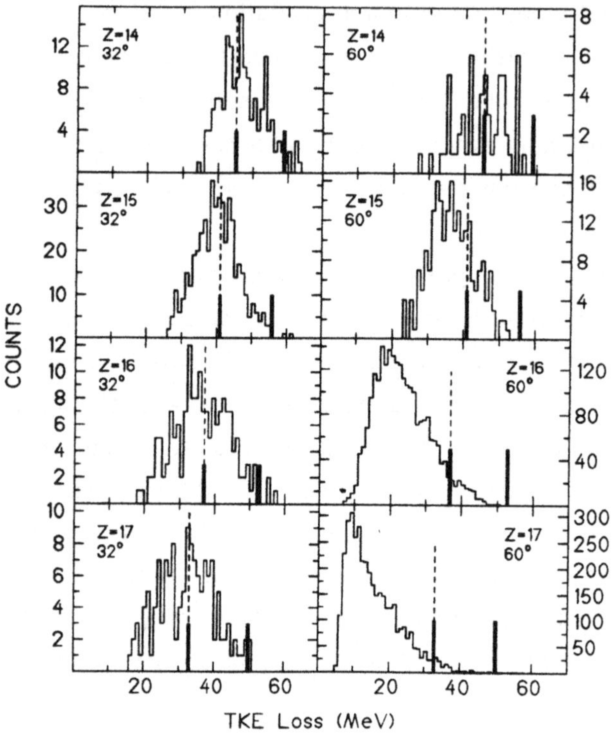

Fig. 5. Total kinetic energy loss spectra for ejectiles of given nuclear charge in the reaction $^{40}Ar + ^{122}Sn$ at 60° (right panels) and 32° (left panels) [11]. The two vertical bars in the spectra represent the expected location of the exit channel fusion barriers and the Viola kinetic energies.

Besides the fact, known from many other studies, that most transfer products are focussed into a narrow angular range, we find that the projectile acts almost exclusively as a *donor* in this reaction (the cross section for $Z = 19$, not shown in fig. 4, is at the level of the $Z = 15$ cross section). These data together with the fusion data of ref. [12] allow us to make a complete reaction balance (for details see ref. [11]). In summary, we find a 'reaction' cross section (not including the first 2^+ levels of Ar and Sn) of about 1000 mb of which 70% are due to fusion. Instead of asking how direct-reaction channels influence fusion, as is done most of the time at this conference, we can ask how fusion influences the observed evolution of the two-nucleus system.

Figure 5 shows the Q-value spectra for ejectiles of given nuclear charge. The right-hand panels show the evolution at an angle close to the grazing angle. One observes an increase of inelasticity as the ejectile's nuclear charge is further removed from the projectile charge. We observe the highest inelasticity at the most forward angle (32°), however (left hand panels). Since our reaction balance is 'complete' (within 50 mb), it is tempting to interprete these spectra as indicative of the last stage of the evolution just prior to absorption into fusion. Two markers in all the panels of fig. 5 indicate the location expected from the outgoing-channel fusion barriers (as obtained from fits to the fusion data [12]) and the location expected from 'completely relaxed' deep inelastic fragments, obtained from Viola's systematics [13]. We expect that kinetic energies of fragments that have passed the barrier and *later* have reseparated should follow fission-like systematics. Figure 5 shows that this is clearly not the case, the Q-value spectra center exactly around the calculated fusion barriers. This is in remarkable contrast to the observations of Rehm et al. [14] for the system $^{48}Ti + ^{208}Pb$ where a considerable cross section for reactions with total kinetic energies well below the fusion barrier was observed at near barrier energies. We conclude that there is no reseparation after passing the barrier for the system $^{40}Ar + ^{122}Sn$, which we know does not exhibit an extrapush phenomenon. We also find that at the 'end of the pre-fusion evolution' (left panels in fig. 5) the Q-value distributions tend to all have the same width, independent on Z. Since most of these reactions are multistep reactions, it is unlikely that this is simply due to Q-matching needs as known [15] from one-step reactions. Instead, we suggest that the widths in the Q-value spectra are strongly correlated with the dynamic fusion barrier fluctuations that have been introduced to understand subbarrier fusion.

Coming now back to the discussion of dynamical hindrance of fusion, we can envision four different scenario's that could cause the failure of two colliding nuclei to fuse, see fig. 6. The simplest suggestion could be the absence of a pocket in the

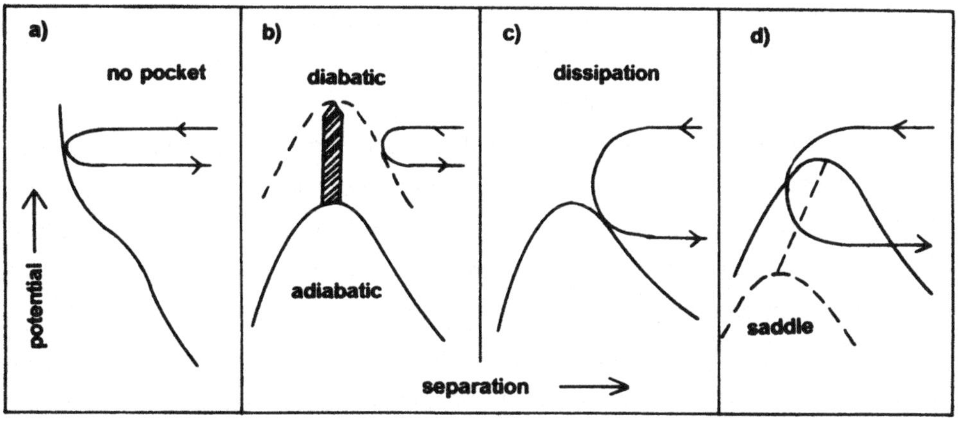

Fig. 6. Various scenario's that could cause hindrance to fusion at near-barrier energies.

nucleus-nucleus potential for sufficiently strong Coulomb interactions (panel a). Simple estimates with theoretical potentials that take into account the saturation of nuclear forces, such as the proximity potential [16] show that this is not likely to be the case for $Z_p \times Z_t < 3000$. The other three, dynamic possibilities (panels b,c,d in fig.6) can be sorted according to the assumed, increasingly 'chaotic' evolution of the dinuclear system.

One suggestion [17] is that the system behaves diabatically, leading to a fast dynamic rise of the fusion barrier, an effect which is predicted to be strongly nuclear stucture dependent, and which in the extreme case is completely reversible, giving rise to a quasielastic reflection of the fragments.

An intermediate possibility (panel c) is that there is sufficient dissipation already in the approach phase, and as a consequence the relative kinetic energy is lowered below the fusion barrier energy. This phenomenon could still be nuclear structure dependent, involving however a fairly large number of strongly coupled channels, or it could be actually so complex that it would be more appropriately described macroscopically [18] in terms of 'surface friction'.

Finally, panel d, we can envision that the decision for fusion occurs at a much later stage, *after* the barrier has been passed [19]. Because of the deeper contact at this later stage, nuclear motion can (or must) be described more 'macroscopically', in terms of the evolution on a multidimensional potential surface along classical trajectories under the influence of friction forces [19].

Experimentally, it is clear that the various scenario's can only be distinguished in *complete* reaction studies, including both the fusion and the reseparation channels. The status of our measurements concerning the latter will be described in a contribution by R.Bellwied [20]. Here I will limit myself to new information we have recently obtained from fusion excitation data [8].

We have used the velocity separator SHIP at GSI, Darmstadt to study evaporation residue cross sections from ^{96}Zr and ^{100}Mo induced reactions at near-barrier energies. SHIP will be described by F.P.Hessberger in a contribution to this conference [21]. Behind SHIP the ejectiles pass a time-of-flight (TOF) device and are stopped in an array of position-sensitive surface-barrier detectors. In general, above the 0.1 mb level the TOF versus energy display (left panel in fig. 7) adds sufficient redundancy for the identification of the fusion products and we obtain integrated evaporation-residue cross sections just by 'ion counting'.

In the case of alpha-active evaporation residues, we can increase our sensitivity down to the sub-nb region by using the position and time correlation between the signals from the incident nuclei and from their subsequent chain of alpha decays. In this way we can identify individual nuclides, as shown in the right panel of fig. 7 ('alpha-spectroscopic method') and establish excitation functions not only for the sum of evaporation residues, but for most of the individual nuclides, see fig. 8.

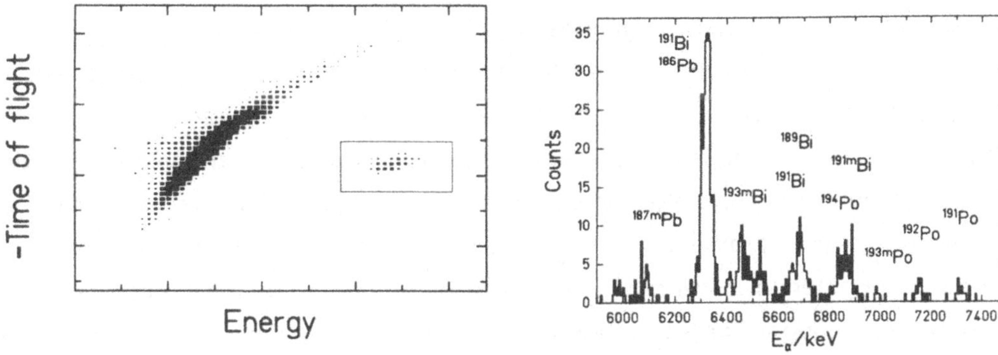

Fig. 7. Left panel: Time-of-flight versus energy spectra of fusion products after passing SHIP. The evaporation residues are inside the rectangle (from ref. [5]). Right panel: Alpha spectrum obtained in the reaction $^{100}Mo + ^{94}Mo$. The isotope ^{191}Po was first identified in this reaction (from ref. [8]).

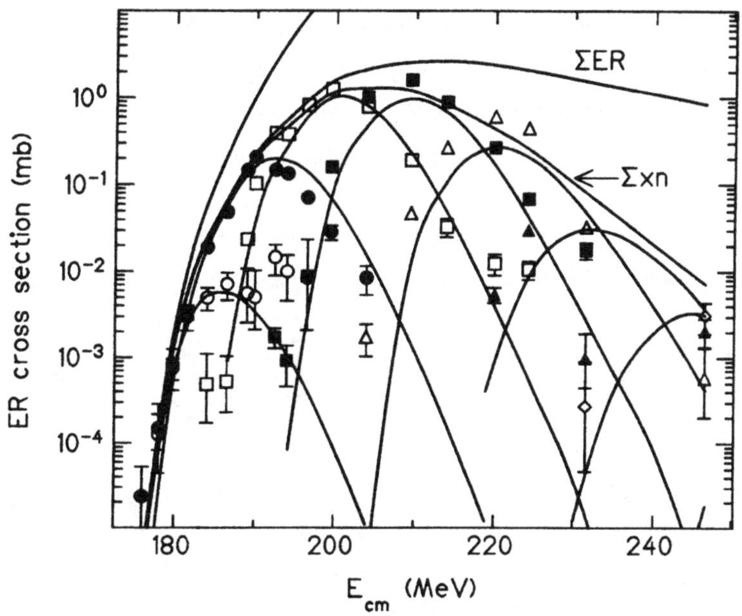

Fig. 8. Measured [8] excitation functions for $^{100}Mo(^{100}Mo,xn)$ reactions, with x = 1 to 7. The smooth curves are from calculations with parameters adjusted to fit the data (see text).

This turns out to be important information for our further data reduction: Indeed, what we actually measure for these systems leading to rather fissile compound nuclei is the product of the fusion probability P_{fus} and the probability P_{er} that a nucleus once formed, will survive the chance to fission during its deexcitation. P_{er} is known from many fissionability studies to depend on the nuclear composition, as well as the excitation energy and the angular momenta.

Actually, for these heavy systems fission acts as a central collision filter: the contribution of partial waves above 30 $\hbar$ to the evaporation residue cross section is negligible. This, together with the fact that we observe only reaction products which by their nuclear composition are unambiguous signatures that *complete* fusion has occurred, allows us to state that we obtain information on truly central fusion probabilities P_{fus}. We parameterize P_{fus} in terms of tunnelling through a dynamical distribution of potential barriers. The parameters are the first and second moment of the barrier distribution. For a symmetric barrier distribution, such as a Gaussian distribution, one can show that $P_{fus} = 1/2$ at the average barrier, just as is the case for transmission through a single barrier. P_{er} is obtained from evaporation calculations using angular momentum dependent level densities and fission barriers. We generally adjust the height of the fission barrier to obtain a better fit in the high energy region where we assume $P_{fus} = 1$ and where we can check experimentally that at least it becomes 'flat' as a function of energy. By doing cross bombardments for some chosen systems, we also check that P_{er} is only dependent on the nuclear composition of the fused system within experimental errors. For more details I refer to [3,4,5].

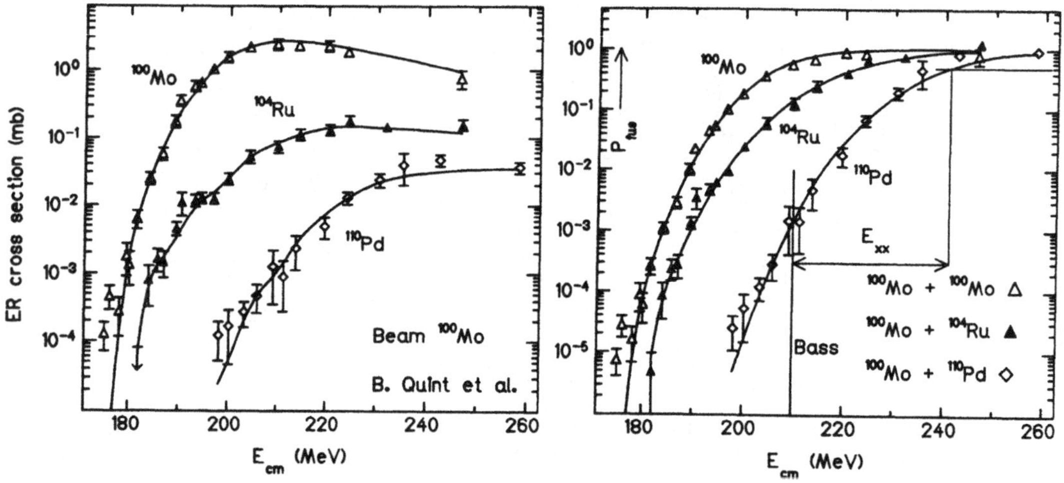

Fig. 9. Left panel: Evaporation residue cross sections for $^{100}Mo + ^{100}Mo, ^{104}Ru, ^{110}Pd$ [8]. The solid lines represent parametric fits to the data taking into account the fission probability. Right panels: Same data after reduction to central fusion probabilities. The average barriers are located at the points where $P_{fus} = 1/2$. For the ^{110}Pd target we have indicated the extrapush shift E_{xx} relative to the Bass barrier.

In fig. 9 we show in the left panel excitation functions for the systems ^{100}Mo + $^{100}Mo, ^{104}Ru, ^{110}Pd$ together with the parametric fits (solid lines). All fits give a rather close description of the data. Not only can we reproduce the total evaporation residue cross sections, but also the excitation functions for individual nuclides (see fig. 8). Finally, on the right hand side of fig. 9 we show how the original excitation functions for evaporation residues (left panel) are affected once the influence of fission is removed yielding what we call central fusion probability excitation functions. The figure also illustrates the barrier shift E_{xx} extracted by comparing the energy at which $P_{fus} = 1/2$ with the energy expected from the Bass barrier systematics. For ^{100}Mo + ^{110}Pd we find a shift as high as 31 MeV. In addition to the ascending Z_t series shown in fig. 9 we have measured various other systems, and in particular the isotopic series ^{100}Mo + ^{92,94,96,98,100}Mo. The new results for the barrier shifts, together with most of the older data, are summarized in the usual 'extrapush' plot shown in fig. 10.

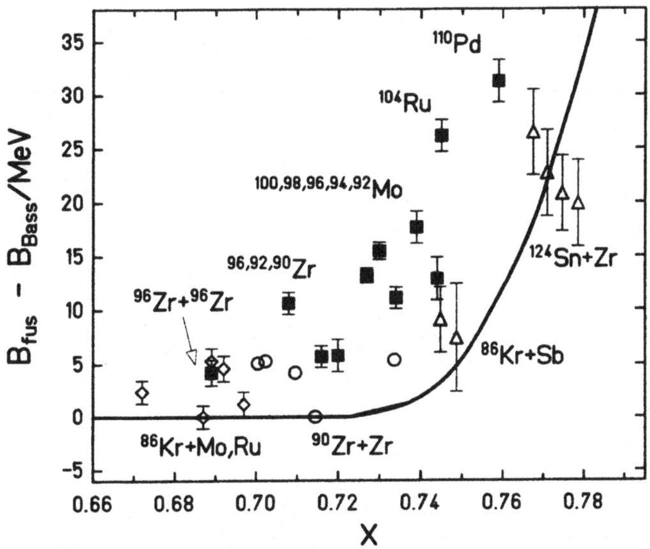

Fig. 10. Summary of extrapush data including the new measurements of ref. [8]. Most of the new data were obtained with a ^{100}Mo beam. The targets are indicated.

At first sight it is obvious from this figure that the effective fissility parameter (abscissa) suggested by the macroscopic extrapush theory [1] is not a very good sorting parameter. More complex influences on the phenomenon must exist that are ignored so far by the macroscopic approach. However it remains true that a steep rise occurs for systems with $x_m > 0.72$. One might argue perhaps that the threshold for this rise is lowered as one moves from nuclei with more magicity (such as ^{124}Sn + Zr) towards nuclei closer to (but not quite at) the midshell region, such as ^{100}Mo + ^{100}Mo, ^{104}Ru, ^{110}Pd. 'Open' shells seem to favour the early onset of the extrapush phenomenon.

There are many ways that one can try to find a better ordering parameter [8]. In figure 11 we display 'asymptotic shifts' D_∞ (in MeV) of the excitation functions. The data are not limited to systems 'with extrapush'. The shifts D_∞ represent the difference between the average barrier and the 'asymptotically' lowest or 'adiabatic' barrier V_{ad} required by the data (see also fig. 2). Whereas the abscissa represents the experimental values for D_∞, the ordinate represents the value one would expect from the coupling of the first 2^+ and 3^- levels (of projectile *and* target) to the fusion process. The simplified coupled-channels approach of ref. [22] was used. While this theoretical shift is too small (and the simplification in the approach would not justify quantitative expectations), it is obvious that it correlates rather well with the experimental data. Essentially, the ordering parameter now is nuclear stucture, and more specifically the collectivity and softness of low-lying states in the colliding nuclei.

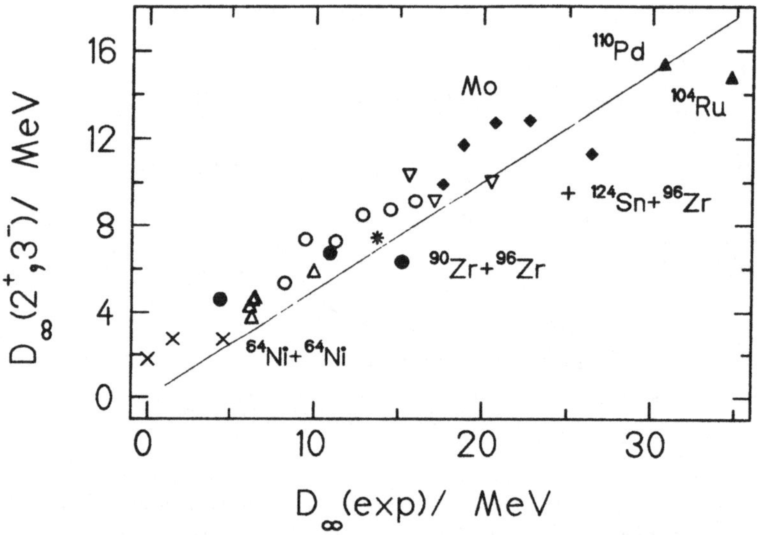

Fig. 11. Abscissa: experimental asymptotic barrier shifts (relative to the adjusted average barrier). Ordinate: theoretical downshifts obtained from the coupling of the first 2^+ and 3^- levels to the fusion process. A few systems are indicated for reference. The points marked Mo (full diamonds, isotopic series), ^{110}Pd and ^{104}Ru were obtained with a ^{100}Mo beam.

While this correlation should not be taken to mean that coupling to the 2^+ and 3^- is all one needs to explain the fluctuation part of the barriers, it suggests that there is a common cause for the vibrational/rotational phenomena in nuclei and fusion in the vicinity of the threshold: the underlying shell structure in nuclei with all its consequences, such as Q-values for transfer and more subtle effects as well. The precise relationship between the shell structure and fusion at the treshold is likely to be a very complex one. In any case, the large barrier shifts shown in fig. 10 require a theory that goes well beyond invoking coupling effects from just a few low-lying states.

In this respect it is remarkable that the diabatic theory of ref. [17] makes barrier shift predictions that are amazingly well correlated with the data as is impressively demonstrated in fig. 12. It is therefore worthwhile to close this contribution by making a few comments on the diabatic theory. For more details we refer of course to the upcoming original paper [17].

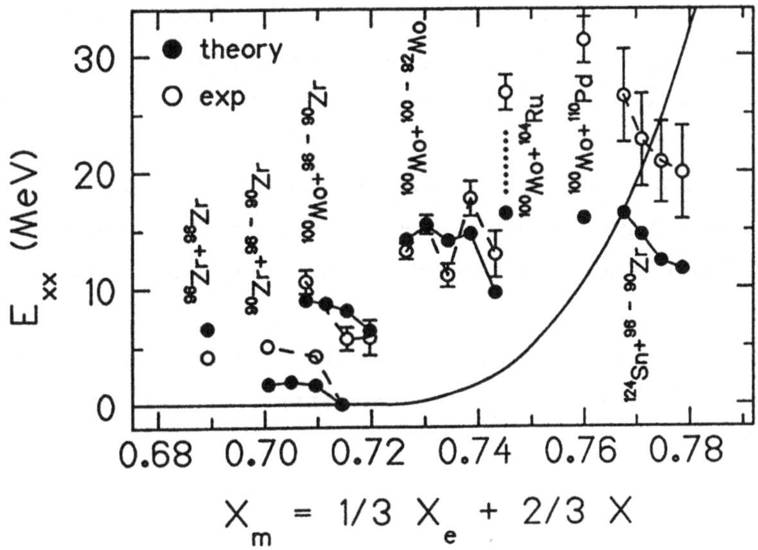

Fig. 12. Comparison of experimental (open circles) and theoretical [17] extrapush data (full circles). The solid line is from ref. [1].

Briefly, the diabatic rise of the fusion barrier is strongly linked to the nuclear shell model. In isolated nuclei, when highly degenerate shell model orbits like the h11/2 orbit are close to the Fermi surface, residual forces such as the pairing force, are known to lead to stronly correlated states. As two such nuclei approach, two-center shell model calculations show that the original degeneracy is removed, the various sub-orbits fanning out. In the diabatic limit no rearrangement of nucleons occurs. In particular, in the one-level approximation one can estimate [17] that on the average the fusion barrier will be raised by an amount $\Delta B = \gamma \frac{n(g-n)}{g}$ and has a variance given by $\sigma_B^2 = 1/3\gamma^2 \frac{n(g-n)}{g-1}$ Here n pairs of nucleons are assumed to be distributed equally (by the residual forces) on g time-conjugate pair states, which at the barrier radius are spread over γ MeV. Typically, in a realistic calculation, one finds $\gamma \simeq 10$ MeV for the h11/2 orbit in Sn at the barrier radius for the system $^{96}Zr + ^{124}Sn$. The effect is largest mid-shell, i.e. when $n = 1/2g$. Since in closed shells no net rearrangement can take take place, the effect is zero there.

The validity of this diabatic and basically *static* model is strongly dependent on the realisation that the approach time to the barrier radius is short compared to the equilibration time, i.e. the time needed to adiabatically rearrange the nucleons. Using results from the work of Bertsch [23] the authors of ref. [17] estimate an equilibration time on the order of 2×10^{-21} s, the approach time being roughly an order of magnitude shorter.

An other important assumption made was that the phase relations between the various sub-orbits are not much changed during the approach time τ_{fus}, i.e. the pair correlation energy is conserved. For this the equation $1/2\gamma\tau_{fus}/\hbar \ll 1$ must be fulfilled, otherwise the pair correlation energy is dissipated in the approach phase. The discrepancy between some of the experimental points and theory in the right hand side of fig. 12 could be explained in this way. For midshell nuclei the pair correlation energy can be as high as 20 MeV. Only dynamical calculations could give a definite answer to this conjecture.

Experimentally, the model should have important consequences for the reseparation channels [20], since it predicts quasielastic reflection at the dynamically raised barrier (panel b in fig. 6).

References

[1] J.P. Blocki, H. Feldmeier and W. Swiatecki, Nucl. Phys. **A459** (1986) 145

[2] R. Bass, Nucl. Phys. **A231** (1974) 45; Proc. Symp. on Deep-Inelastic and Fusion Reactions with Heavy-Ions, Berlin 1979 (Springer, Berlin 1980)

[3] C.-C. Sahm, H. G. Clerc, K.-H. Schmidt, W. Reisdorf, P. Armbruster, F. P. Hessberger, J. G. Keller, G. Münzenberg and D. Vermeulen, Nucl. Phys. **A441** (1985) 316

[4] J.G. Keller, K.-H. Schmidt, F.P. Hessberger, G. Münzenberg, W. Reisdorf, H.-G. Clerc and C.-C. Sahm, Nucl. Phys. **A452** (1986) 173

[5] W. Reisdorf, F.P. Hessberger, K.D. Hildenbrand, S. Hofmann, G. Münzenberg, K.-H. Schmidt, W.F.W. Schneider, K. Sümmerer, G. Wirth, J.V. Kratz, K. Schlitt and C.-C. Sahm, Nucl. Phys. **A444** (1985) 154

[6] F. P. Hessberger, G. Münzenberg, S. Hofmann, W. Reisdorf, K.-H. Schmidt, H. J. Schött, P. Armbruster, R. Hingmann, B. Thuma and D. Vermeulen, Z. Phys. **A321** (1985) 317

[7] M. Beckerman, M. Salomaa, A. Sperduto, H. Enge, J. Ball, A. Dirienzo, S. Gazes, Yan Chen, J.D. Molitoris and Mao Nai-Feng, Phys. Rev. **45** (1980) 1472

[8] A.B. Quint, thesis, Technische Hochschule Darmstadt; A.B. Quint, K.-H. Schmidt, W. Reisdorf, P. Armbruster, F.P. Hessberger, S. Hofmann, G. Münzenberg, H. Stelzer, H.-G. Clerc, C.-C. Sahm, to be published

[9] W. Reisdorf, Proc. Int. Conf. on Nuclear Physics, Harrogate, 25-30 august 1986

[10] J.S. Lilley, M.A. Nagarajan, D.W. Banes, B.R. Fulton, I.J. Thompson, Nucl. Phys. **A463** (1987) 710

[11] W. Reisdorf, E.A. Bakkum, P. Decowski, K.A. Griffioen, R. Kamermans, R.J. Meijer, H.W. Wilschut, GSI Scientific Rep. 1987,41
[12] W. Reisdorf, F.P. Hessberger, K.D. Hildenbrand, S. Hofmann, G. Münzenberg, K.-H. Schmidt, J.H.R. Schneider, W.F.W. Schneider, K. Sümmerer, G. Wirth, J.V. Kratz, K. Schlitt, Nucl. Phys. **A438** (1985) 212
[13] V.E. Viola, K. Kwiatkowski and M. Walker, Phys. Rev. **C31** (1985) 1550
[14] K.E. Rehm, A.M. van den Berg, J.J. Kolata, D.G. Kovar, W. Kutschera, G. Rosner, G.S.F. Stephans, J.L. Yntema, preprint PHY-5085-HI-87,ANL (1987)
[15] P.J.A. Buttle and L.J.B. Goldfarb, Nucl. Phys. **A176** (1971) 299
[16] J. Blocki, J. Randrup, W.J. Swiatecki and C.F. Tsang, Ann. of Phys. **105** (1977) 427
[17] D. Berdichevsky, A. Lukasiak, W. Nörenberg, and P. Rozmej, GSI Scientific Report 1987,100 and to be published
[18] P. Fröbrich, contribution to this conference
[19] W.J. Swiatecki, Nucl. Phys. **A376** (1982) 471
[20] R. Bellwied, contribution to this conference
[21] P.F. Hessberger, contribution to this conference
[22] R.A. Broglia, C.H. Dasso, S. Landowne and G. Polarolo, Phys. Lett. **133B** (1983) 34
[23] G.F. Bertsch, Z. Phys. **A289** (1978) 103

INFLUENCE OF INELASTIC COUPLINGS ON $^{32}S + ^{24}Mg$ SUB-BARRIER FUSION

J. Díaz, J.L. Ferrero and F. Sánchez

IFIC, Centro Mixto Universidad de Valencia–CSIC, Burjasot, Valencia, Spain

and

B. Bilwes and R. Bilwes

Centre de Recherches nucléaires and Université Louis Pasteur, Strasbourg, France

abstract: Elastic, inelastic and mutual excitation angular distributions have been measured for $^{32}S + ^{24}Mg$ at 65, 75, 86.3, 95 and 110 MeV of ^{32}S incident energy. The data have been analysed by optical model and coupled channel calculations. Energy-dependent optical potentials have been obtained which have been used to describe the $^{32}S + ^{24}Mg$ fusion data of Berkowitz *etal* [2]). Barrier penetration and coupled channel models are compared. It is found that the energy-dependence of the empirical potentials is essential if a consistent simultaneous description of elastic, inelastic an fusion data is required.

1. Introduction

The sub-barrier fusion cross section is underestimated by the barrier penetration model for many systems[1]). One particular case is the system $^{32}S + ^{24}Mg$, measured by Berkowitz *et al* [2]).

In order to explain the observed enhancement several models have been proposed, in which the inelastic degrees of freedom are included, often making adiabatic or sudden approximations[1]). The main characteristic shared by all these models is the use of a two body potential which in principle should be the potential to be employed in optical model or coupled channel calculations. However, for most of the systems studied, elastic and inelastic angular distributions have not been measured. Thus, there is no certainty that the potential used to describe the fusion data describes also the elastic and inelastic angular distributions. Thus, in order to have a severe test to the models proposed it would be very useful to have data for elastic, the relevant inelastic and transfer channels and fusion cross sections for some systems. For this reason we have measured elastic and inelastic angular distributions for $^{32}S + ^{24}Mg$ at 65, 75, 86.3, 95 and 110 MeV.

2. Experimental

The experiments were done at the MP Tandem of the CRN of Strasbourg. The kinematical coincidence method was used as described previosly [3,4]). The energy resolution was about 300 keV. The events were gathered in 1° CM steps. The angular resolution was about 0.4° CM. Targets were made of $20\mu g/cm^2$ of enriched ^{24}Mg evaporated onto a $20\mu g/cm^2$ carbon foil. Angular distributions were normalized to the Rutherford cross section at low angles. The normalization error is estimated to be less than 2%. The 2_1^+ levels in both ^{32}S and ^{24}Mg and the mutual excitation of these two levels could be well resolved. Concerning the transfer reactions, one, two and three nucleon transfers give a negligible yield. Only alpha transfer reactions are clearly seen in the spectra, although with yields considerably smaller than the inelastic reactions. The total integrated alpha transfer cross section is estimated to be about 3 mb. In Fig. 1 are shown the data at 95 and 110 MeV.

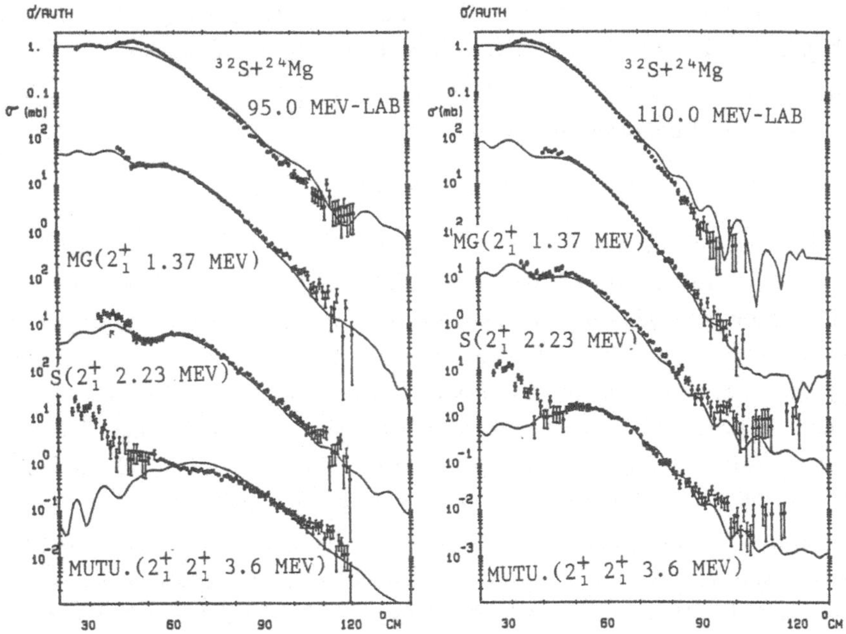

Fig. 1. Elastic, inelastic and mutual excitation data for ^{32}S + ^{24}Mg at 95 and 110 MeV. The curves have been obtained from coupled channel fits.

3. Analysis of elastic and inelastic data

In the first place, the elastic angular distributions were analysed by fitting Woods-Saxon potentials. In this way, the sensitivity radius, defined as the crossing radius of different potentials fitted to the experimental data, is determined. The barrier height and position extracted from the data analysis will be well determined only if they are placed near the sensitivity radius. In the energy range studied, the sensitivity radius found by us is $R_s = 9.7 \pm 0.1$ fm which in fact agrees with the barrier position $R_B = 9.545$ fm. In the second place the data were analyzed by the folding model potential. The M3Y effective nucleon-nucleon interaction of Bertch *et al* [5])

$$v_{M3Y} = 9999\frac{e^{-4r}}{4r} - 2134\frac{e^{-2.5r}}{2.5r} - 262\delta(r)$$

was used.

The nucleon densities were calculated by unfolding the nucleon charge distributions in the standard way[6]). The nuclear charge distributions were taken from electron scattering experiments[7]), parametrized in the Fermi-parabolic form. The parameters used were $\omega = -0.213$, $c = 3.441$ fm, $a = 0.624$ fm for ^{32}S and $\omega = -0.249$, $c = 3.192$ fm and $a = 0.604$ fm for ^{24}Mg. The imaginary potential was taken of Woods-Saxon shape.

A renormalization coefficient was allowed for the folding potential in order to fit the data. Thus the optical potential used was

$$V_{opt}(r) = NV_F(r) + iW(r) + V_C(r)$$

The use of a renormalization coefficient is essential in order to reproduce the data at low energy. It accounts for the polarization potential, which is assumed of the same shape that the bare potential. This polarization potential is linked to the imaginary potential by a dispersion relation derived by Mahaux *et al* [8]). According to this dispersion relation, the energy dependence in the real potential at low energies, also known as the 'threshold anomaly', is produced by the increase in absorption due to the fact that more and more channels start to be open as energy increases. Thus, it can be said that the energy dependence in the real part is produced by the nonelastic couplings.

The inelastic data were analyzed by the coupled channel method. The code ECIS 79 [9]) was used in all the calculations. The rotational model with prolate shape was assumed for the 2_1^+ levels in both ^{24}Mg and ^{32}S with $\beta_2(^{24}\text{Mg}) = 0.48$ and $\beta_2(^{32}\text{S}) = 0.29$ deduced from time-life measurements[10]). The real transition form factors were obtained by deforming the transition densities as described in Bilwes *et al* [11]). The imaginary transition form factors were calculated in the standard way[12]). The multipolar expansion of the potential was carried out up to second order in the deformation parameters. The renormalization coefficients found in both optical model and coupled channel analyses are consistent with the imaginary potentials at the sensitivity radius in the sense that they fulfil the dispersion relation

$$\Delta V_{E_S}(R_S, E) = (E - E_S)\frac{P}{\pi}\int_{-\infty}^{+\infty} \frac{W(R_S, E')}{(E' - E_S)(E' - E)} dE'$$

where

$$\Delta V_{E_S}(R_S, E) = V(R_S, E) - V(R_S, E_S)$$

The energy of reference was taken as $E_S = 110$ MeV and $W(R_S, E)$ was approximated by a polygonal line with the empirical points as vertices and vanishing at E= 55 MeV-Lab.

In Fig. 2 it is shown the empirical real and imaginary potentials at the sensitivity radius and the curve obtained from the dispersion relation for the optical model analysis of the data. In the coupled channel calculations a similar curve was found although with a less pronounced energy dependence.

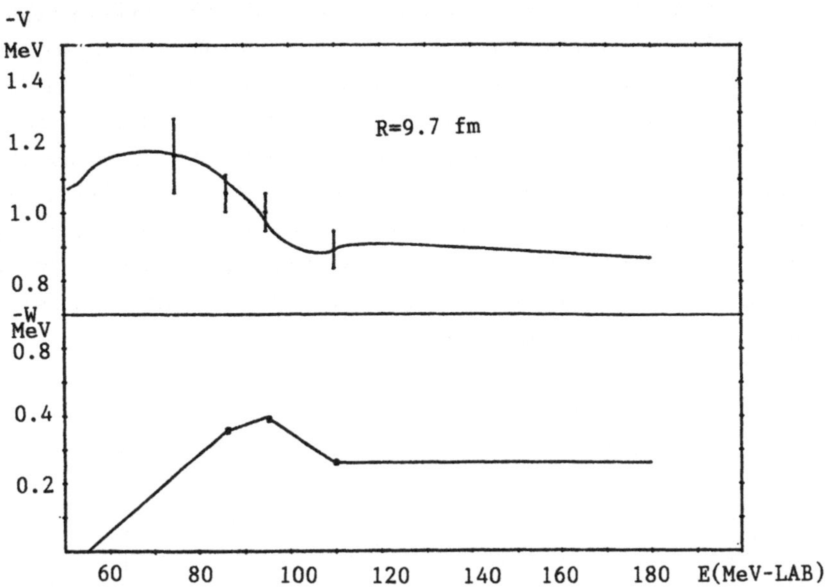

Fig. 2. Empirical real and imaginary potentials obtained from optical model fits and curves obtained from the dispersion relation.

4. Analysis of the fusion data

In the first place, calculations were done with the barrier penetration model. The bare folding potential understimates the fusion data even at the highest energy measured by Berkowitz *et al* [2]). When the folding potential is multiplied by 1.38, the highest energy fusion data are correctly reproduced but the lower energy data are still grossly underestimated as seen in Fig. 3a. In the second place, the BPM was used with the energy dependent potential obtained by fitting the elastic data. This procedure has been suggested recently by Nagarajan and Satchler[13]). It is found that the barrier penetration model using the empirical energy-dependent potential gives a much better description of the fusion data than the energy-independent potential, although the empirical potential overestimates slightly the data at the higher energies as it is shown in Fig. 3a. Thus, it seems that we have to go beyond the barrier penetration model if a good description of the fusion data is required.The inclusion of inelastic and transfer degrees of freedom is one of the most widely used extensions of the barrier penetration model. One way of treating the inelastic degrees of freedom is to solve the full coupled channel equations with an imaginary potential restricted to the inner barrier region and to calculate the fusion cross section as the difference between the reaction and inelastic cross sections[14]) $\sigma_F = \sigma_R - \Sigma\sigma_I$. This procedure has the uncertainty associated to the choice of the imaginary potential. If the empirical imaginary potential is taken, then σ_F contains the reaction cross section of all the channels not included explicitely in the coupled equations. If on the contrary an imaginary potential restricted to the inner part of the barrier is used, the model wave fuction is larger than the true one, because the former has not been damped by the external part of the imaginary potential. We have made calculations under each of these two assumptions. It is found that none of them gives a good description of the fusion data in the whole measured energy range. The results are shown in Fig. 3b in which it is seen that a short imaginary potential works relatively well at the higher energies while the empirical imaginary potential gives reasonable results only at the lower energies where the nonelastic processes not included in the coupled equations are negligible.

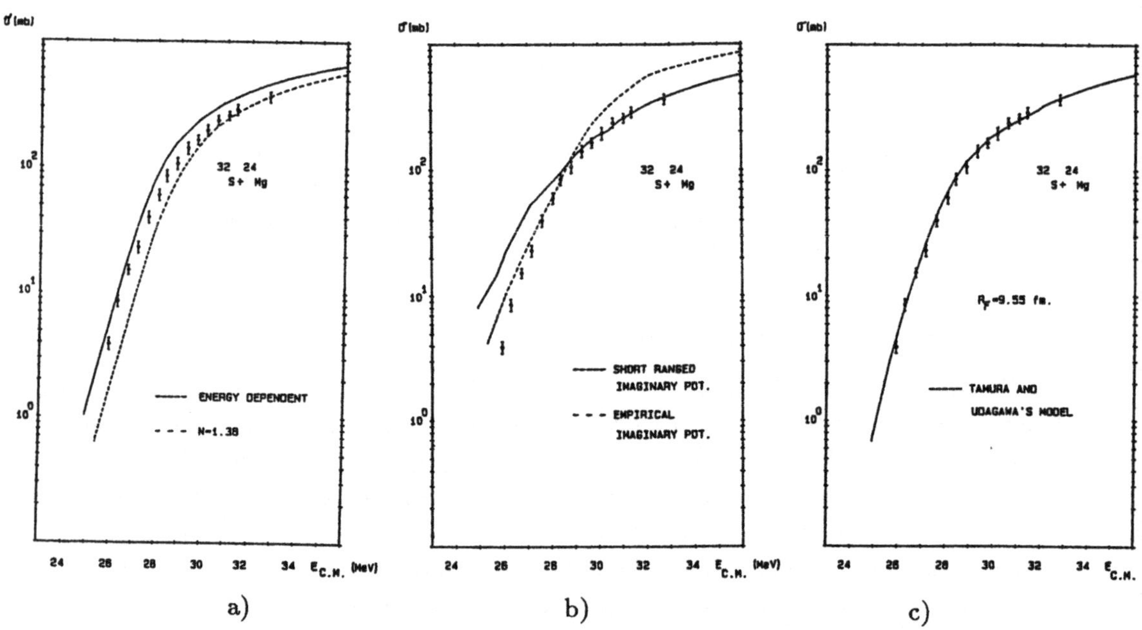

Fig. 3. Experimental fusion data for $^{32}S+^{24}Mg$ compared to calculations by a)barrier penetration model, b)coupled channel model and c)Udagawa and Tamura's model.

Finally, the model proposed by Tamura and Udagawa[15]) was employed. In this model the fusion cross section is calculated as the absortion due to the imaginary potential in the inner part of the barrier but the wave function used is obtained with the full empirical potential. The imaginary potential is divided into a peripherical and a fusion part and the fusion cross section is calculated from

$$\sigma_F = \frac{\pi}{k^2} \sum_\ell (2\ell + 1) \frac{8}{\hbar v} \int_0^\infty |\chi(r)|^2 W_F(r) dr$$

where $W_F(r)$ is the imaginary potential associated to the fusion process. The simplest way of splitting the imaginary potential into a peripherical and a fusion part is to define a fusion radius R_F, such that the fusion potential is the empirical imaginary potential for $r < R_F$. For our system we have found a fusion radius of $R_F = 9.55$ fm which is placed just at the barrier top position. Thus, in this case the Udagawa and Tamura's model is compatible with fusion taking place by a tunneling mechanism. The description of the fusion data obtained in this model is in fact very good, as seen in Fig. 3c, and what is more important, is consistent with the empirical energy-dependent potential obtained from the analysis of the elastic data.

Acknowledgements

This work has been supported in part by CAYCIT(Spain) under projet PR84-0208. One of us (F. S.) is grateful to Ministerio de Educación y Ciencia for a fellowship.

5. References

1) M. Beckerman, Phys. Report **C129** (1985) 145

2) G. M. Berkowitz *et al* , Phys. Rev. **C28** (1983) 667

3) B. Bilwes *et al* , Nucl. Phys. **A408** (1983) 173

4) A. Baeza, B. Bilwes, R. Bilwes, J. Díaz J. L. Ferrero, Nucl. Phys. **A419** (1984) 412

5) G. Bertch *et al* ,Nucl. Phys. **A284** (1977) 399

6) G. R. Satchler and W. G. Love, Phys. Report **55** (1979) 183

7) G. C. Li, M. R. Yearian and I. Sick, Phys. Rev. **C9** (1974) 1861

8) M. A. Nagarajan, C. Mahaux and G. R. Satchler, Phys. Rev. Lett. **54** (1985) 1136
C. Mahaux, H. Ngô and G. R. Satchler, Nucl. Phys. **A449** (1986) 354

9) J. Raynal, Phys. Rev. **C23** (1981) 2571

10) P. M. Endt and C. Van der Leun, Nucl. Phys. **A310** (1978) 1

11) B. Bilwes, R. Bilwes, J. Díaz, J. L. Ferrero, Nucl. Phys. **A449** (1986) 519

12) T. Tamura, Rev. Mod. Phys. **37** (1965) 679

13) M. A. Nagarajan and G. R. Satchler, Phys. Lett. **173B** (1986) 29

14) M. J. Rhoades-Brown and P. Braun-Munzinger, Phys. Lett. **136B** (1984) 19

15) T. Udagawa and T. Tamura Phys. Rev. **C29** (1984) 1922
T. Udagawa, S. W. Hong and T. Tamura, Phys. Rev. **C32** (1985) 1435

NUCLEAR DEFORMATION EFFECTS IN SUB-BARRIER FUSION OF $^{16}O+^{147,149}Sm$

D.E. DiGregorio, M. diTada, D. Abriola, M. Elgue, A. Etchegoyen,
M.C. Etchegoyen, J.O. Fernández Niello, A.M.J. Ferrero, S. Gil,
A.O. Macchiavelli, A.J. Pacheco, J.E. Testoni
Departamento de Física, Comisión Nacional de Energía Atómica
Avda. del Libertador 8250, 1429 Buenos Aires, Argentina

P.R. Silveira Gomes
Instituto de Fisica da Universidade Federal Fluminense
Caixa Postal 296, Niteroi, Brazil

V.R. Vanin, R. Liguori Neto, E. Crema
Instituto de Fisica da Universidade de Sao Paulo
Caixa Postal 20516, Sao Paulo, Brazil

R.G. Stokstad
Nuclear Science Division, Lawrence Berkeley Laboratory
University of California, Berkeley, CA 94720, U.S.A.

Introduction

Sub-barrier fusion reactions and their relation with the structure of the projectile and target nuclei have been the subject of intensive investigation, both from the experimental and theoretical point of views. In particular, one of the aspects that has drawn special attention is the influence of the shape degrees of freedom on the enhancement of the fusion cross sections at energies below the barrier, beyond what could be expected from simple models based on the quantal penetration of one-dimensional barriers.

From an experimental standpoint, an important indication of the role played by the shape degrees of freedom in the sub-barrier enhancement came from the study of the fusion of ^{16}O with various even samarium isotopes [1]. The samarium nuclei are known to exhibit a wide range of deformation, from the spherical, semimagic ^{144}Sm to the well deformed ^{154}Sm. The use of a doubly magic projectile such as ^{16}O was considered to be important in order to be able to isolate effects that could in principle be attributed to the different deformations of the different target nuclei. The results obtained in Ref. 1 clearly showed that larger deformation (i.e., heavier isotopes of samarium) corresponded to larger sub-barrier enhancements of the fusion cross sections.

The study of the $^{16}O + ^{144}Sm$ [2] not only confirmed the trend previously observed for the rest of the even samarium isotopes, it also provided the *basic* barrier parameters needed to undertake a more quantitative description of the phenomenon.

In the present work the study of fusion cross sections at energies in the vicinity of the interaction barrier was extended to include the reactions of ^{16}O with the two stable odd samarium isotopes, ^{147}Sm and ^{149}Sm. The purpose of this investigation was two-fold: On the one hand it might prove possible to observe effects due to the presence of valence nucleons, in excess of those expected from the deformation of the core. On the other hand, the inclusion of these two nuclei may be expected to provide information regarding the behaviour of the extracted nuclear deformation parameters in the whole mass range. This might be particularly important for the case of ^{147}Sm, in order to bridge the change in structure between ^{144}Sm and ^{148}Sm, which from a structural point of view is probably larger than for any other pair of succesive isotopes.

Experimental

The experiment was performed at the TANDAR laboratory in Buenos Aires. Beams of ^{16}O delivered by the 20UD tandem accelerator with laboratory energies ranging from 61 to 75 MeV were used to bombard isotopically enriched ^{147}Sm and ^{149}Sm targets of about $100 \mu g/cm^2$.

Fusion cross sections were extracted from the observation of delayed X-rays emitted following the radioactive decay of the evaporation residues produced in the reaction and trapped in an aluminum catcher foil positioned 2 mm behind the samarium target. The thickness of the Al catcher foils was sufficient to stop the evaporation residues of interest, while letting through projectile-like fragments and evaporation residues originating from light target contaminants and from the catcher foil itself.

The method of analysis determines the evaporation-residue cross sections for the separate isotopes. These are taken as adjustable parameters in a chi-square minimization procedure (incorporating the theoretical expressions for the time depedence of the X-ray activities) and are fit to the experimental data. From considerations based on the systematics of reactions with other samarium isotopes, the 2n, 3n, 4n and α-xn evaporation channels were considered. The normalization required for the determination of absolute cross sections was obtained using the Rutherford scattering from the target, detected by two solid-state detectors positioned at $\pm 30^o$ from the beam, and by monitoring the beam intensity as a function of time. In order to check the whole method, various points of the previously known fusion cross sections for the ^{16}O + ^{148}Sm system were also measured.

Results and Discussion

The relative values obtained for the evaporation-residue cross sections as a function of the bombarding energy were in good agreement with the expectations of statistical model calculations using the codes CASCADE and PACE. The fusion cross sections were calculated as the sum of the contributions from the different evaporation channels, since for these masses and bombarding energies, fusion-fission can be neglected.

Fig. 1 summarizes the experimental results. Fig. 1 a) shows *reduced* fusion cross sections versus *reduced* center of mass energies. These reduced quantities are defined in such a way that geometric effects due to small variations in nuclear radii and masses are removed (see Ref. 2). The squares and triangles correspond to ^{147}Sm and to ^{149}Sm, respectively. The circles indicate fusion cross sections for ^{148}Sm whereas the rhombs represent data on the same isotope taken from Ref. 1. The comparison between cross sections of ^{148}Sm and ^{147,149}Sm is crucial if one desires to evaluate, at least qualitatively, any observed difference that might be attributed to the presence of the odd nucleons, in excess of what could be expected from the systematic behaviour as a function of deformation. In addition, the points for ^{148}Sm are in excellent agreement with those of Ref. 1. Fig. 1b) shows the same data of 1a), although processed in such a manner that the differences below the barrier can be clearly appreciated. The abcissa represents reduced cross sections divided by the ^{148}Sm data. The solid curves are fits to the experimental points of each of the three samarium isotopes under consideration, performed with a one-dimensional model that includes deformation effects [3].

Qualitatively, the pattern observed in Fig. 1 indicates that the odd isotopes ^{147}Sm and ^{149}Sm do not exhibit any significant departure from the known behaviour of the even isotopes. However, a more sensitive test in order to check possible effects due to the unpaired nucleons could be provided by a detailed analysis of the behaviour of the quadrupole deformation parameter β_2.

The basic idea is that if one uses the one-dimensional model of Ref. 3, the deduced values of β_2 will depend on any effects that produce a sub-barrier enhancement, even those effects that are not directly related to shape degrees of freedom. In that case, one could expect a discrepancy between the deformation parameters deduced from the fusion data using the one-dimensional model and the corresponding values obtained from other sources, such as B(E2) values. Since the deformation relevant to fusion enhancement and that relevant to the determination of the B(E2) are not necessarily the same, the above mentioned discrepancy should be apparent in the systematic behaviour of β_2 rather than in the absolute values for a given isotope.

Fig. 2 shows the results for the quadrupole deformation parameter β_2 extracted in this work from the fits using the model of Ref. 3, together with those obtained in Ref. 2. These values of β_2 are plotted versus the corresponding values obtained from the experimental B(E2) values, which were calculated for each isotope using the prescription given in Ref. 4. In the case of the odd isotopes, it was assumed that the extra neutron occupies a given Nilsson state. States with $K = \frac{3}{2}$ and $K = \frac{5}{2}$ were taken for ^{147}Sm and ^{149}Sm, respectively.

This comparison shows that although the relation between the parameters extracted using the one-dimensional model and those obtained from B(E2) values is not linear, all the isotopes, including the odd ones, fall in a rather smooth curve . This indicates that within the sensitivity of this method the odd nucleons do not have a pronounced effect on fusion cross sections below the barrier. Regarding this dependence, the β_2 values obtained from fusion seem to saturate for the heaviest samarium isotopes, in contrast with the rapid rise observed starting from ^{144}Sm. Note that the existence of this initial slope is confirmed by the inclusion of the new point corresponding to ^{147}Sm. Preliminary calculations performed with the coupled-channel code CCFUS [5] seem to reproduce this behaviour.

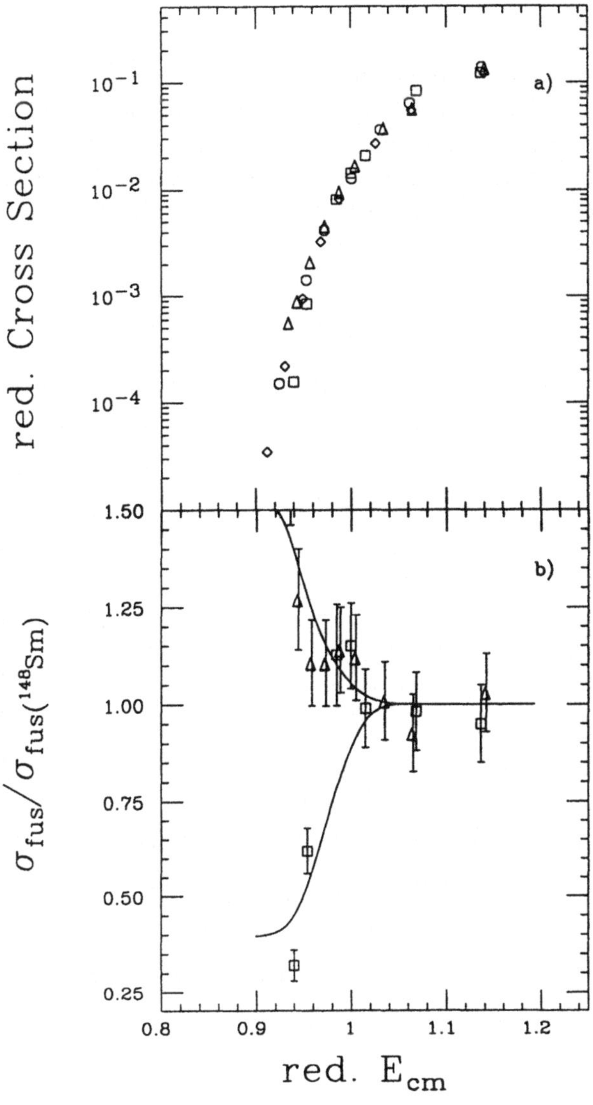

Fig. 1: a) Reduced fusion cross sections vs. reduced center-of- mass energies. The squares and triangles correspond to ^{147}Sm and to ^{149}Sm, respectively. The data corresponding to ^{148}Sm are indicated by circles (this measurement) and rhombs (taken from Ref. 1) b) Experimental and theoretical reduced fusion cross sections for the odd isotopes divided by the corresponding cross sections for ^{148}Sm, as a function of reduced center-of- mass energies.

Conclusions

The excitation functions measured for the odd $^{147,149}Sm$ do not show unusual effects related to the unpaired neutrons. The behaviour at below-barrier energies follows the systematics of the even isotopes, i.e., the quadrupole deformation parameters obtained for the odd isotopes confirm the dependence of β_2 as a function of target deformation. This dependence indicates an important difference between ^{144}Sm and ^{148}Sm and a smoother behaviour for the heaviest isotopes.

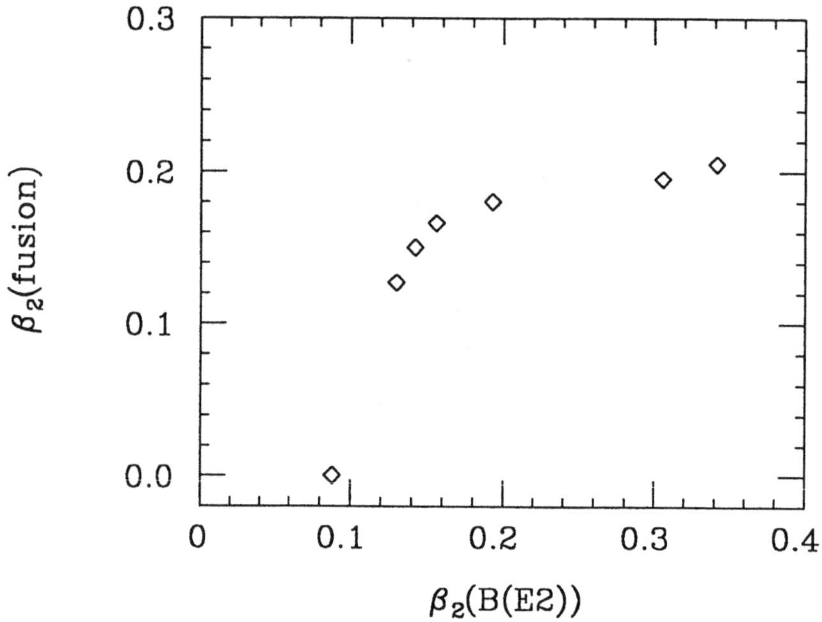

Fig. 2: β_2 values deduced from fusion vs. β_2 values deduced from experimental B(E2) transition probabilities.

References

[1] R.G. Stokstad, Y. Eisen, S. Kaplanis, D. Pelte, U. Smilansky and I. Tserruya, Phys. Rev. Lett. **41**, (1978) 465; Phys. Rev. **C21**, (1980) 2427.

[2] D.E. DiGregorio, J.O. Fernández Niello, A.J. Pacheco, D. Abriola, S. Gil, A.O. Ma- cchiavelli, J.E. Testoni, P.R. Pascholati, V.R. Vanin, R. Liguori Neto, N. Carlin Filho, M.M. Coimbra, P.R.S. Gomes and R.G. Stokstad, Phys. Lett. **B176**, (1986) 322.

[3] C.Y.Wong, Phys.Rev.Lett. **31** (1973) 766.

[4] A. Bohr, B.R. Mottelson, "Nuclear Structure" Vol.II, W.A. Benjamin Inc, New York, 1974.

[5] C.H. Dasso and S. Landowne, Comp. Phys. Comm. **46** (1987) 187.

Sub - Coulomb Fusion of $^{24,26}Mg$ with $^{90,92,94,96}Zr$ *

R.-D. Ecker, J. Friese, A. Gillitzer, H.-J. Körner,
L. Müller, G. Rosner** and K. Zeitelhack
Physik-Department, Technische Universität München
James-Franck-Straße, D-8046 Garching

Since a couple of years it is a well known fact that for medium heavy nuclei systems fusion cross sections below the classical Coulomb barrier are observed to be enhanced by orders of magnitude compared with 1-dimensional tunneling calculations [1]. As a reason for this phenomenon it is believed that additional degrees of freedom like inelastic excitations or transfer processes have to be taken into account as doorway states towards fusion.

Investigating the systems $^{24,26}Mg$ + $^{90,92,94,96}Zr$ we examined in particular the role of two nucleon transfer (2n, 2p). For the systems ^{24}Mg + ^{A}Zr ground- state Q-values for 2n-pickup reactions increase with increasing target mass (cf. table 1). If 2n transfer is an important process, fusion cross section should increase for this systems with increasing target mass. On the other hand the heavier projectile allows the examination of the influence of 2p transfer. For these systems the ground- state Q- value for 2p-pickup reactions decrease with increasing target mass (cf. table 1). Therefore one expects decreasing cross sections with increasing target mass if the latter process plays a dominant role.

Table 1: Ground-state Q-values for two nucleon transfer [MeV]

	(^{24}Mg, ^{26}Mg)	(^{24}Mg, ^{26}Si)	(^{26}Mg, ^{28}Mg)	(^{26}Mg, ^{28}Si)
^{90}Zr	- 2.86	- 7.64	- 6.34	+ 4.42
^{92}Zr	+ 2.59	- 9.31	- 0.88	+ 2.76
^{94}Zr	+ 3.47	- 11.11	- 0.01	+ 0.95
^{96}Zr	+ 4.11	- 13.39	+ 0.63	- 1.33

Fusion cross sections were measured in the energy range of 56 MeV $\leq E_{CM} \leq$ 72 MeV at the Munich tandem accelerator. An electrostatic deflector placed shortly downstream from the target was used to separate evaporation residues from the direct beam (fig. 1). The length of the deflector was 27 cm, the two plates were separated by a 3.5 cm gap. The high voltage applied ranged up to 100 kV. The detector system consisted of 4 surface barrier detectors mounted in an ionization chamber. The charge of the evaporation residues was identified by measuring energy loss and residual energy. Mass resolution was achieved via time- of- flight and total energy measurements. The array of 4 silicon surface barrier detectors was mounted perpendicular to the deflection plane, thus allowing to sample the evaporation residue angular distribution from zero to 4.5 degrees.

Since these angular distributions in the deflection plane are changed by the deflector according to mass, energy and charge states, one has to determine the ratio $\eta = \Delta\Omega_{eff} / \Delta\Omega_{geo}$ of the effective solid angle to the geometric solid angle of the detectors. The ratio η was determined to η = 80- 90% employing a Monte Carlo simulation program. The overall efficiency of the detectorsystem is ε = 0.5%.

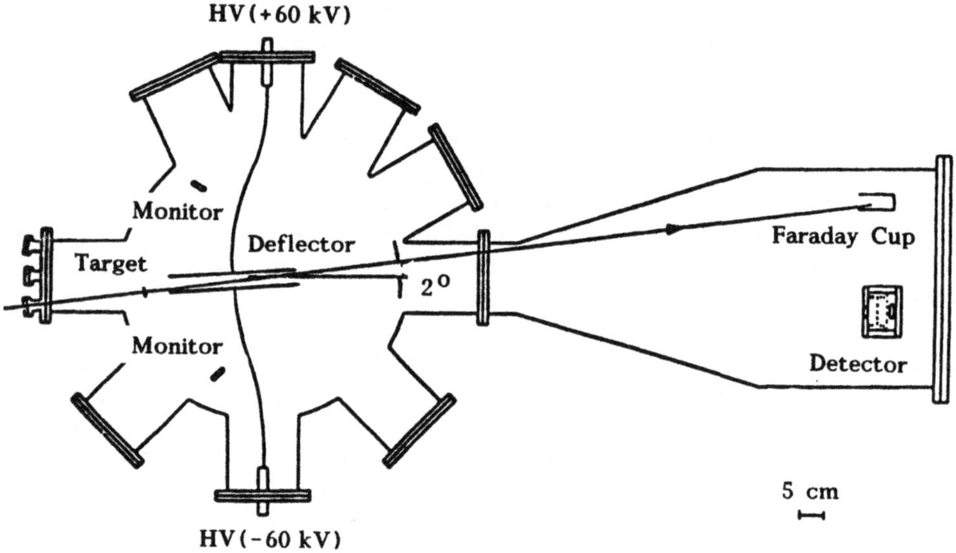

Fig. 1: Experimental setup

Fig. 2 shows reduced fusion cross sections vs. reduced center of mass energy for the reactions $^{24}Mg + {}^{A}Zr$. The normalization was done in order to account for trivial radius dependences of both cross section and barrier, assuming $r \propto A^{1/3}$. The data show a significant enhancement of cross sections below the fusion barrier compared to a one dimensional tunneling calculation (marked "uncoupled" in fig. 2), using the proximity potential of Blocki and Randrup [2]. The enhancement increases with increasing target mass. Difference in experimental fusion cross sections amounts to about two orders of magnitude comparing ^{90}Zr and ^{96}Zr at about 57 MeV.

Coupled- channels calculations were performed within the limits of constant coupling employing the code CCFUS [3]. First, the inelastic excitation of collective states (first 2^+ state in ^{24}Mg, lowest 2^+ and 3^- states in the zirconium isotopes) was taken into account. The coupling strengths were calculated using the relationship given by Broglia and Barranco in [4] with B(E2) values taken from the compilation of [5]. The effect of inelastic excitation is a large increase in cross section (marked "coupled" in fig. 2) which, however, still underestimates the data. In addition, the relative discrepancies between the different target isotopes cannot be explained.

The consideration of the various transfer degrees of freedom is a difficult task and no consistent treatment of this problem exists so far. In order to make a first approach, we only allowed for 2n transfer. Each of the ground- state Q- values concerned (cf. table 1) has been employed additional in CCFUS calculations. Then an effective coupling strengh was fitted, which turned out to be 3 to 5 ± 1 MeV (cf. table 2). The fact that we yield values lying so close together is a strong hint for the dominant influence of neutron pair transfer.

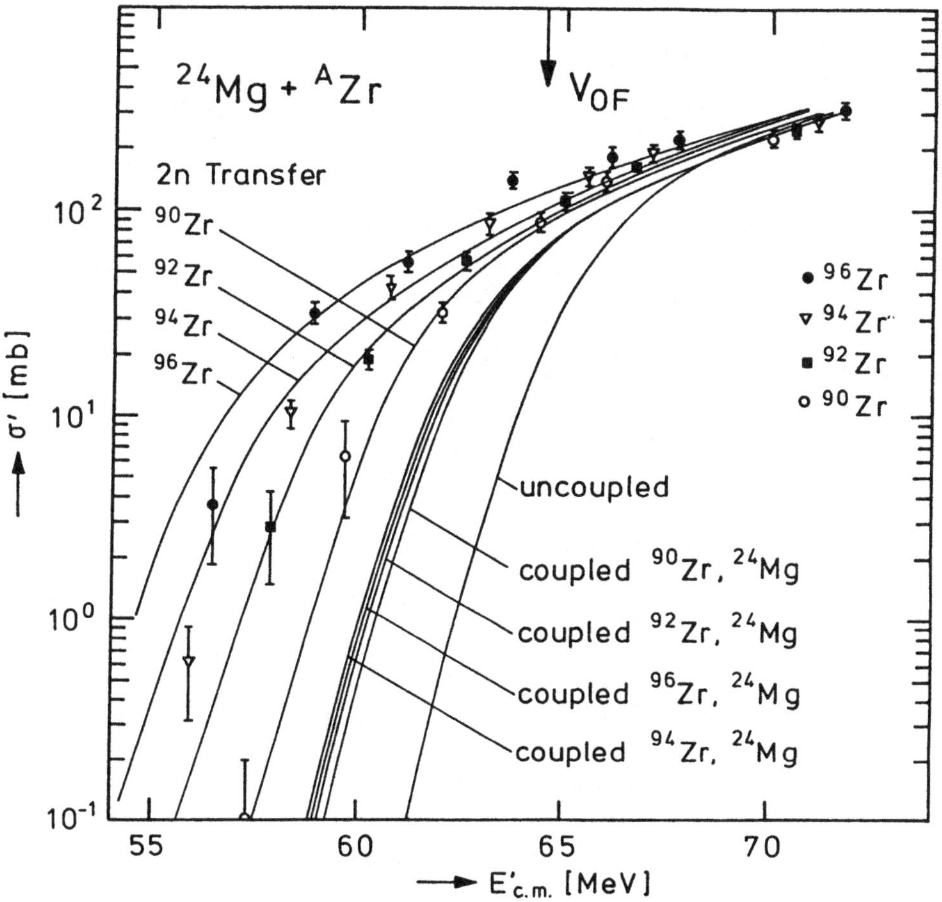

Fig. 2: Reduced fusion cross sections σ' vs. reduced center of mass energy E'_{CM} for the reactions ^{24}Mg + ^{90,92,94,96}Zr. Curves represent calculations:

uncoupled	1-dim tunneling calculation
coupled	coupling of inelastic excitation
2n Transfer	2n-pickup

Fig. 3 shows reduced fusion cross sections vs. reduced center- of- mass energies for the reactions ^{26}Mg + ^{A}Zr. Again, strong enhancement is seen compared to the 1-dim. calculation. Like observed in the ^{24}Mg case, this enhancement increases with increasing target mass. This is, as it was just argued above, exactly the opposite trend one would expect for dominance of 2p transfer. The observed shift for the heaviest isotope ^{96}Zr is, however, not so pronounced as compared to the ^{24}Mg- induced reaction.

Good agreement with the experimental data was achieved by coupling the 2n transfer in the same way as it was done for ^{24}Mg, and the coupling strengths turned out to be 4 ± 1 MeV (cf. table 2). It should be emphasized that the relatively small difference in cross sections observed between ^{90}Zr and ^{96}Zr is reproduced by the calculations in an excellent manner, because it corresponds to a similar shift for the Q-values for 2n transfer.

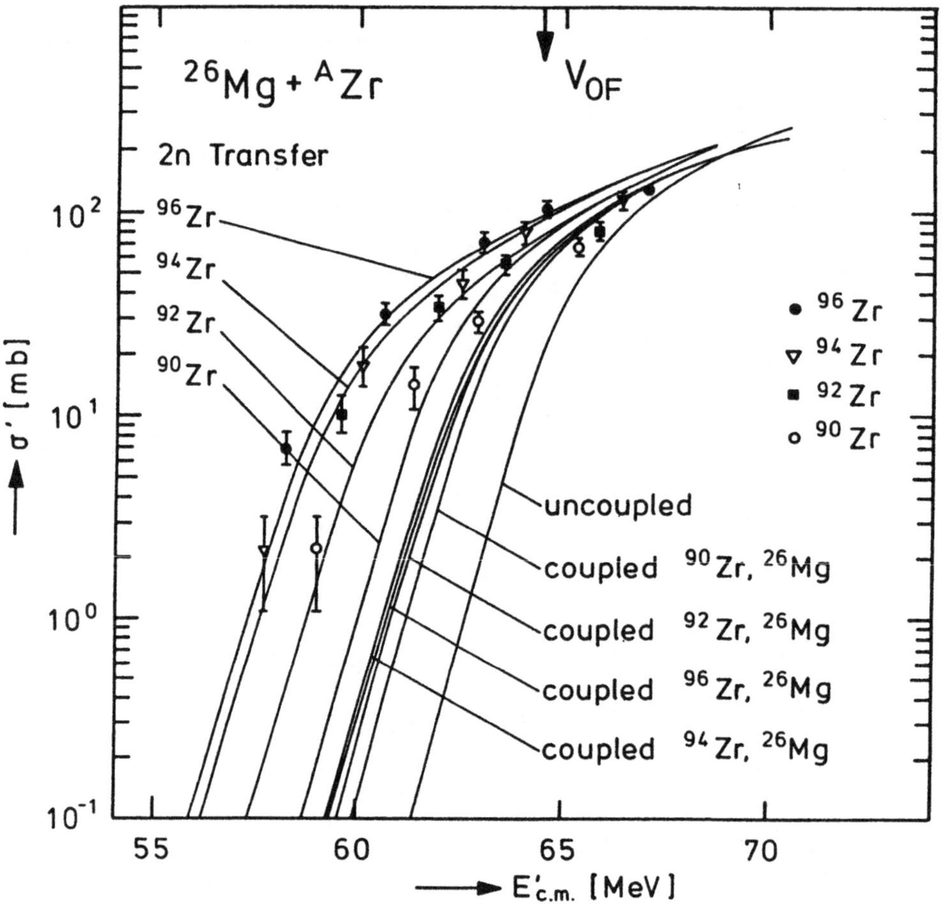

Fig. 3: Same as fig. 2 for ^{26}Mg + ^{A}Zr

To summarize, the results of our experiments give evidence that 2n transfer is the dominant process leading to fusion whereas the proton transfer channels seem to be of minor importance. More experimental information, however, is needed for a deeper understanding of the various degrees of freedom relevant to fusion. In a forthcoming experiment, we will examine transfer cross sections for our systems close to the Coulomb barrier.

Table 2: Coupling strength of the first collective 2^+ and 3^- states and of the 2n pickup

	Inelastic excitation E^* [MeV]	I^π	F_I [MeV]	$(^{24}Mg, ^{26}Mg)$ F_{2n} [MeV]	$(^{24}Mg, ^{26}Mg)$ F_{2n} [MeV]
^{24}Mg	1.37	2^+	3.27		
^{26}Mg	1.81	2^+	2.76		
^{90}Zr	2.19	2^+	0.45	4.0	4.0
	2.75	3^-	0.95		
^{92}Zr	0.93	2^+	0.73	3.0	4.0
	2.34	3^-	1.32		
^{94}Zr	0.92	2^+	0.81	4.0	4.0
	2.06	3^-	1.58		
^{96}Zr	1.74	2^+	0.54	5.0	4.0
	1.89	3^-	1.45		

[1] M. Beckermann, Phys. Reports 129 (1985) 145

[2] J. Blocki, J. Randrup, W.J. Swiatecki and C.F. Tsang, Ann. of Physics 105 (1977) 427

[3] C.H. Dasso, S. Landowne, Comp. Phys. Comm. 46 (1987) 187

[4] R.A. Broglia and F. Barranco, Lecture Notes in Physics, Springer Verlag, Berlin (1984) 147

[5] S. Raman, C.H. Malarkey, W.T. Milner, C.W. Nestor Jr. and P.H. Stelson, Atomic Data and Nuclear Data Tables 36, 1 - 96 (1987)

* Supported by: Bundesministerium für Forschung und Technologie, Federal Republic of Germany

** Permanent address: Institut für Kernphysik, Joh. Gutenberg Universität, 6500 Mainz

Influence of Hexadecapole Deformations of the Nuclear Shape on the Enhancement of Subbarrier Fusion Cross Sections

J. Fernández Niello
Departamento de Física, Comisión Nacional de Energía Atómica
Avda. del Libertador 8250, 1429 Buenos Aires, Argentina

C.H. Dasso
The Niels Bohr Institute, University of Copenhagen
Blegdamsvej 17, DK-2100, Copenhagen Ø, Denmark

In this work we review the arguments which can be advanced to characterize the role of positive and negative hexadecapole shapes in the enhancement of subbarrier fusion cross sections. We also investigate the possibilities that result from combining the hexadecapole distortions with characteristic prolate quadrupole deformations. This survey was done employing the coupled channel code CCDEF [1] which allows for static deformations in both projectile and target. The influence of the deformed nuclear shapes is taken into account within the sudden approximation. In this limit the relevant cross sections can be constructed by averaging over the results of calculations performed for all the possible relative orientations of the projectile and target [2].

We consider a spherical projectile impinging on a deformed target; in this way we isolate the shape effects which are to be attributed to only one of the colliding ions. For the deformed partner we considered only axially symmetric shapes, characterized by nuclear quadrupole and hexadecapole deformation parameters β_2 and β_4. In terms of these, the radius of a deformed system is written as

$$R(\chi) = R_o \left[1 + \beta_2 Y_{20}(\chi) + \beta_4 Y_{40}(\chi)\right] , \tag{1}$$

where χ is an angle measured with respect to the symmetry axis of the deformed nucleus and R_o is given by

$$R_o = 1.233\, A^{1/3} - 0.98\, A^{-1/3} . \tag{2}$$

Fig. 1 shows a cut of the nuclear profile for a prolate quadrupole deformation of $\beta_2 = 0.2$ and hexadecapole deformations of $\beta_4 = +0.1$ and $\beta_4 = -0.1$. At the poles, a positive hexadecapole

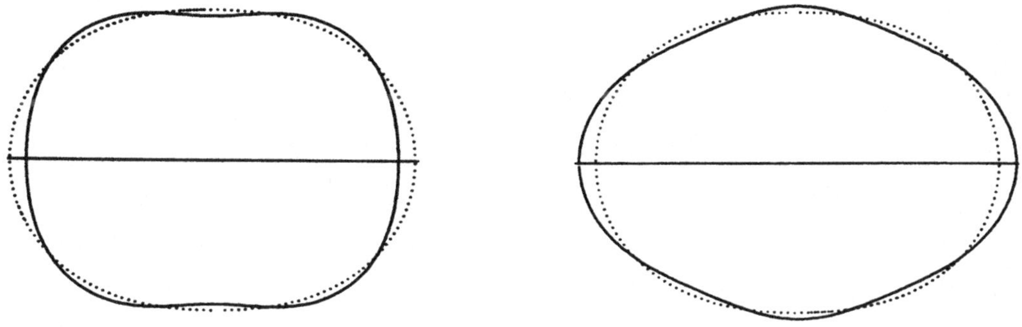

Figure 1: *Cut of the nuclear profile for $\beta_2 = 0.2$ and $\beta_4 = -0.1$ (left) and $\beta_4 = +0.1$ (right). In these plots the dotted curve indicates the case with only prolate deformation, i.e. $\beta_4 = 0$.*

deformation increases the radius respect to the one obtained with $\beta_4 = 0$. Negative hexadecapole deformations, on the other hand, lead to a flattening of the nucleus.

There are two ingredients which govern the outcome of this survey. On the one hand we have the effective modulation of the barrier height associated with a given spatial orientation. This effect can be seen in fig. 2, where the effective barriers are plotted as a function of the angle χ. The modulation of the barrier is essentially given by the shift of the nuclear surfaces measured with respect to a spherical (or prolate-deformed) reference. The other ingredient is the statistical weight of the different orientations, which clearly favor hitting the equator of the deformed nucleus rather than its poles. This is also shown in fig. 2. To represent the combined effect of these factors we use the quantity

$$\mathcal{F}(\chi) = \sin\chi \, \exp\left(\frac{E - \tilde{V}_B(\chi)}{\epsilon(\chi)}\right)\Bigg|_{\beta_2,\beta_4} - \sin\chi \, \exp\left(\frac{E - \tilde{V}_B(\chi)}{\epsilon(\chi)}\right)\Bigg|_{\beta_2,\beta_4=0} , \tag{3}$$

where $\tilde{V}_B(\chi)$ is the height of the effective barrier, E is any bombarding energy satisfying $E \ll \tilde{V}_B$ and $\epsilon(\chi)$ measures the thickness of the barrier at that particular orientation.

In fig. 3 we display the quantity $\mathcal{F}$ as a function of the orientation angle χ for the same cases presented in fig. 2. Fig. 3 shows that the range of latitudes that enhance the fusion in the case of positive hexadecapole deformations is about 20° wide. This identifies the pole-orientation of the deformed nuclei as being indeed the most favourable configuration to increase the fusion rates. For negative deformations the enhancements arise from relative orientations around 45° and they are not very large in spite of the larger statistical weights. Moreover, these effects are almost cancelled by the lost of fusion cross section resulting from the flattening of the nucleus at the poles. From this analysis one concludes that positive hexadecapole deformations should always contribute to further enhancements of the fusion cross sections. In this statement

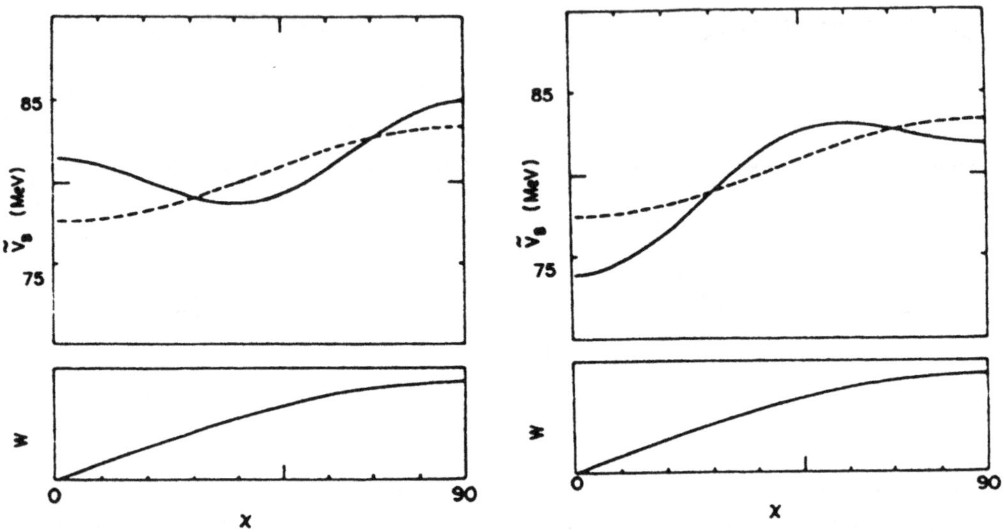

Figure 2: *Effective barriers $\tilde{V}_B$ and statistical weights W as a function of the angle χ between the symmetry axis of the deformed target and the vector $\vec{r}$ connecting the centers of the two ions. The plots on the left half of the figure correspond to a negative hexadecapole deformation $\beta_4 = -0.1$, while the right half corresponds to a positive deformation of the same magnitude, $\beta_4 = +0.1$, for the system $^{16}O+^{232}Th$. In both cases a quadrupole deformation of $\beta_2 = 0.2$ was considered. The dashed curve indicates the reference, obtained for $\beta_4 = 0$.*

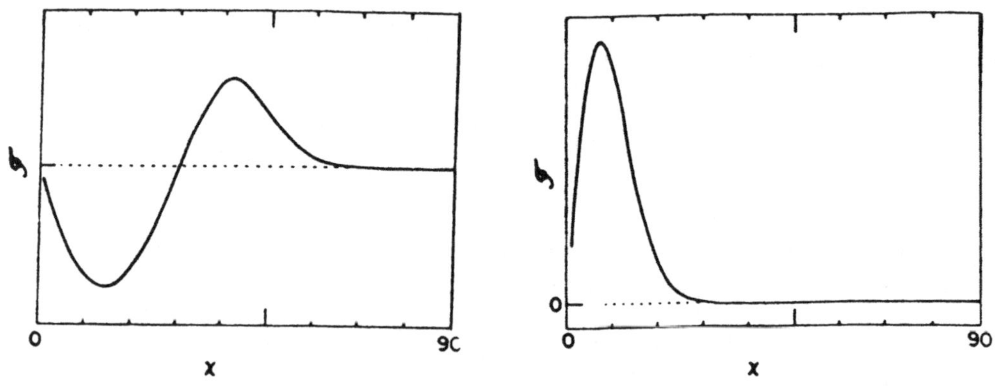

Figure 3: *Indicator $\mathcal{F}$ as a function of the angle χ for the same cases presented in fig. 2. The left half of the figure corresponds to a negative hexadecapole deformation $\beta_4 = -0.1$ and the right half to a positive deformation $\beta_4 = +0.1$.*

it is understood that one speaks about increases which are measured with respect to the ones already generated by the prolate nucleus. From these qualitative arguments on may expect, on the other hand, situations in which the addition of a negative hexadecapole degree of freedom may reduce the enhancement of the fusion cross section.

We would like to illustrate the range of possibilities that arises from the combination of the different quadrupole and hexadecapole deformations. To this end we define the enhancement $\mathcal{E}$ as the ratio of the cross sections expected for a nucleus with a given deformation β_2, β_4 respect to the cross section expected for a spherical nucleus,

$$\mathcal{E} = \frac{\sigma(E \ll \tilde{V}_B)_{\beta_2,\beta_4}}{\sigma(E \ll \tilde{V}_B)_{\beta_2=0,\beta_4=0}} \,. \tag{4}$$

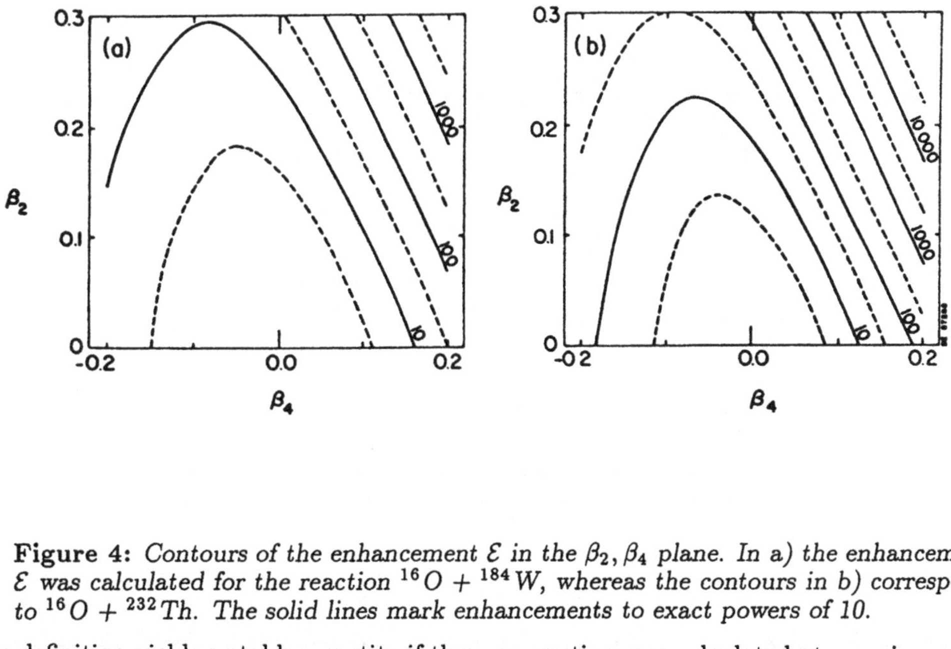

Figure 4: *Contours of the enhancement $\mathcal{E}$ in the β_2, β_4 plane. In a) the enhancement $\mathcal{E}$ was calculated for the reaction ${}^{16}O + {}^{184}W$, whereas the contours in b) correspond to ${}^{16}O + {}^{232}Th$. The solid lines mark enhancements to exact powers of 10.*

This definition yields a stable quantity if the cross sections are calculated at energies well below the lowest effective barrier $\tilde{V}_B$. In fig. 4 we present contours of the enhancement for two systems with the same atomic numbers as in the systems ^{16}O + ^{184}W, ^{16}O + ^{232}Th studied in refs. [3,4]. These publications stressed the importance of hexadecapole deformations but differed in the emphasis they placed on the sign of the β_4 parameter. The range of the parameters β_2, β_4 used for this illustration has been chosen to cover the typical situations encountered in heavy ions.

The real values of the enhancement depend, of course, on the size of the nuclear systems considered. However, the two plots in fig. 4 are qualitatively very similar. In both situations the minimum of the enhancements for a given value of β_2 follows a ridge that *does not* coincide with the condition $\beta_4 = 0$. Moreover, the enhancements obtained with a negative value of β_4 are larger than the corresponding ones for $\beta_4 = 0$ only for large values of the hexadecapole deformation. This situation can be better appreciated if one cuts the enhancement surface at a constant value of β_2. In fig. 5 we show such a cut for the system $^{16}O + ^{232}Th$ and corresponding to a value of $\beta_2 = 0.2$. The enhancements are normalized to the value obtained for $\beta_4 = 0$. This curve confirms previous considerations, as it shows that small negative deformations do indeed decrease the fusion cross sections. For positive β_4 the fusion rates are, on the other hand, always improved.

Systematic measurements in reactions involving nuclei with hexadecapole deformations are needed to assess the role of this shape degree of freedom. Suitable candidates targets can be found in the rare earth region, where β_4-parameters conveniently range from positive to negative values.

This work was performed in part at the Sektion Physik der Universtität München. It was supported by a grant (06 LM 177 II) from the Bundesministerium für Forschung und Technology, Federal Republic of Germany.

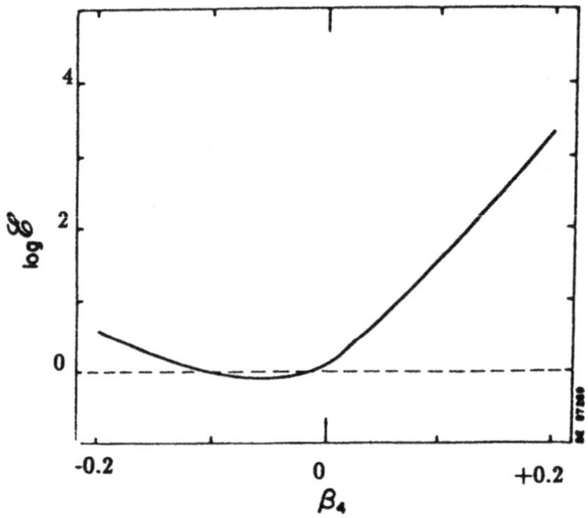

Figure 5: *Enhancement $\mathcal{E}$ as a function of β_4 for the reaction ^{16}O+^{232}Th for a fixed value of $\beta_2 = 0.2$. The values are normalized to the enhancement obtained for $\beta_4 = 0$*

References:

1.- J. Fernández-Niello, C.H. Dasso and S. Landowne, Comp. Phys. Comm. in press
2.- C.Y. Wong, Phys. Rev. Lett. **31** (1973) 766
3.- M.J.Rhoades-Brown and V.E. Oberacker, Phys. Rev. Lett. **50** (1983) 1435
4.- H. Esbensen and S. Landowne, Nucl. Phys. **A467** (1987) 136

SUBBARRIER FUSION IN THE SURFACE FRICTION MODEL

P. Fröbrich
Hahn-Meitner-Institut, Berlin-West, Germany
and
Freie Universität Berlin, Fachbereich Physik (WE3)

1. Motivation

Motivated by the fact that coupled-channel calculations [1], which include as many channels as technically possible, fail to describe subbarrier fusion for heavier systems quantitatively we investigate the problem in a more global way by using the concept of friction also at subbarrier energies. This is done in the framework of the surface friction model [2] which describes the global trends of fusion and capture excitation functions above the barrier throughout the periodic table. Moreover the barrier systematics for heavier systems has been reproduced superior to the extra-push parametrization [4]. More recent examples are shown in fig. 1, where the excess energy over the Bass barrier has been calculated within the surface friction model. A comparison is made to the dynamically calculated excess energies of Blocki et al. [5] and to recent experimental data [6]. The average trend of the measured barriers is reproduced.

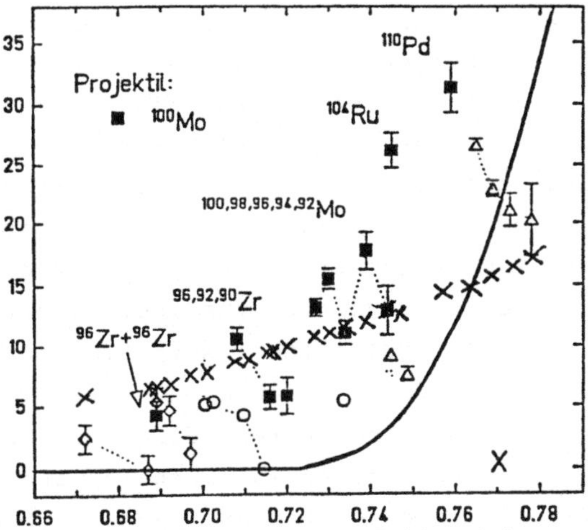

Fig. 1: Data [6] and dynamical extra-push calculations [5, solid line] are compared with results of the surface friction model [crosses]. The excess energy over the Bass barrier is plotted versus the fissility parameter.

However, the scattering of the data around the predictions of the surface friction model indicates clearly the need for taking into account more detailed nuclear structure effects than are contained in the present macroscopic friction models.

We attack the problem of using friction models for subbarrier fusion in two different ways, each involving a number of approximations. The main idea of the first one is to follow frictional trajectories until the barrier and then tunnel through the barrier with the local energy and angular momentum neglecting dissipation under the barrier. The second approach is a purely classical one which is based on the fact that because of statistical fluctuations one is dealing with distributions of trajectories such that also at subbarrier energies a part of the distribution function can surmount the barrier and contributes to the subbarrier fusion cross section. Of course one would like to overcome the shortcomings of these approaches in a proper theory of dissipative tunneling which not yet exists. Because our model is of global nature one cannot expect to describe situations where selective channels are of importance. But general trends should be describable. For another global approach see ref. 7, where a fusion absorptive potential is introduced.

2. Frictional trajectories and subsequent tunneling [8]

Using the full surface friction Hamiltonian would lead to a too complicated multi-dimensional tunneling problem. In order to deal with a manageable effective one-dimensional problem we introduce the following simplifications. We construct an adiabatic potential by minimizing the linearized potential energy of the Hamiltonian with respect to the deformation parameters for each radial position. As improvement over the global model [2] we take into account nuclear structure effects in using instead of the liquid drop model experimental energies and BEλ values to determine mass and stiffness parameters of the surface modes. Further we use in constructing the folding potential mass densities from Hartree Fock calculations [9] instead of a global parametrization. A one-dimensional tunneling calculation with this adiabatic potential then yields results above the data. Friction is taken into account in the following way. We follow trajectories including frictional forces until the barrier and tunnel then with the local energy and angular momentum through the barrier, neglecting dissipation below the barrier. The inclusion of friction leads to a reduction of the cross section as compared to the tunneling through the adiabatic barrier. This is because due to friction a trajectory loses energy and reaches the barrier at a lower than the incoming energy. Because the barrier is thicker at this point as in the no friction limit a corresponding tunneling with the local energy reduces the probabilty for fusion. One should note that the approach above treats fusion from the adiabatic side where additional degrees of freedom (the friction) reduce the fusion probabilities. This is conceptually different from the usual approach where one starts with a sudden potential and additional degrees of freedom lead to a subbarrier enhancement of the fusion cross section. In figs. 2 and 3 we show a few results of the calculations. The $^{58}Ni+^{64}Ni$ system, where 2n-transfer channels are supposed to be of importance, is not described in detail, whereas the experimentally observed isotopic behaviour for the Ca systems is reproduced.

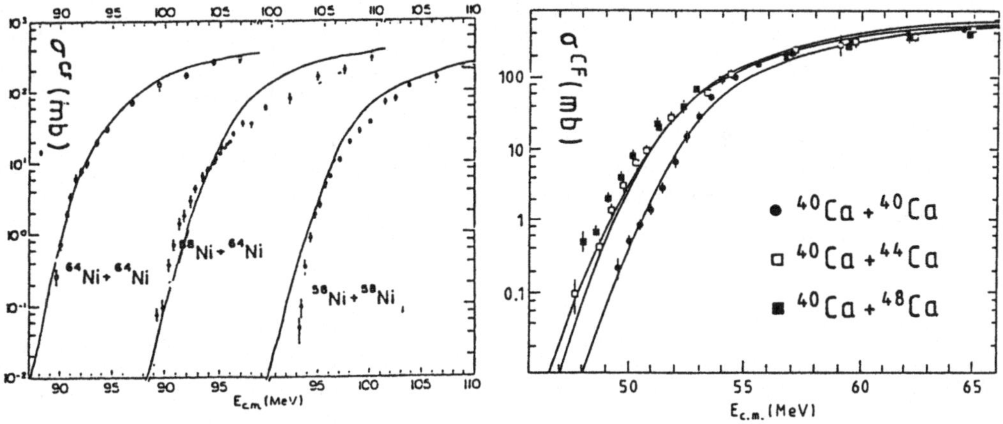

Fig. 2 + 3: Data [ref. 16] are compared to the results of the adiabatic surface friction tunneling model (solid lines).

3. Langevin treatment of subbarrier fusion

It has been shown that a proper treatment of statistical fluctuations within a Langevin approach [10] instead of solving a Fokker-Planck equation by a moment expansion not only leads to an improved description of deep-inelastic collisions but also to a smearing out of the triangular spin distributions of the sharp cut-off model and has consequences for the fusion excitation function above the barrier. Moreover, the statistical fluctuations yield a finite fusion probability at subbarrier energies. This is due to the spreading of the classical distribution function (obtained from solving the Langevin equations) which for a given impact parameter also at subbarrier energies splits into two parts, one that surmounts the barrier giving the probability for fusion and one that is reflected from the barrier representing the probability for deep-inelastic scattering. We calculate the subbarrier fusion completely classically by using the same equations of motion as in the preceding section adding to them random forces whose strengths are determined via the fluctuation-dissipation theorem. Then we obtain the fusion cross sections by sampling fusing trajectories. (For more details of the method see ref. 10).

In fig. 4 we show the result of a Langevin calculation and of the standard surface friction model [2] (without fluctuations) as compared to the data of ref. 11. One obtains similar agreement as with the coupled-channel calculations of this reference. Fig. 5 shows the corresponding spin distributions at a subbarrier energy and for an energy above the barrier.

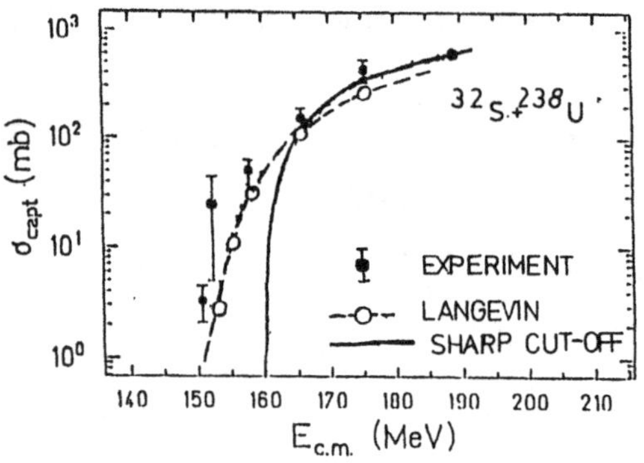

Fig. 4: Data [11] are compared to a Langevin calculation (---o---) and to the standard sharp-cut off surface friction model.

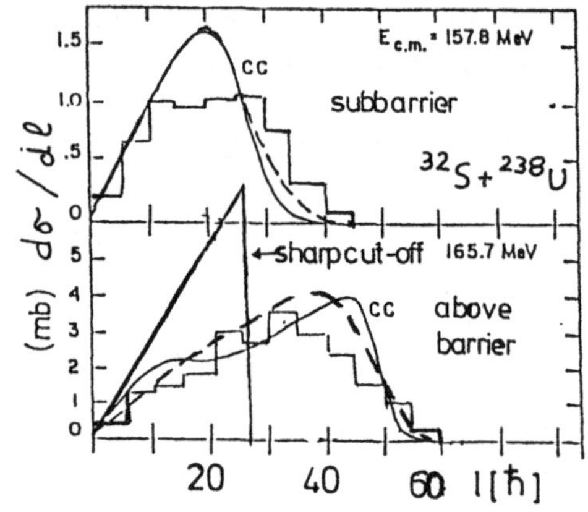

Fig. 5: Spin distributions are compared for coupled channel [11, solid lines] Langevin (histograms) and the model of the last section (dashed lines).

Analysing fusion-fission data one finds for a number of systems [12] at energies around the barrier too large $\langle J^2\rangle$ values which are not compatible with the usual analysis within transition state theory. Because our calculations yield broad spin distributions we investigated these systems. As example the $^{16}O+^{232}Th$ system is shown in fig. 6 and 7.

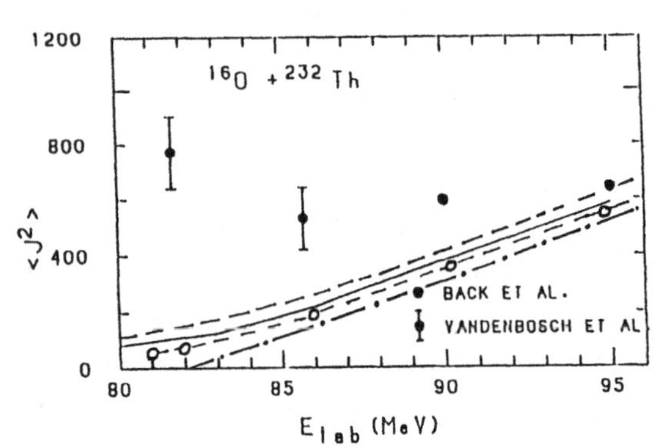

Fig. 6: Experimental [12] $\langle J^2\rangle$ values are compared to coupled-channel [13] (solid line), a Langevin calculation (-o-o-), adiabatic tunneling (-----) and sharp cut-off (-.-.-) surface friction model.

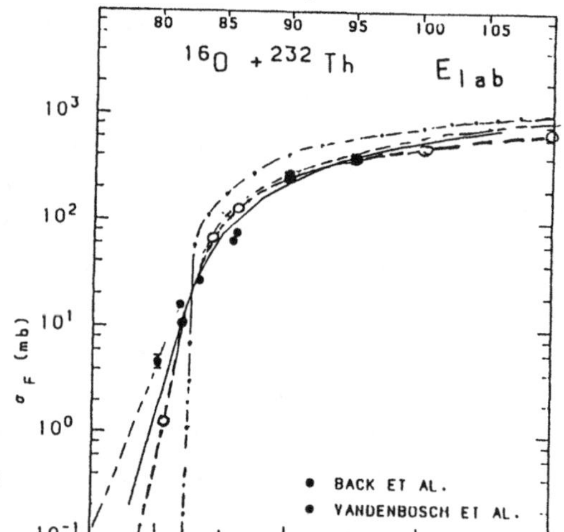

Fig. 7: The corresponding fusion cross sections are shown for $^{16}O+^{232}Th$.

The enhanced $\langle J^2 \rangle$ values at energies around the barrier cannot be explained in any of these models, which give reasonable fusion excitation functions. This we consider as a hint that the formation process seems not to be responsible for the observed large $\langle J^2 \rangle$ values but that the dynamics of the fission exit channel behaves different from the assumptions of transition state theory.

Finally we investigate the influence of thermal fluctuations on fusion probabilities $T_{\ell=0}(E)$, which have been measured for heavy systems [14]. The fusion probability might be a more sensitive quantity to check fusion models than the fusion excitation function. The inclusion of statistical fluctuations, calculated with the usual Einstein relation, yields a somewhat too strong (as compared to the data) smearing out of the sharp cut-off result of the full surface friction model [2]. (see fig. 8). It is observed that an artificial reduction of the strength of the fluctuating force by a factor of two gives results consistent with the data. Such a reduction, however, leads to difficulties in explaining the widths of cross sections for deep-inelastic collisions. In any case, the statistical (thermal) fluctuations seem to play a major role for an explanation of the measured broad energy distribution of $T_{\ell=0}(E)$. However one should not forget that additional quantal effects exist.

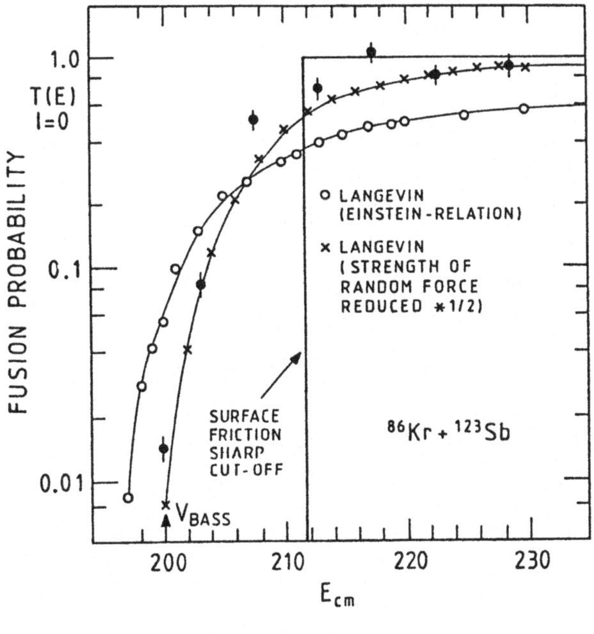

Fig. 8: A measured [14] fusion probability $T_\ell(E)$ for $^{86}Kr+^{123}Sb$ is compared to a Langevin calculation (-o-) with the standard Einstein relation and with a calculation where the strength of the random force is reduced by a factor of 2 (x). Also shown is the result of the sharp cut off surface friction result [2] which shows an extra push of 11.8 MeV over the Bass barrier.

4. Conclusions

Although the models discussed above show some appealing features one has to be aware of the fact that they involve several approximations. Up to now our attempts to unify the approaches above and to overcome their shortcomings (the neglect of dissipation under the barrier in one case and the neglect of tunneling in the other case) have not been satisfactory. Trying to use the path integral influence functional formalism [15] for imaginary times leads to equations which up to now we were not able to solve except in simple limits.

References

1. M.J. Rhoades-Brown et al., Lecture Notes in Phys. 219 (1985) 162
2. P. Fröbrich, Phys. Rep. C 116 (1984) 337
3. P. Fröbrich, Proc. Hirschegg 1985, p. 53 and
   Proc. Tsukuba 1984, World Scientific, Singapore, p. 386
4. W.J. Swiatecki, Nucl. Phys. A 376 (1982) 51
5. J.P. Blocki et al., Nucl. Phys. A 459 (1986) 145
6. B. Quint, private communication
7. T. Udagawa et al., Phys. Rev. C 32 (1985) 124 and 1435
8. P. Fröbrich, K. Möhring, Annual Report 1985, Hahn-Meitner-Institut, p. 25
   P. Fröbrich, Invited Lectures, Mikolajki 1987, to be published in the Gordon and Breach Conference Series
9. J. Friedrich, P.G. Reinhard, Phys. Rev. C 33 (1986) 335
10. P. Fröbrich, S.Y. Xu, Nucl. Phys. A 477 (1988) 143
    P. Fröbrich, 3rd Int. Conf. on Nucleus Nucleus Collisions, Saint-Malo, June 1988, Contributed papers, p. 124
11. R. Freifelder et al., Phys. Rev. C 35 (1987) 2097
12. R. Vandenbosch et al., Phys. Rev. Lett. 56 (1986) 1234
    H. Rossner et al., this symposium
13. O. Tanimura, Z. Physik A 327 (1987) 413
14. C.C. Sahm et al., Nucl. Phys. A 441 (1985) 316
15. K. Möhring, U. Smilansky,Nucl. Phys. A 338 (1980) 227
    A.O. Caldeira, A.J. Leggett, Ann. of Phys. 149 (1983) 374
16. M. Beckerman, Phys. Rep. C 129 (1985) 145

SPECTROSCOPIC STUDY OF SUB-BARRIER QUASI-ELASTIC NUCLEAR REACTIONS

C. N. Pass, P. M. Evans, A. E. Smith, L. Stuttgé†
Department of Nuclear Physics, Oxford University,
Oxford OX1 3RH. U.K.

R. R. Betts
Physics Division, Argonne National Laboratory,
Argonne, Illinois 60439. U.S.A.

J. S. Lilley, K.A. Connell, J. Simpson, J. R. Smith
Science and Engineering Research Council,
Daresbury Laboratory, Warrington WA4 4AD. U.K.

A. N. James, P. A. Butler
Physics Department, University of Liverpool,
Liverpool L69 3BX. U.K.

B. R. Fulton
Physics Department, University of Birmingham,
Birmingham B15 2TT. U.K.

Recent studies have revealed anomalous behaviour in the low energy reactions of colliding heavy nuclei [1]. Examples of intriguing phenomena have appeared in the fusion and elastic scattering of heavy nuclei occuring at energies approaching their classical interaction barrier. Fusion, traditionally understood to occur through the single channel tunnelling of the imposed potential barrier, is found to be orders of magnitude stronger than expected. Elastic scattering measurements, conventionally described by folding model prescriptions of the nucleus-nucleus interaction, indicate the real part of the collision potential to be considerably deeper than such simple treatments would suggest. It would thus appear that toward lower energies the reaction processes become more sensitive to the details of the collision mechanism and hence the structure of the partners begins to play a dominant role. This realization indicates the potential importance of surface degrees of freedom of the colliding nuclei and it is these elements of the inclusive reaction mechanism which must be addressed to develop a coherent description of the collision process.

One may expect the character of the surface degrees of freedom to be reflected in the quasi-elastic reaction channels. For the purpose of experiment these may be divided into three classes; inelastic excitation, nucleon transfer and the combination of these two. At energies below the Coulomb barrier inelastic excitation is expected to be well described by pure electromagnetic interaction and is thus calculable. However, nucleon transfer and its interplay with inelastic excitation is less well determined and the physics of both must be explored by experiment.

Such studies are faced with considerable experimental difficulties which explain the lack of data currently available [2]. At energies near and below the Coulomb barrier transfer reactions are controlled by the classical distance of closest approach of the collision partners. As a consequence the reactions are strongly backward peaked with the ejectile recoiling with a small fraction of the incident collision energy. Positive identification of the fragment is, therefore, no longer possible. However, associated with the low energy ejectile is a heavy, energetic target–like recoil which, providing it can be separated from the beam and elastic flux, may be used to analyze the reaction process [3]. Such a separation is achieved using a recoil mass separator which is capable of efficiently rejecting beam and unwanted products so as to isolate the complementary recoil nucleus. Studies of the inclusive neutron transfer strength have been made using this technique [4] and suggest the process to be strongly influenced by the structure of the collision partners [5].

Information concerning the detailed mechanism of nucleon transfer may be obtained through a spectroscopic study of the the recoil product. However, once again, conventional techniques are no longer applicable. Charged particle spectroscopy, successful in analyzing higher energy reactions induced by light projectiles, is not suited to systems in which the nuclei are heavy. At energies close to the Coulomb barrier transfer angular distributions have little structure. Furthermore, the considerable straggling associated with the passage of a heavy fragment through target and window material precludes the the resolution of close–lying residual nuclear states. Alternatively, one may study the reaction spectroscopy through the γ–radiation associated with the decay of the residual nuclei. In this manner the detailed distribution of the inclusive reaction strength across the spectrum of available nuclear states may be determined and the reaction mechanism subsequently studied.

The arrangement used in these studies comprised the recoil mass separator (RMS) [6] at Daresbury Laboratory in conjunction with the POLYTESSA array of sixteen Compton suppressed intrinsic germanium γ–ray detectors [7]. The unit mass resolution of the RMS was used to identify the reaction channel of interest and prompt γ–radiation, detected in delayed coincidence with the recoiling nucleus, provided spectroscopic information. Of the many collision systems which may be studied with this technique we here principally address a single example: ^{124}Sn(^{58}Ni,^{59}Ni)^{123}Sn at 220 MeV. In this system the target is a spherical–vibrational nucleus and one would expect the transfer to follow a relatively simple mechanism. A mass spectrum of the three groups accepted by the separator focal plane is presented in Figure 1 and illustrates the clean separation of elastic and inelastic (124), neutron pickup (123) and two neutron pickup (122) reaction channels. Spectra of γ–rays coincident with the recoil products are presented in Figure 2. In this figure the upper ((a), (b)) and lower ((c), (d)) quadrants respectively show γ-rays coincident with recoil products of all masses observed in the focal plane and with the mass 123 fragments alone. The left hand section of the figure ((a), (c)) comprises spectra corrected for the Doppler shift induced by the motion of the target-like recoil and, similarly, the right hand spectra are corrected for the motion of the ejectile fragment. Lines corresponding to inelastic excitation and single neutron transfer are clearly evident, rising cleanly out of the background.

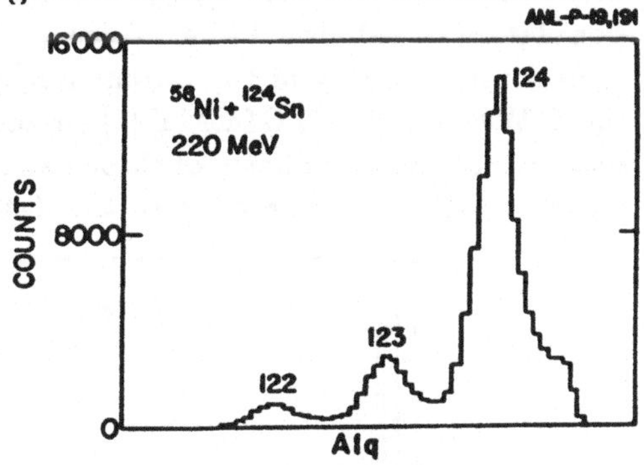

Figure 1

Focal plane mass spectrum of the quasi-elastic recoil products corresponding to the reaction $^{124}Sn(^{58}Ni,\ ^{58,59,60}Ni)^{124,123,122}Sn$ at 220 MeV

Yields extracted from these spectra have been corrected for both the energy dependent efficiency of the γ-ray spectrometers and feeding between connected individual levels to obtain estimates of the relative population of residual nuclear states following transfer.

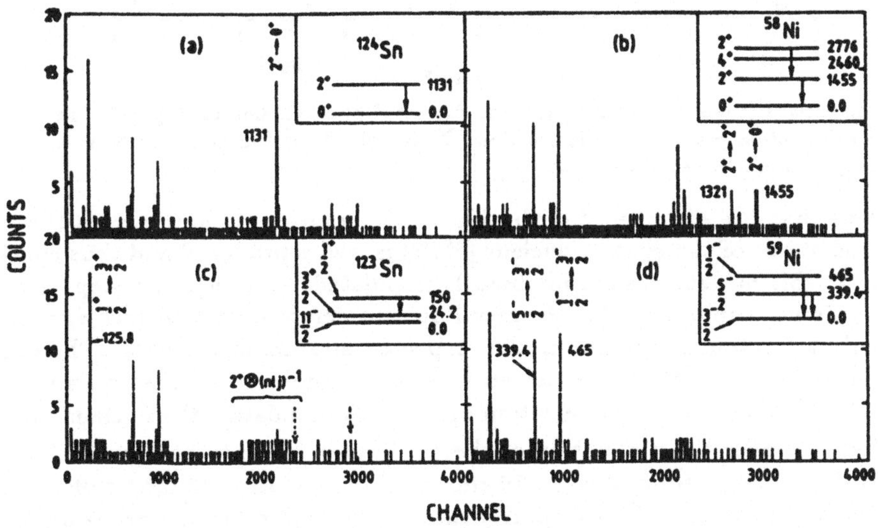

Figure 2

Spectra of total γ-rays detected in delayed coincidence with the recoil nucleus. Figures (a) and (b) present spectra gated by masses all recoil masses observed in the focal plane with (c) and (d) gated by the neutron pickup channel (123) alone. The left hand quadrants are corrected for the Doppler shift induced by the motion of the recoil and similarly those on the right for that of the ejectile nucleus. The dashed arrows indicate the expected position of unobserved lines as discussed in the text.

These estimated yields have then been scaled relative to the calculated Coulomb excitation of the first 2^+ level in the ^{124}Sn target nucleus to obtain cross sections for the population of individual final states and these are presented as solid bars as a function of excitation energy of the parent nucleus in Figure 3. Calculations performed using the DWBA code PTOLEMY [8] successfully reproduced the observed inclusive cross section and the breakdown of these calculations for the direct population of component nuclear states is compared with the data as hollow bars in Figure 3.

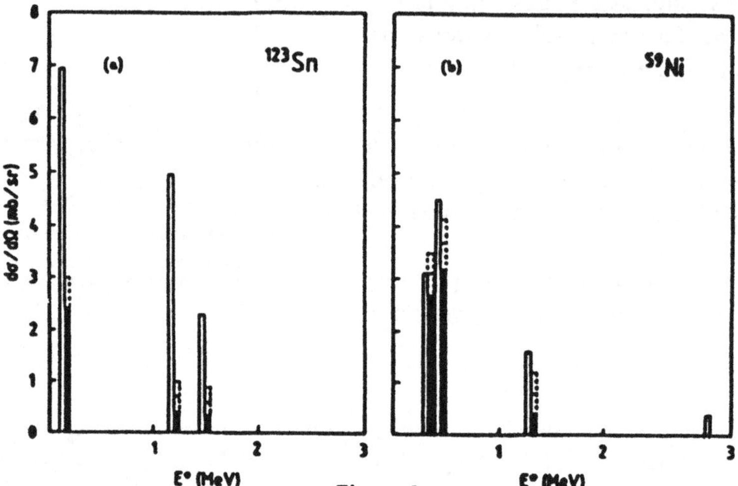

Figure 3

Comparison of observed (solid bars) and predicted direct (hollow bars) population of the residual products of the reaction $^{124}Sn(^{58}Ni,\ ^{59}Ni)^{123}Sn$ at 220 MeV as a function of nuclear excitation energy.

We find that the observed distribution of the transfer strength across the spectrum of final states of the ejectile nucleus (^{59}Ni) is well reproduced and thus conclude that states in this nucleus are predominantly populated via direct, one–step, transfer. However, a similar comparison for the recoil nucleus demonstrates no such agreement. We find only a fraction of the population expected appears in the data. The dashed lines in Figure 2(c) indicate the expected location of unobserved lines and an examination of Figure 3(a) suggests a depletion of strength populating the nucleus directly. Indeed, if we consider the population of the $\frac{1}{2}^{+}_{1}$ level in ^{123}Sn to be representative of the overall discrepancy, we find some 14 mb/sr of the inclusive 22 mb/sr of the 180° cross section to be missing. This disparity may be accounted for by a re–examination of Figure 2(a) in which a clump of strength is indicated at around 1 MeV. This may potentially correspond to the population of unresolved hole states comprising the weak coupling multiplet built on the single phonon nuclear core excitation. Indeed, following efficiency correction, this yield corresponds to some 10 mb/sr which may suitably account for the remaining fraction of the total inclusive reaction cross section.

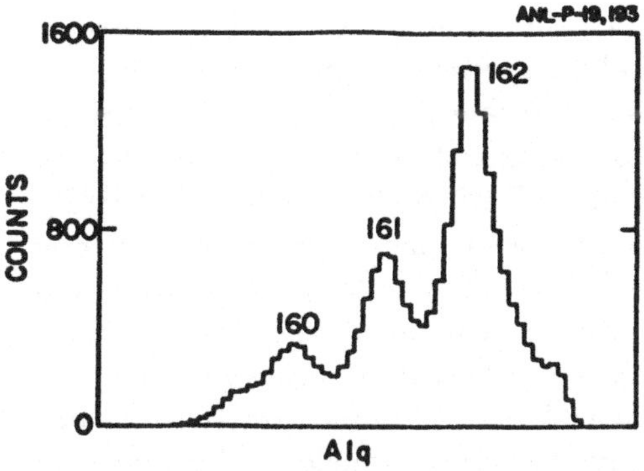

Figure 4

Focal plane mass spectrum of the quasi–elastic recoil products corresponding to the reaction $^{162}Dy(^{58}Ni, ^{58,59,60}Ni)^{162,161,160}Dy$ at 275 MeV

Despite the limited statistics upon which these measurements are based, the data clearly suggest that as much as half of the neutron transfer is occuring via vibrational core excitation. It is interesting to note that the probability of Coulomb excitation of the 2_1^+ level of the target is only around 10%. In view of the multistep transfer being comparable in strength to that proceeding via a direct mechanism this would suggest a considerably higher probability of neutron transfer from the excited core.

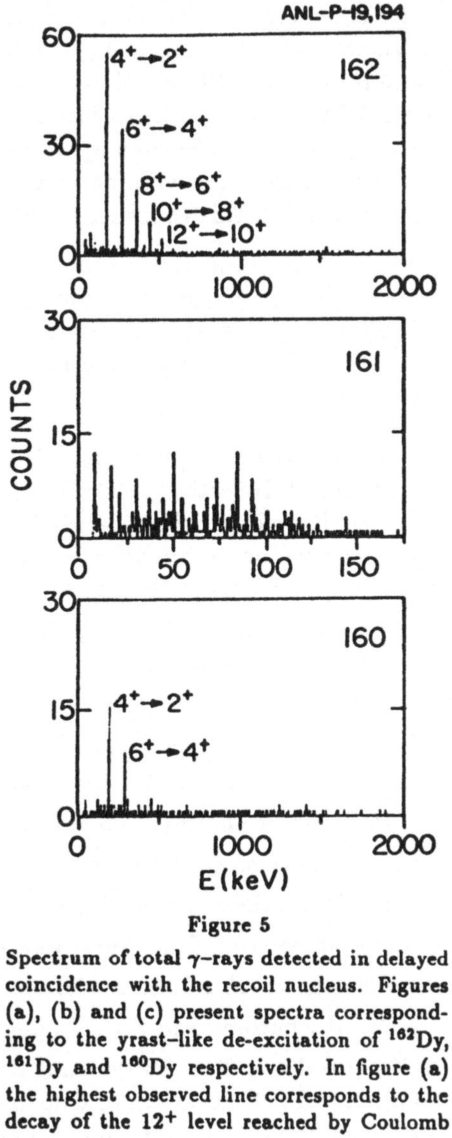

Figure 5

Spectrum of total γ–rays detected in delayed coincidence with the recoil nucleus. Figures (a), (b) and (c) present spectra corresponding to the yrast–like de-excitation of ^{162}Dy, ^{161}Dy and ^{160}Dy respectively. In figure (a) the highest observed line corresponds to the decay of the 12^+ level reached by Coulomb excitation.

In summary, the technique developed in this paper is particularly well suited to the detailed spectroscopic study of low energy quasi–elastic nuclear reactions and, by overcoming the limitations of conventional procedure, the prospect of detailed studies of inclusive reaction mechanism may be realised. With only limited statistics we find evidence for strong multistep character in the transfer of a single nucleon from spherical vibrational target to spherical projectile nuclei. The suggestive measurements reported here may be made definitive through extended runs based on this technique and experiments planned for the future offer the real prospect of developing a quantified interpretation of the reaction process.

The technique discussed in this paper

may be extended to study rapidly rotating cold nuclei [9]. A heavy nuclear projectile incident on a collectively deformed nucleus at energies approaching the Coulomb barrier will quasi–elastically populate near yrast states to moderately high spin in the residual recoil nucleus and the opportunity is presented to examine the influence of rotation on its structure. Preliminary results for the reactions ^{162}Dy(^{58}Ni,^{58,59,60}Ni)^{162,161,160}Dy at 275 MeV are presented in Figures 4 and 5. The clean mass separation of strong transfer channels is observed and lines corresponding to the decay of levels populated through Coulomb excitation (a), single (b) and two (c) neutron transfer may be studied simultaneously. As only five hours of experimental beam time would indicate, the prospect of good, quantifiable data is realistic. Already, with limited statistics, we observe the population of states up to spins of 14$\hbar$ in the residual nuclei and a tentative examination of the spectrum corresponding to the decay of ^{161}Dy reveals new aspects of the structure of this complex nucleus.

With many laboratories around the world presently developing recoil separators and sophisticated arrays of γ–ray spectrometers, the opportunity will soon become widely available to study in detail not only low energy nuclear reaction mechanism but also the structure of the exotic nuclei produced.

We acknowledge support from NATO Collaborative Research Grant No.0076/87. Work at the Universities of Oxford, Liverpool and Birmingham was supported by the United Kingdom Science and Engineering Research Council. Work at Argonne National Laboratory was supported by the U.S. Department of Energy, Nuclear Physics Division, under contract No. W–31–109–Eng–38.

† Present address: Centre de Recherches Nucléaires, Strasbourg, France

REFERENCES

[1] M. Beckerman, Phys. Rep. **C129**, 145 (1985) and refs. therein

[2] See, for example, J. Wiggins *et al.* Phys. Rev. **C31**, 1315 (1985)

[3] R.R. Betts *et al.* Phys. Rev. Letts. **59**, 978 (1988)

[4] C.N. Pass *et al.* (submitted to Nuc. Phys.)

[5] R.R. Betts (these proceedings)

[6] A.N. James *et al.* Nuc. Inst. and Meths. **A267**, 144 (1988)

[7] P.J. Nolan *et al.* Meth. **A236**, 95 (1985)

[8] M.H. MacFarlane and S.C. Pieper, Argonne rep. no. ANL–76–11. Unpublished

[9] C. Lauterbach, P.A. Butler and H.J. Wollersheim, in **Proc. of the Niels Bohr Centennial Conference**, Copenhagen 1985 eds. R.A. Broglia, G. Hagemann, B. Herskind (North Holland, Amsterdam, 1985) p. 405

FUSION OF ^{14}N, $^{16}O+^{59}Co$ AT NEAR BARRIER ENERGIES

T.J.P.Penna, P.R.S.Gomes, E.F.Chagas
Departamento de Física - Universidade Federal Fluminense
Outeiro de São João Batista s/n - Niterói, RJ - 24020 - Brasil

R.Liguori Neto, J.C.Acquadro, E.Crema, N.Carlim F., M.M.Coimbra
Instituto de Física - Universidade de São Paulo
C.P. 20516 - S.Paulo - SP - Brasil

ABSTRACT

Fusion and transfer cross sections have been measured for the ^{14}N and ^{16}O + ^{59}Co systems at near and sub-barrier energies. There is no larger enhancement of the sub-barrier fusion for the system, which has a large transfer cross section, than for the system with no measurable transfer cross section. The value of the effective deformation parameter of the target which fit the fusion excitation functions is 20% higher than the predicted one. The fusion enhancement can be explained by the CCFUS code through the coupling of transfer channels, but there is a need for a better understanding of the form factors used.

INTRODUCTION

Among the different degrees of freedom, which may enhance the fusion cross section below the Coulomb barrier,deformations and the superficial vibrations of the nuclei. The dependence of the Coulomb barrier height of the system on the nuclei orientation and surface vibration has the effect of splitting the barrier, and a net result of increasing the fusion probability. The coupling of the fusion with transfer channels may have the same net effetc of lowering the effective Coulomb barrier. There has been an increasing interest in the last few years in the investigation of how the transfer reactions influence the sub-barrier fusion cross section, but there are, at the moment, conflicting evidences about the relations among transfer cross sections, transfer Q-values and fusion cross section enhancement.

In order to contribute to the understanding of the influence of transfer reactions and deformation of the colliding nuclei on the sub-barrier fusion, we have measured the fusion and transfer reactions for the ^{14}N + ^{59}Co and ^{16}O + ^{59}Co systems. Some transfer reactions induced by the ^{14}N projectile have positive transfer Q-value whereas all transfer channels for the ^{16}O induced reaction have negative Q-values. The measurement of the fusion and transfer channels of both systems could therefore give information about the influence of transfer Q-values and transfer cross sections on the fusion enhancement. On the other hand, as the two systems roughly have the same effective deformation, due to the ^{59}Co target, we might be able to investigate the influence of the effective deformation on the fusion.

EXPERIMENTAL RESULTS

The experiments were performed with the 8Ud Pelletron accelerator at the Universidade de S. Paulo, and the experimental technique used was the in-beam and off-beam gamma ray spectroscopy method. This method allows us to measure the fusion cross section by adding all the evaporation channels identified and simultaneously to measure the inelastic scattering and transfer cross section for different channels. The irradiation times lasted from 3 to 10 hours. After each irradiation, decay spectra were accumulated during periods of 1-2 hours and there were backing irradiations in order to identify and correct the lines of the spectra coming from target contaminants and background. For each spectrum all lines were identified and the existance of the residual nuclei was confirmed not just by the energy idenfitication of the transitions but also by their excitation functions.

The fusion cross sections were obtained, adding up to 13 evaporation channels. For the ^{14}N + ^{59}Co system, two transfer cross sections were measured: stripping of one proton (Q = +2 MeV) and stripping of one neutron (Q = -2 MeV). The pickup of one neutron (Q = 0.4 MeV) was identified but the cross section was not measured due to the low energy gamma ray transitions envolved. There are no evidences of any other transfer channel, neither for the ^{14}N + ^{59}Co nor for ^{16}O + ^{59}Co. The transfer cross section for the stripping of one proton is roughly the same as for the stripping of one neutron at energies right above the barrier, but when near or below it, the transfer cross section of proton becomes up to five times larger. The inelastic scattering for both systems are either below the measurable level or have the gamma transitions contaminated by other reactions and cannot be measured.

In figure 1 it is shown the measured reduced cross section for both systems. One can notice that the reduced fusion cross sections are quite similar although the transfer cross section is big for one of the systems while for the other it stood below the measurable level. This fact shows that there is no clear relation between the fusion enhancement and the transfer cross section.

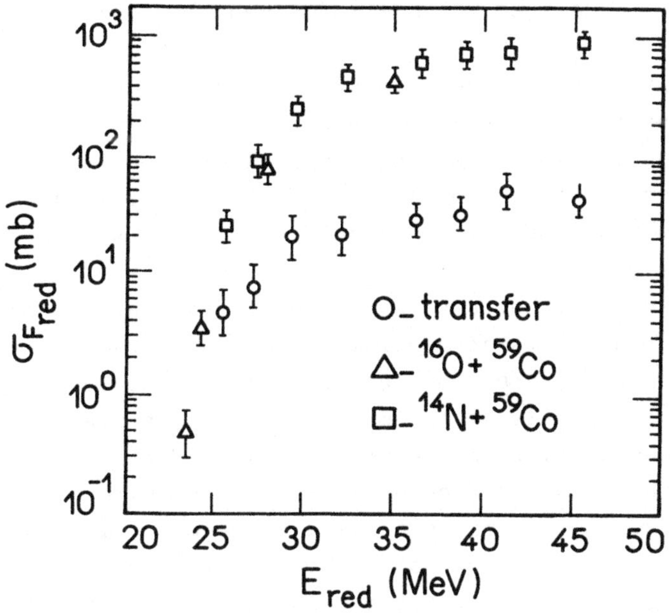

Figure 1 - Reduced excitation functions of the fusion of $^{14}N + ^{59}Co$(□), $^{16}O + ^{59}Co$ (Δ) and total transfer for $^{14}N + ^{59}Co$ (O).

FIT OF THE FUSION DATA ABOVE THE BARRIER BY UNIDIMENSIONAL POTENTIALS

As a first step, to explain the experimental fusion excitation functions, only the data at energies above the Coulomb barrier are considered, because at this energy region the fusion cross section is well understood and therefore the three parameters which define the interaction barrier can be determined by the fit of the data through one dimensional potentials such as the Krappe-Nix-Sierk(1).

The three parameters are the height V_B, the radial position R_B and the curvature $\hbar w$ of the interaction barrier. The values obtained are compared with the systematics and they are used as reference for the fit of the whole fusion excitation function, including the sub-barrier energy region.

Another interesting unidimensional model was proposed by Wong(2), where it is used the parabolic approximation for the barrier and it is assumed that the barrier parameters are independent of the angular momentum of the partial wave, In this model the fusion cross section is expressed by an analytical form.

THE WONG MODEL WITH DEFORMATION

The first approach used in this work to explain the fusion excitation functions was the simple Wong model with deformation(2), where the effective quadrupole deformations of the colliding nuclei are explicitly taken into account in the definition of the interaction radius.

Although the Wong model is derived from the basis of static deformation, the quadrupole deformation β_2 can be interpreted as an effective deformation caused by surface vibrations (dynamical deformation) and changing orientation (static deformation).

In this model there are four parameters: V_B, R_B, $\hbar w$ and β_2. From the unidimensional analysis of the data above the barrier, the initial values of the three barrier parameters are given and the fit of the fusion excitation functions is done by least square, with only one free parameter. The other three parameters are allowed to vary only near the values derived from high energy data. The effective deformation parameter which fits the excitation function must then be compared with the predictions from other methods such as Coulomb excitation and scattering. Particularly for the ^{59}Co nucleus, the predicted deformation parameter was obtained by averaging the values of the deformations of the even-even neighbours ^{60}Ni and ^{58}Fe(3). The effective deformation obtained for the ^{59}Co was $\beta_2 = 0.22 \pm 0.02$.

Table 1 shows the values of the parameters which fit the data. The deformation value needed to fit both fusion excitation functions is the same, showing that two systems with the same deformation have the same behaviour for the sub-barrier fusion. However, the value which fits the data is 20% higher than the predicted one, a fact that could indicate that there is another degree of freedom responsible for the fusion enhancement.

It is important to notice that when one uses this method it is essential that the value of the barrier curvature ($\hbar w$) which fits the data is compatible with the predicted one by realistic unidimensional potentials, because if one allows the two parameters which govern the fusion below the barrier ($\hbar w$ and β_2) be free, it will always be possible to find a "good fit of the data", as can be seen in figure 2 for the $^{14}N + {}^{59}Co$ system. In this figure, there is a visual coincidence of the two fits: (a) $\beta_2 = 0$, $\hbar w = 5.19$ MeV (no deformation but a very narrow barrier) (b): $\beta_2 = 0.294$, $\hbar w = 2.90$ MeV (a large deformation but a wide barrier). There will be no meaning in these fits if one does not use realistic $\hbar w$ values. The correct fit with the parameters of table 1 has the same visual in this figure.

Table 1 - Values of the parameters which fit the experimental data

System	V_b (MeV)	R_B (fm)	$\hbar w$ (MeV)	β_2
$^{14}N + ^{59}Co$	26.8	9.18	3.15	0.263 + 0.017
$^{16}O + ^{59}Co$	30.4	9.25	3.35	0.266 $\pm$ 0.025

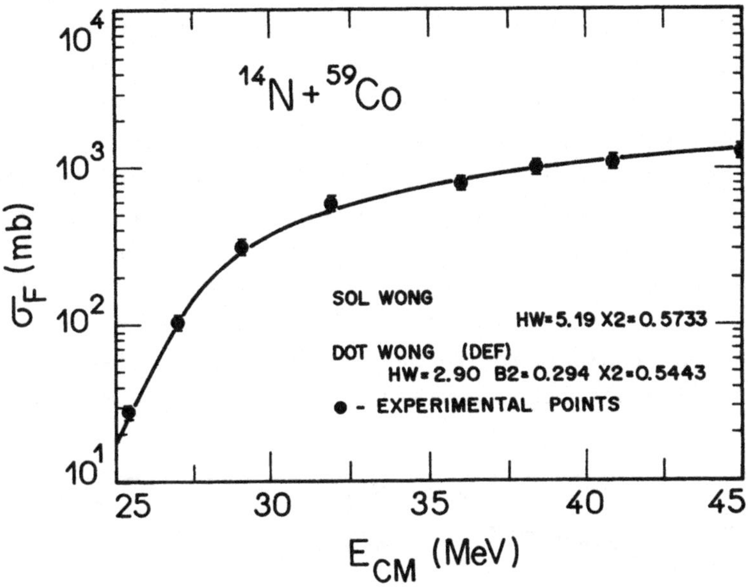

Figure 2: Fits of the excitation function of $^{14}N + ^{59}Co$ by the Wong model: (a) without deformation; (b) with a large deformation.

COUPLING TRANSFER CHANNELS

Another approach used to explain the fusion excitation functions was the use of the simplified coupled channels code CCFUS[4,5]. In a first step, inelastic channels were coupled to the fusion for both systems, but these couplings gave very small fusion enhancements, showing the need of coupling transfer channels.

In order to couple the transfer channels, effective Q-values have been used by taking into account the variation of the Coulom barrier due to the transfer. For the transfer strengths of all transfer channels it was used the value of 3 MeV, giving rise to form factors around 0.7, from the expression given by Landowne[6]. The reliability of this kind of analysis depends on the possibility of testing the coupling strenghts with measured transfer cross sections. Nevertheless, this approach is quite useful to have a qualitative feeling of the coupling effect of transfer channels to fusion.

For the $^{14}N + ^{59}Co$, we have coupled the three transfer channels measured or identified: stripping of one proton, stripping of one neutron and pickup of one neutron. The single transfer coupling which gives the largest enhancement is the stripping of one proton, which is also the channel with the highest cross section. Figure 3 shows the predictions of the fusion for the $^{14}N + ^{59}Co$ system for different couplings.

For the $^{16}O + ^{59}Co$, where there were no measurable transfer cross sections, we have coupled the two transfer channels with higher Q_{eff} values: stripping of one alpha (Q_{eff} = 4.5 MeV) and stripping of one proton (Q_{eff} = 0.4 MeV). Using the same transfer strength and form factor expression for the α transfer as for p and n transfer, the results of the coupling are the ones shown in figure 3. One could say that the fit is quite reasonable although a more reliable conclusion should be drawn only after a better understanding of the form factors to be used. A very interesting fact about the $^{16}O + ^{59}Co$ fusion is that the stripping of one alpha particle seems to be responsible for most of the fusion enhancement although its cross section is below the measurable level.

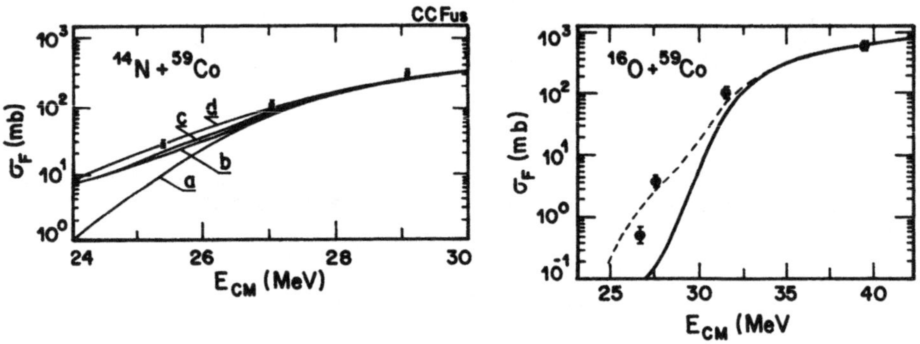

Figure 3: Coupling of transfer channels by the CCFUS code. Left: $^{14}N + ^{59}Co$ - (a) inelastic coupling; (b): (a) + stripping of 1 p; (c): (b) + stripping of 1 n (d): (c) + pickup of 1 n - Right: $^{16}O + ^{59}Co$ - full line: inelastic coupling; dashed line: inelastic + stripping of α and 1 p couplings.

THIS WORK WAS PARTIALLY SUPPORTED BY THE CNPq AND FAPERJ.

REFERENCES

1) H.J. Krappe, J.R.Nix, A.J.Sierk, Phys.Rev.C 20 (1979) 992
2) C.Y. Wong, Phys.Rev.Lett 31 (1973) 766
3) Nuclear Data Sheets 42 (1984) 457 and 48 (1986) 251
4) C.H.Dasso, S,Landowne, A. Winther, Nucl.Phys. A 405 (1983) 381
5) C.H.Dasso, S.Landowne, Phys Lett B 183 (1987) 141
6) S.Landowne, S.C.Pieper, F.Videbaeck, Phys.Rev C 35 (1987) 597

FUSION REACTIONS OF ^{58}Ni + ^{90,91,94}Zr AROUND THE COULOMB BARRIER

F.Scarlassara

Dipartimento di Fisica dell'Università e INFN-Sezione di Padova

Fusion cross section enhancements at sub-barrier energies have been the subject of many studies and are still not fully understood. In our experimental work we started studying the reactions ^{58}Ni + ^{90,91,94}Zr for which an interesting situation exists as for transfer Q-values that may affect fusion at sub-barrier energies.

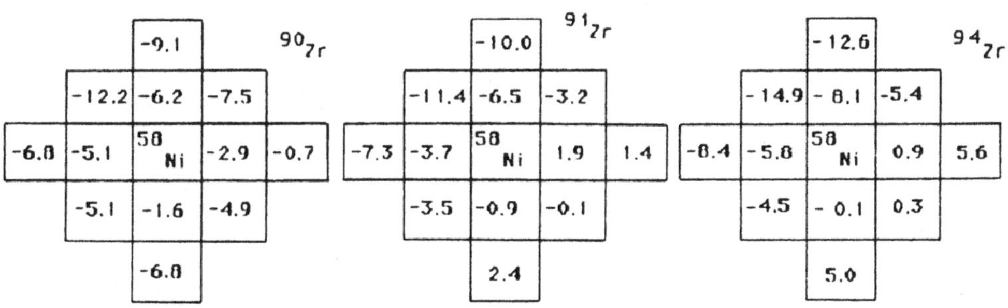

Fig 1 - Effective Q-values (in MeV) for transfer of up to two nucleons

As can be seen in fig 1, transfer reactions are energetically unfavoured for the ^{90}Zr target whereas very large positive Q-values are calculated for neutron pick-up and proton stripping with ^{94}Zr. So it could be possible to check the effect of positive Q-values on the enhancement of sub-barrier fusion.

A beam of ^{58}Ni was delivered by the Legnaro XTU Tandem on a set of ^{90,91,94}Zr enriched targets 30-120μg/cm^2 thick (a thicker ^{91}Zr target (300μg/cm^2) was used in a second run) at laboratory energies ranging from 199 to 280 MeV. These limits actually correspond to about 20% below and 30% above the barrier respectively.

As the evaporation residues mainly recoil around 0° they had to be separated from the beam particles by means of an electrostatic deflector [1]. Finally the evaporation residues (ER) were detected with a Time of Flight / Energy telescope consisting of a MCP and a SSBD Si detector. Another Si detector placed at 16° to the beam direction was used as monitor for an absolute normalization of the data. The lower cross section sensitivity is set by noise coming from scattered beam particles; this was particularly critical in this measurement due to the fact that ER electric rigidities were not so different from beam rigidities as we used to have in previous studies.[2] Therefore in a second run also an ionisation chamber was used in front of the Si detector to further discriminate between ER and beam through ΔE signals.

ER angular distributions were measured in the 0°-7° angular range and turned out to be quite narrow (FWHM~ 3°). Target thickness and isotopic corrections were taken into account.

In the isotopic correction, cross sections due to ^{92}Zr and ^{96}Zr contaminants for which no measurements were available have been inter- and extrapolated from the known cases; anyway the percentages are small and this has little effect on the results. Also, transmission efficiency of the electrostatic deflector was calculated with a Monte Carlo program which, together with possible small misalignment of the device gives an overall normalization uncertainty of about 30%. The resulting ER cross sections are shown in figs 2A,2B,2C where only statistical errors are shown.

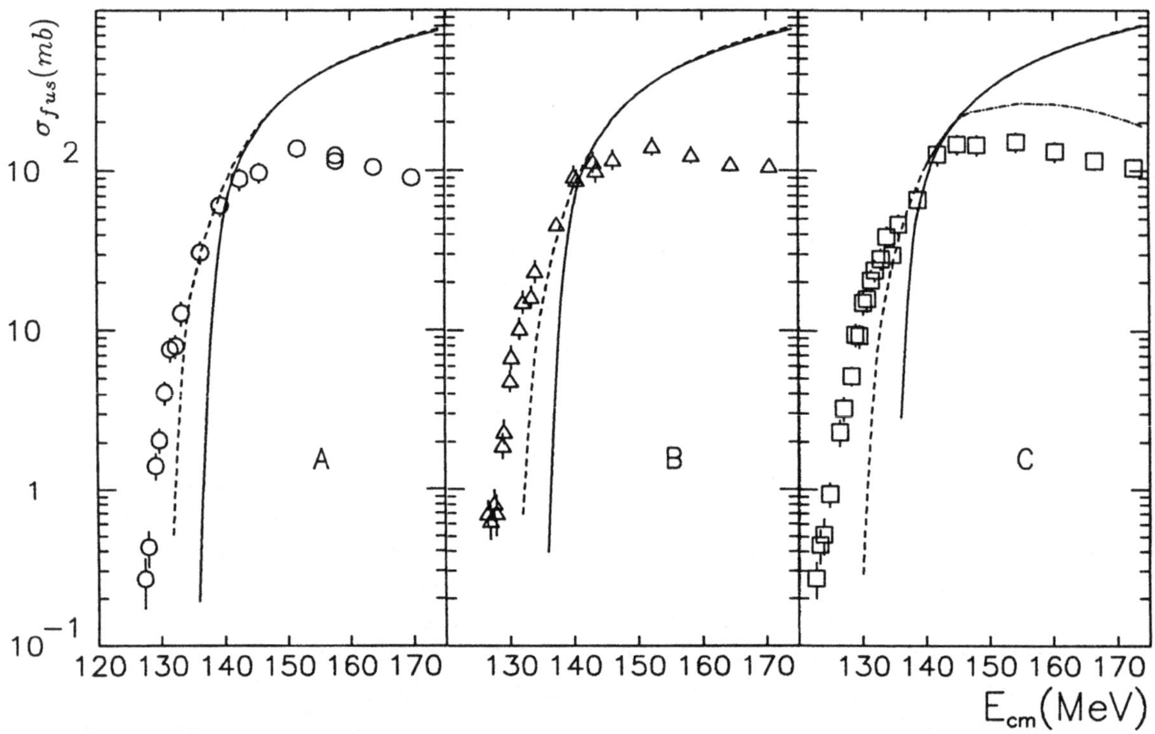

Fig 2 - ER cross sections for $^{58}Ni + ^{90,91,94}Zr$. CCFUS calculations with (dashed lines) and without (solid lines) coupling to inelastic excitations are also shown.In fig 2C also a result from CASCADE calculation (dashed-dotted line) is shown for energies above the barrier. CCFUS results have been normalized to the classical expectation values $\sigma_{FUS} = \pi R^2(1 - V_C/E)$ for $E_{CM} > 155MeV$

One can notice the large enhancement of sub-barrier cross sections which increase with increasing target mass. Also apparent in the figure is a strong hindrance of ER production immediately above the barrier, probably due to fission competition.

In figs 2A,2B,2C, we also compare our data to calculations with CCFUS both with no coupling and with coupling to inelastic channels of projectile and target, namely the first 2^+ and 3^- levels of the ^{58}Ni projectile and of the even-even targets.

As for ^{91}Zr, which is an odd nucleus with non-zero ground state spin, all B(E2) strengths up to 2.13 MeV excitation energy have been considered. The $B(E\lambda)$'s were taken from gamma lifetime data when available and are listed in table 1 together with the extracted β_λ parameters used in the calculation.

Table 1 - *Adopted $B(E\lambda\downarrow)$'s and deformation parameters β_λ of the inelastic channels coupled to the ground state of projectile and targets.*

Nucleus	trans.	$E_\gamma(MeV)$	$J^\pi_{gs} \to J^\pi_{ex}$	$B(E\lambda\downarrow)_{WU}$	β_λ
$^{58}Ni^{a)}$	E2	1.45	$0^+ \to 2^+$	10	0.18
	E3	4.47	$0^+ \to 3^-$	16	0.22
$^{90}Zr^{a)}$	E2	2.19	$0^+ \to 2^+$	5.5	0.09
	E3	2.75	$0^+ \to 3^-$	25	0.19
$^{91}Zr^{b,c)}$	E2	1.21	$5/2^+ \to 1/2^+$	53	0.07
	E2	1.45	$5/2^+ \to 5/2^+$	3.6	0.03
	E2	1.88	$5/2^+ \to 7/2^+$	4.2	0.04
	E2	2.04	$5/2^+ \to 3/2^+$	8.3	0.04
	E2	2.13	$5/2^+ \to 9/2^+$	2.3	0.035
	E3	2.17	$5/2^+ \to 11/2^-$	84	0.19
$^{94}Zr^{b,d)}$	E2	0.92	$0^+ \to 2^+$	8.4	0.11
	E3	2.06	$0^+ \to 3^-$	25	0.19

a) P.M.Endt At.Data and Nuc.Data Tab. 23(1979)547

b) P.M.Endt At.Data and Nuc.Data Tab. 26(1981)48

c) L.B.Horodynski-Matsushigue et al. Phys. Rev. C33 (1986)1594

d) E.Prota-Pessoa et al. Nuovo Cim. A 96 (1987) 347

We see that a sizable difference still remains between data and curves, which is an indirect evidence of the importance of other coupled reaction channels such as transfer and we can also see that this difference is much larger with ^{94}Zr where positive Q-values might favour coupling of transfer reactions to fusion. Transfer cross sections for these systems have been measured too but data have not been analysed yet.

In the high energy side of the plots, a comparison of these cross sections with simple computations of the total fusion cross section above the barrier, implies that about 80% of the compound nuclei undergo fission at the maximum energies used in our experiment. A statistical calculation (CASCADE code) was made for $^{58}Ni + ^{94}Zr$ that yields a compound nucleus (^{152}Er) for which experimental information on fission, obtained from different reactions, already exists[3]. The relevant parameters used were choosen as follows: the level density parameter "a" which seems to be ininfluent within a large range of variation was fixed to A/8; the level density ratio a_f/a_n was assumed to be 1.0 as in agreement with results from fission analysis in this nuclear region[3,4,5] and the fission barrier was taken to be the liquid drop one reduced by a factor 0.69 [3].

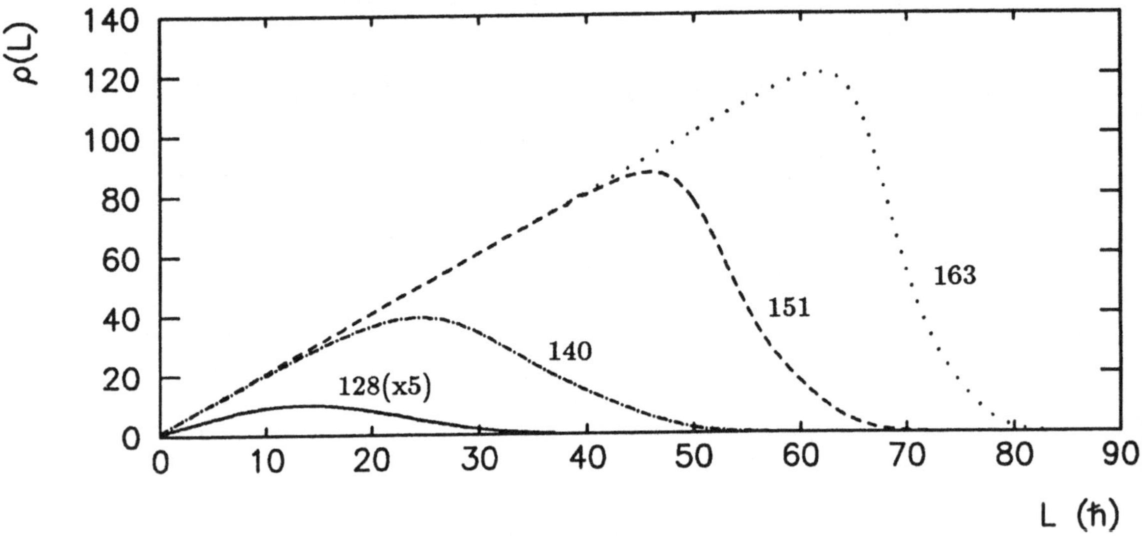

Fig 3 - Angular momentum distributions calculated from σ_{ER} at CM energies of indicated.

The angular momentum distribution of the compound nucleus was obtained from the measured ER cross section assuming[3] that the total fusion cross section above the barrier is reproduced by semiclassical calculations (dashed line in fig 2C) and that the transmission coefficients T(L,E) depend on the angular momentum L according to

$$T(L, E) = T(0, E - E_{rot})$$

One can thus deduce that

$$\rho(L) = \frac{(2L+1)}{\pi R^2}\left[(E - E_{rot})\left(\frac{d\sigma}{dE}\right)_{E-E_{rot}} + \sigma(E - E_{rot})\right]$$

where $\rho(L) = (2L+1) * T(L)$ and σ is the compound nucleus cross section. At high energies where fission is large the calculated fusion cross section was assumed (solid lines in fig 2).

After fitting the data in fig 2 with an arbitrary function we could thus calculate the angular momentum distributions shown in fig 3 whose maximum angular momenta and diffusenesses are listed in table 2 for $^{58}Ni + ^{94}Zr$.

Table 2 - Maximum angular momenta and diffusenesses obtained by fitting the data in fig 3 with the function $\rho(L)=(2L+1)/(1+exp((L-L_{MAX})/\Delta L)$.

E_{cm}(MeV)	E/V_{coul}	L_{MAX}	ΔL
128	0.93	0	> 10
140	1.02	31	5.4
151	1.10	54	3.2
163	1.18	69	2.5
174	1.26	82	2.1

Using these parameters CASCADE calculations yield the dashed-dotted line shown in fig 2C. The behaviour of our data is reproduced reasonably well, if we consider the overall uncertainty in the experimental cross section.

This experiment was performed in collaboration with S.Beghini, G.Montagnoli, Y.Nagashima, C.Signorini, F.Soramel A.M.Stefanini, A.Tivelli whom I wish to thank here.

1) - S.Beghini et al., Nucl. Instr. and Meth. A239(1985)585

2) - A.M.Stefanini et al., Nucl. Phys. A456 (1986)509

3) - W. Reisdorf et al, Nuc. Phys. A438 (1985)212

4) - Van der Plicht et al., Phys. Rev. C28 (1983)2022

5) - M. Beckermann, M. Blann, Phys. Rev. Lett. 45 (1978)333

NUCLEAR REACTIONS IN COLLISIONS OF VERY HEAVY IONS AT ENERGIES BELOW AND NEAR THE BARRIER.

G. Wirth, W. Brüchle, Fan Wo, and K. Sümmerer
GSI Darmstadt
Postfach 110552, D-6100 Darmstadt 11, Fed. Rep. Germany

F. Funke, J.V. Kratz, and N. Trautmann
Institut für Kernchemie, Universität Mainz
Postfach 3980, D-6500 Mainz, Fed. Rep. Germany

Introduction

Nuclear reactions in heavy ion collisions at energies around the Coulomb barrier are of current interest. An increasing amount of experimental data becomes available for medium heavy ion collisions but only little is known on the interaction of very heavy ions as U + U at low energies. The interaction potential for these very heavy ions is dominated by the strong repulsive Coulomb term. Contrary to light and medium heavy ion reactions the potential exhibits no stable minimum and fusion is not possible. The nominal Coulomb barrier $V_C = Z_1Z_2e^2/R_{INT}$ calculated e.g. for U + U with the experimentally determined interaction radius of $R_{INT} = 16.9$ fm is $E_{LAB}/A = 6.1$ MeV/u. But the ion-ion potential clearly depends on the relative orientation of the two deformed U-nuclei. Several theoretical calculations of the potential including deformation and orientation degrees of freedom were published [1]. Some of these predict shoulders or pockets in the interaction potential which depend on the relative orientation of the deformed actinide nuclei. This could give rise to delayed collisions with deviations from Rutherford trajectories; even the existence of a long-lived dinuclear system was suggested. Nuclear reactions should be influenced by the shape of the potential and may therefore serve as an experimental probe.

In the present work nuclear reactions in collisions of U + U, U + Au and Au + Au were investigated at energies well below and near the barrier. Originally, this work was motivated by the observation of positron line structures in U + U collisions, which have attracted considerable interest [1]. The source of the positron lines is still unknown and the various processes proposed depend strongly on the collision dynamics, in particular on the question whether nuclear resonance scattering is involved or not.

Experimental

The experimental investigation of the nuclear reactions in very heavy ion collisions at subbarrier energies is difficult because one needs excellent mass and charge resolution and in addition a very high sensitivity to detect low cross sections. The direct detection of outgoing particles by counters or magnetic spectrometers does not provide the necessary mass and charge resolution for these very heavy ions especially at forward angles in the presence of a large background of elastic scattered beam particles. In-beam γ-spectroscopy with coincident particle identification could give the necessary resolution, but it does not allow the measurement of very small cross sections of rare reaction channels. Therefore we have used an activation technique measuring off-line the radiation from the decay of reaction products which are unstable in their ground state. We reach a high sensitivity down to the μb-level together with unique isotope identification which is necessary to get an overview on nuclear reactions that occur in very heavy ion collisions at beam energies below the barrier and to look for macroscopic effects of the interaction potential. But we loose information on the final states. For one-nucleon transfer e.g. we sum over all transfers of the nucleon to different excited levels including transfer to the ground state.

One-neutron transfer

It turns out from our measurements that one-neutron transfer is by far the most probable direct reaction in these very heavy ion collisions at energies below the barrier and cross sections of several mb are observed still at beam energies 20% below the barrier. The cross sections of all other transfer reactions are smaller by more than an order of magnitude. Recently we have reported on the excitation function and angular distributions for the one-neutron transfer product ^{239}U in $^{238}U + ^{238}U$ collisions [2]. The integral excitation function for the one-neutron transfer product was measured over a wide energy range from the Coulomb barrier V_C down to 73% of the barrier. The exponential dependence as well as the slope of the excitation function can be understood within a semiclassical description of sub-Coulomb transfer. Angular distributions were measured at various energies below the barrier. The experimental data in Fig. 1 are taken from [2]. The reduced distance of closest

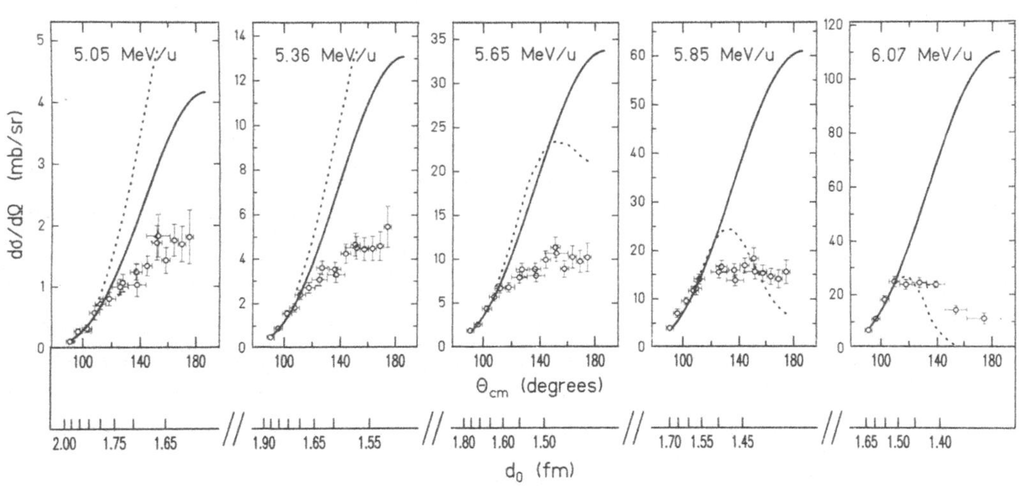

Fig.1: Angular distributions of ^{239}U in the system $^{238}U + ^{238}U$ at various indicated beam energies. Solid lines are semiclassical calculations. Dotted lines are DWBA calculations.

approach $d_0 = R_{MIN}/(A_1^{1/3} + A_2^{1/3})$ calculated from the scattering angle θ_{CM} assuming Rutherford trajectories is also given. For the more peripheral collisions the transfer probabilities are also in excellent agreement with the semiclassical description (solid lines). But deviations of the one-neutron transfer angular distributions from the semiclassical description and also from DWBA calculations (dotted lines) are observed for the more central collisions. This cannot be explained by sequential fission of the one-neutron transfer product [3].

A quite different picture compared to $^{238}U + {}^{238}U$ results from a preliminary data analysis of the analogue one-neutron transfer measurements in the reaction $^{238}U + {}^{197}Au$. Angular distributions for the one-neutron transfer product ^{198}Au were measured at various beam energies near and well below the barrier. One-neutron transfer was measured still at a very large distance of 27 fm between the two colliding nuclei; this is 11 fm more than the interaction radius as illustrated in Fig. 2. We don't

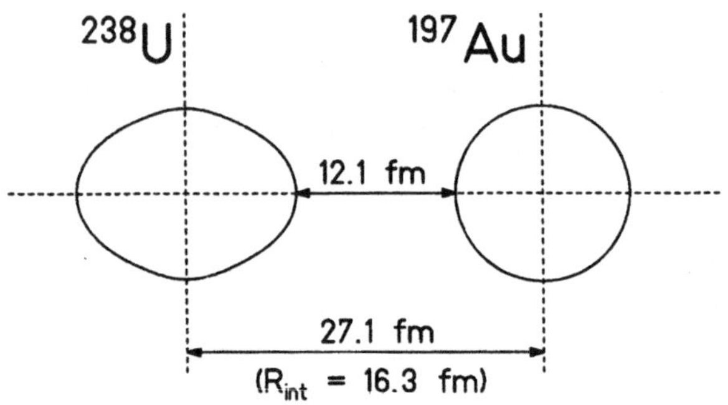

Fig.2. One-neutron transfer was observed still at the shown very large distance of closest approach which is 11 fm more than the interaction radius.

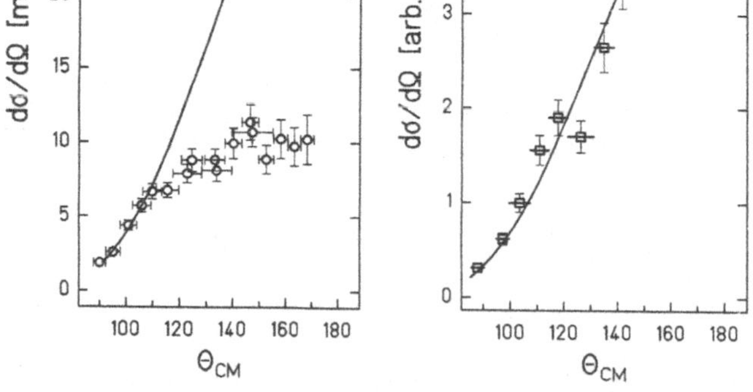

Fig.3. Angular distributions of the one-neutron transfer products from the reaction $^{238}U + {}^{238}U$ and $^{238}U + {}^{197}Au$ respectively. The solid lines are semiclassical calculations.

know of any other experiment where mass transfer was shown to occur between nuclei so far from each other. More than seventy transfer probabilities deduced from the differential cross sections and ranging over four orders of magnitude depend exponentially on the distance of closest approach as expected from the semiclassical transfer theory. Deviations from the theory are observed only for collisions with the colliding nuclei coming closer than $d_0 = R_{MIN}/(A_1^{1/3} + A_2^{1/3}) = 1.5$ fm. Systematic lower transfer probabilities for these close collisions are explained by absorption, i.e. the loss of elastic flux by the opening of many complex reaction channels when approaching the barrier. Fig. 3 demonstrates the different shapes of the one-neutron transfer angular distributions in the reaction U + U and U + Au. Both angular distributions were measured at the same beam energy relative to the corresponding barrier V_C. Good agreement is obtained between experiment and semiclassical theory (solid lines) for the reaction U + Au whereas strong deviations are observed for the more central collisions in U + U.

What makes the difference between U + Au and U + U ? Why can the simple sub-Coulomb transfer theory explain the one-neutron transfer probabilities for the reaction U + Au whereas it does not for U + U. The reason for the unusual behaviour is unknown but there are a few possibilities to be discussed. In the case of U + U both partners are strongly deformed. The transfer probability may depend on the relative orientation of the deformed nuclei at the turning point but the experiment with unpolarized particles averages over all initial relative orientations. On the other hand, dynamical alignment by the strong Coulomb forces between very heavy deformed nuclei may favour some relative orientations at the turning point. It is not clear whether the deformation of the colliding nuclei should show up in the angular distributions of one-neutron transfer, averaging over several final states. Very recently the theory of sub-Coulomb transfer of one neutron was extended to include deformation and orientation degrees of freedom treated by classical equations of motions [4]. Deformation was shown to be important for central collisions, but the calculated effect is not strong enough to explain the differences between experiment and sub-Coulomb transfer theory. Another reason for the observed discrepancies could be multi-step processes which are not treated in the DWBA formalism and in the semiclassical transfer theory. Using a simplified model it was shown schematically that channel coupling in principle may reduce cross sections at large angles [5]. To see whether Pauli blocking influences the transfer probability in the symmetric reaction ^{238}U + ^{238}U we have measured one-neutron transfer angular distributions in a second symmetric reaction ^{197}Au + ^{197}Au; data analysis is still in progress.

Possibly, nucleon transfer might provide a useful experimental tool to search with high sensitivity for the formation of long-lived giant nuclear molecules as mentioned in the introduction. In the asymmetric reaction U+Au target-like and projectile-like transfer products can be distinguished. At backward angles, practically no target-like transfer products are expected because of very large distances of closest approach of the corresponding Rutherford trajectories and the low transfer probabilities for these large distances at the turning point. The expected cross section for the one-neutron transfer product ^{198}Au integrated in the angular cone $137^\circ \leq \theta_{CM} \leq 174^\circ$ at a beam energy of 5.7 Mev/u is three orders of magnitude lower than the experimental detection limit. Therefore the observation of one-neutron transfer products ^{198}Au at these backward angles would give evidence for collisions not following Rutherford trajectories. In order to see resonance scattering the beam energy must fit very precisely the unknown position of such a possible small pocket in the potential and one must search in a certain energy window without any gap. As an advantage our experiment can use targets of 360 $\mu g/cm^2$ thickness thus integrating in each irradiation a reasonable energy bin by the energy loss of the beam in the target without increasing the background. By now a few irradiations were performed between 5.4 MeV/u and 6.5 MeV/u. Data analysis is not yet completed, but at all energies used so far no no ^{198}Au-activity with a limit of approximately $\sigma \leq 10$ μb was found in the angular range $137^0 \leq \theta_{CM} \leq 174^0$. To estimate this limit a constant cross section was assumed within the energy bin covered by the target thickness. Under the assumption of an isotropic distribution of the reseparating nuclei after the formation of a long-lived dinuclear system, the cross section limit for the formation of a long-lived dinuclear system with the transfer of one neutron during the lifetime is approximately $\sigma \leq 0.1$ mb in ^{238}U + ^{197}Au collisions for the beam energies used so far.

Fission and multi-nucleon transfer

Besides one-neutron transfer, fission is another important reaction channel in the collision of very heavy ions at energies below the barrier. Two processes may contribute: Coulomb fission and sequential fission after nucleon transfer. Integral fission cross sections for the system ^{238}U + ^{238}U were measured and scattering angle dependent probabilities and integral cross sections for Coulomb fission were calculated [3]. The experimental fission excitation function increases steeper than the excitation function for Coulomb fission already at beam energies below the barrier due to a growing contribution from sequential fission. But only a small fraction of sequential fission comes from sequential fission after one-neutron transfer. It is concluded that in U+U collisions multi-nucleon transfer induced fission takes place already at low energies $(0.90 \pm 0.02) \times V_C$. This agrees with the observation of surviving reaction products from the transfer of many nucleons such as the 2p9n-transfer product ^{227}Th in U+U collisions down to energies $E_{LAB}/A = 5.6$ MeV/u, although with much lower cross sections.

The detection of complex transfer reactions at bombarding energies well below the Coulomb barrier V_C appears to be plausible if one considers the static deformation of the U-nucleus. The calculated barrier for a central collision of the deformed U + U system in a tip-to-tip configuration at the distance of closest approach corresponds to $E_{LAB}/A \simeq 5.7$ MeV/u. Therefore, at energies below the nominal barrier $V_C = 6.1$ MeV/u, but higher than $\simeq 5.7$ MeV/u, a fraction of collisions with small impact parameter and favourable relative orientation at the distance of closest approach is actually above the barrier and reaches nuclear contact with the well known opening of many inelastic transfer channels. The fraction of collisions reaching nuclear contact may depend on the potential and on the collision dynamics. It is of interest how that fraction develops as a function of the bombarding energy; the excitation function for the complex transfer products indicates a strong energy dependence. Fig. 4 shows that the cross section for the 2p9n-transfer product changes in U+U collisions by nearly two orders of magnitude between 5.6 MeV/u and 5.9 MeV/u, whereas the cross sections of the most dominant nuclear reaction, one-neutron transfer, and the sum of all fission contributions change only by a factor 2 or 3 in the same energy range.

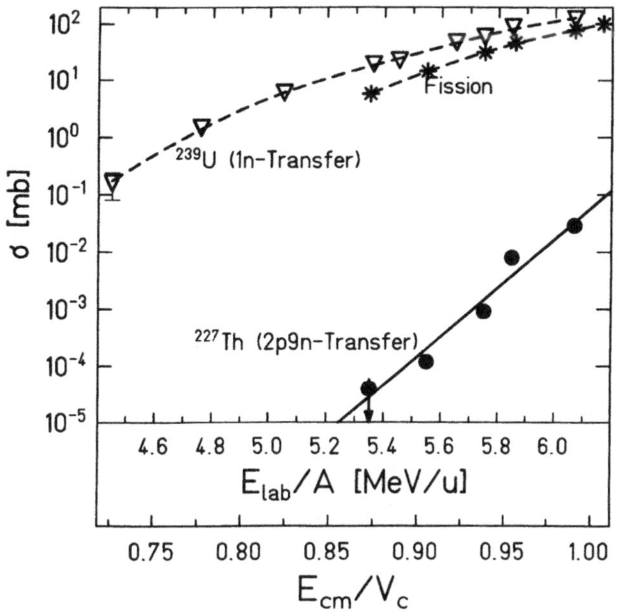

Fig.4: Excitation functions for different reaction channels in the reaction ^{238}U + ^{238}U at subbarrier energies.

References

1. See references given in Refs. 2 - 3
2. Wirth, G., Brüchle, W., Brügger, M., Wo, Fan, Sümmerer, K., Funke, F., Kratz, J.V., Lerch, M., Trautmann, N.: Phys. Lett. **177B**, 282 (1986)
3. Wirth, G., Brüchle, W., Wo, Fan, Sümmerer, K., Funke, F., Kratz, J.V., Lerch, M., Trautmann, N., Oberacker, V.E.: Z. Phys. **A 330**, 87 (1988)
4. Gupta, Raj K., Maass, R., Scheid, W., Phys. Rev. **C 37**, 1502 (1988)
5. Pinkston, W.T., Russell, D.P., Oberacker, V.E.: In: Physics of strong fields, Greiner, W. (ed.), p. 501, New York : Plenum Press 1987

FUSION AND QUASI-ELASTIC REACTIONS

SUB-BARRIER REACTIONS MEASURED USING A RECOIL MASS SEPARATOR

R. R. Betts
Physics Division, Argonne National Laboratory
Argonne, IL 60439, U.S.A.

Much recent effort [1] has been directed to the study of the interaction of heavy nuclei at energies close to and below their mutual classical interaction barrier. This interest is motivated by the observation of characteristics of nuclear reactions at these energies which cannot be understood in terms of simple first order models. The most graphic example of this being, of course, the observed enhancement of the cross-section for fusion of heavy ions at sub-barrier energies. Our current understanding of this phenomenon involves the influence of the coupling between elastic, quasi-elastic and fusion channels which leads to an interplay of nuclear structure and dynamics resulting in the observed enhancements to the fusion channel. In principle, therefore, a framework exists for a self-consistent description of all features of nuclear interactions in the sub-barrier region. In addition, the understanding of transfer between heavy ions is important in its own right. The hope being that new nuclear structure information might be obtained from such reactions. The sub-barrier region, in which the number of competing reaction channels is minimal, should therefore be the first testing ground of this understanding.

Despite these strong motivations, few data exist in the sub-barrier region for reaction channels other than fusion. In particular, our experimental knowledge of quasi-elastic transfer reactions is sparse, despite the belief that this particular channel may be dominant in determining some features of the sub-barrier fusion enhancement. This deficiency is due in part to experimental difficulties. Transfer reactions are governed primarily by the closest approach of the colliding nuclei which, at low energies, results in a strong backward peaking of the angular distribution in the center-of-mass frame. For situations where the projectile has a significant fraction of the target mass, as is so in most cases of interest, the backscattered projectile-like fragment has such low energy that the usual techniques of measurement and identification become invalid. Here, we report on a solution to this problem which allows a systematic study of many aspects of transfer reactions in the energy regime of interest. We exploit the fact that associated with the low-energy backscattered projectile-like fragment is a complementary target-like fragment which recoils to forward angles with a large fraction of the incident beam energy. These target-like fragments were detected and identified using the Daresbury Recoil

Mass Separator (RMS) thus allowing the measurement of quasi-elastic transfer over a hitherto inaccessible energy range from the vicinity of the barrier to several tens of MeV below.

The Daresbury RMS has been described in the literature [2] and recent developments are discussed elsewhere in these proceedings. [3] Briefly, it consists of a velocity filter followed by momentum selection giving an image in the focal plane dispersed in A/q. The position of the transmitted ions in the focal plane is measured using a position sensitive channel plate detector and their specific energy loss and total energy are then measured using an ionization chamber. The experiments described here used ^{58}Ni beams of energies ranging from 180 to 260 MeV provided by the Daresbury Laboratory Nuclear Structure Facility tandem accelerator. Data on sub-barrier transfer for targets of ^{116,118,120,122,124}Sn [4,5] and ^{144,148,150,152,154}Sm were obtained.

A spectrum of energy loss versus total energy for recoils detected in the bombardment of ^{122}Sn with a 214 MeV beam of ^{58}Ni is shown in Fig. 1. The Sn-

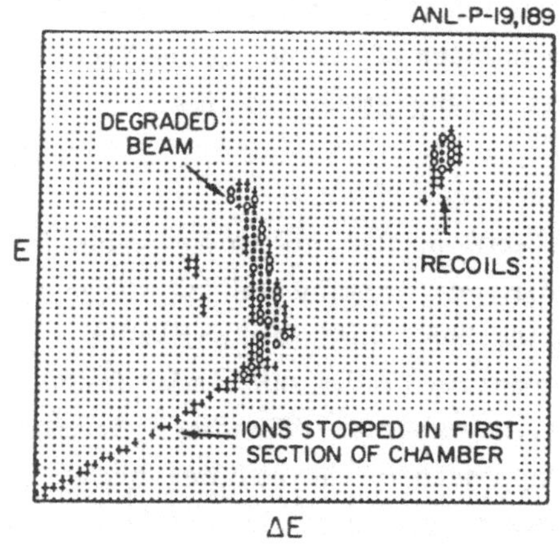

Fig. 1. Two-dimensional spectrum of energy loss versus total energy for recoils from 214 MeV ^{58}Ni + ^{122}Sn.

like recoils are identified by their large energy loss compared to a tail of degraded beam particles produced by scattering from the electrostatic plates in the velocity filter. The distribution of the Sn-like events as a function of energy and focal-plane position is shown in Fig. 2 together with the projections

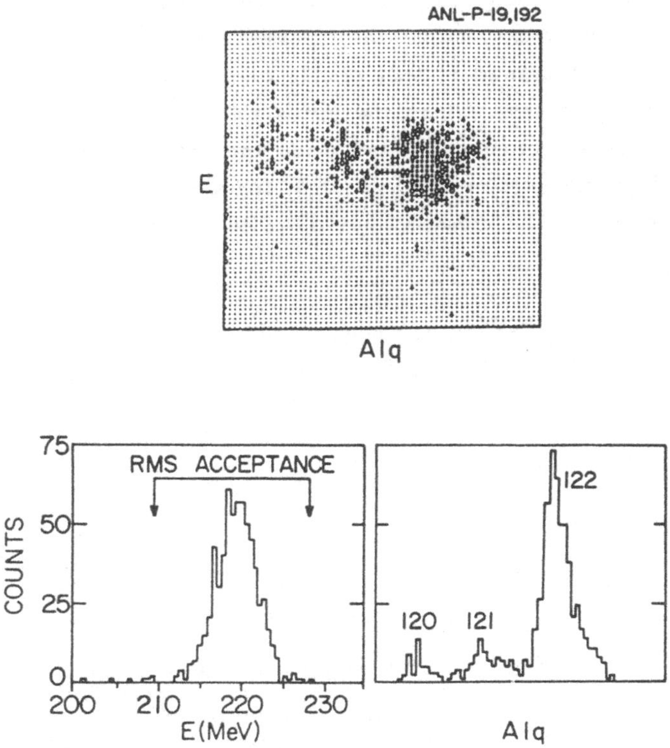

Fig. 2. Two-dimensional spectrum of position versus total energy for Sn-like events. The projections on to the position and energy axes are also shown.

of these events onto the two axes. The better than unit mass resolution of the RMS is evident from the position spectrum which, in this case, shows peaks corresponding to masses 122, 121 and 120 all of charge-state 28^+. The energy spectrum shows a broad peak which, nevertheless, falls well within the acceptance of the recoil separator as indicated on the spectrum. This observation is further confirmed by a scan of the yield as a function of the central velocity setting of the Wien filter which gives a distribution of width equal to the expected velocity acceptance ($\Delta v/v \simeq 4\%$) consistent with the velocity spread of the detected fragments being narrower than the RMS acceptance.

On the basis of Q-values and the expected fall off of form factors at large radius, as well as the results of DWBA calculations, we expect that the observed yields correspond to neutron, rather than proton, transfer. This expectation is confirmed by the results of a measurement of coincident gamma rays in the case of

$^{58}Ni + {}^{124}Sn$ at a bombarding energy of 220 MeV. In this case the one nucleon transfer channel is conclusively identified as neutron transfer. These results are discussed in more detail in another paper [6].

Cross sections were obtained by assuming that the total yield observed corresponds to the Rutherford scattering cross-section; the cross-sections for individual masses then being given by the ratio of their yields to the total. The validity of this procedure was verified in cases where the yields were measured for all the strong charge states and a cross-section calculated using the ratios of the solid angles of the RMS and a monitor detector fixed at 36.5° together with the measured yield in the monitor counter. The cross-sections obtained by the two methods agreed to within 15%.

Cross sections for $^{58}Ni + {}^{116,118,120,122,124}Sn$ one-neutron pickup reactions are shown plotted as a function of center-of-mass bombarding energy in Fig. 3. The data for each isotope fall with decreasing bombarding energy with

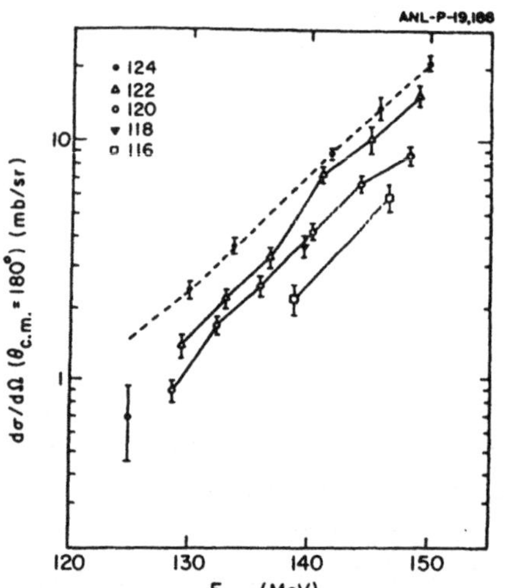

Fig. 3. Cross-sections for one-nucleon pickup in ^{58}Ni + Sn reactions plotted versus center-of mass bombarding energy. The dashed curve shows the DWBA prediction for ^{124}Sn. The solid lines join the data points.

slopes which are independent of the target. The absolute magnitudes, however, depend strongly on target mass, varying by a factor of three from the heaviest to the lightest isotope. The general features of these data can be understood as a result of the monotonic change in the ground-state neutron binding energy across the isotopic sequence. This, coupled with the almost unchanging collective and single-particle structure of the Sn isotopes results in the rather simple behavior observed. Table I lists some of the relevant properties of the Sn isotopes studied.

Table I

Properties of the Even Sn Isotopes

A	B_n MeV	Q_n MeV	κ_n fm^{-1}	β_2 [a]	β_3 [a]
116	9.57	-0.56	0.681	0.120	0.046
118	9.33	-0.33	0.672	0.110	0.035
120	9.10	-0.11	0.664	0.108	0.031
122	8.80	0.18	0.653	0.104	0.035
124	8.49	0.51	0.641	0.095	0.041

[a] Deformation associated with the collective inelastic excitation of the lowest 2^+ and 3^- states

Within the framework of a simple semi-classical treatment of neutron transfer, we expect the dependence of the $\theta_{cm}=180^\circ$ cross-section to be given by;

$$\frac{d\sigma}{d\Omega} \propto \frac{e^{-2\kappa_n d}}{\kappa_n} \left.\frac{d\sigma}{d\Omega}\right|_{Rutherford} \quad \text{where} \quad d = \frac{Z_1 Z_2 e^2}{E_{cm}} ,$$

the distance of closest approach for a head-on collision. The cross-sections calculated with this expression are shown in Fig. 4 together with the data. The

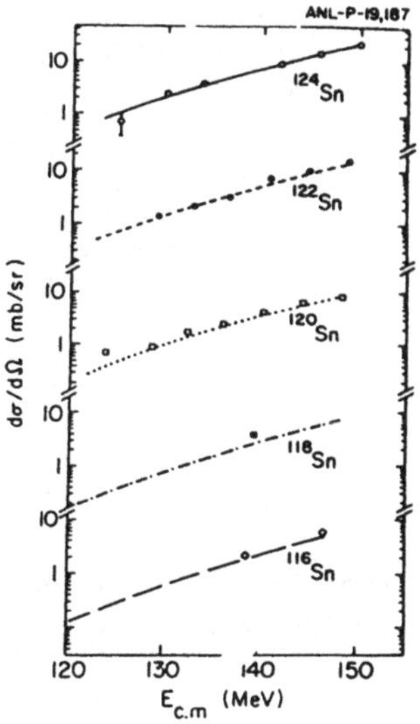

Fig. 4. Cross-sections for one-nucleon pickup in ^{58}Ni + Sn together with cross-sections calculated using the semi-classical expression given in the text. The curves are normalized to the highest energy point for ^{124}Sn.

calculated curves were normalized to the data at one point, that for ^{124}Sn at the highest energy. The overall agreement is quite remarkable. The near constancy of the observed slopes can be seen from the form of the semi-classical expression which has a logarithmic slope which depends only linearly on κ_n - and remains constant within a few percent across the isotopic sequence. The magnitude of the cross-section is, however, exponentially dependent on the product of κ_n and d which leads to the observed strong dependence. It would not have been expected a priori that this simple treatment give such a good account of the variation of the absolute magnitudes of the cross-sections. This has the additional requirement that the dependence of the overall magnitude of the valence neutron wave function at the nuclear surface vary only weakly from isotope to isotope such that its magnitude at large distances is governed solely by κ_n the binding energy parameter.

A more detailed analysis of these data has been carried out in DWBA using spectroscopic information taken from light ion studies [7,8] in this mass region together with the code PTOLEMY. [9] The resultant energy dependence is in similarly good agreement with the data, as is the isotopic dependence which is shown together with the data and the semi-classical results in Fig. 5. The

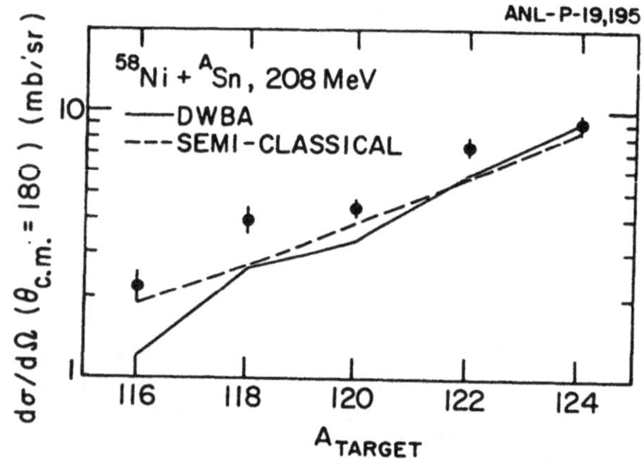

Fig. 5. Cross-sections for one-nucleon pickup in ^{58}Ni + Sn at 208 MeV together with the predictions of the semi-classical model and the results of DWBA calculations.

conclusion is therefore that the sub-barrier neutron transfer of ^{58}Ni + Sn is well understood within the framework of simple semi-classical and quantum mechanical theories of the transfer process.

The influence of the transfer process on the sub-barrier fusion of these systems has been investigated using the model of Dasso et al., [10] which provides a transparent connection between the strengths of the quasi-elastic reaction channels and fusion. The results of these calculations are shown in Fig. 6 which shows, successively from left to right, the cross-sections

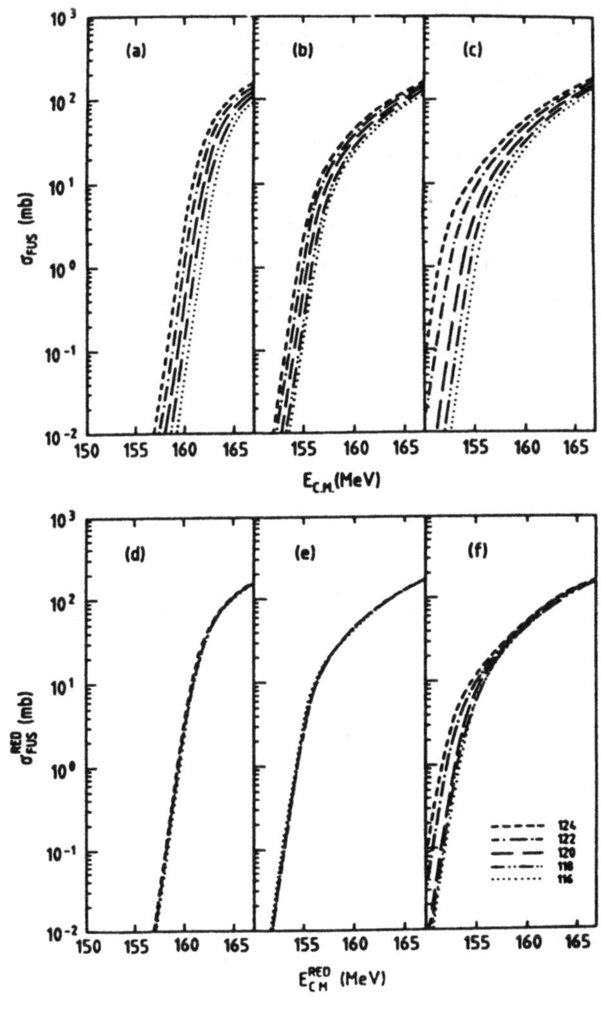

Fig. 6. Calculated fusion cross-sections for Ni + Sn. Figures a)-c) show the results using the bare potential, bare + inelastic and bare + inelastic + transfer. Figures d)-f) show the same results corrected for geometric effects as discussed in the text.

calculated with no-coupling, inelastic coupling and, inelastic + transfer coupling. The relative strengths of the transfer coupling were taken from the present measured cross-sections. The lower portion of the figure shows the effect of removing geometric effects on the isotopic dependence through defining the appropriately scaled quantities

$$\sigma_{FUS}^{RED} = \sigma_{FUS}\left(\frac{R_o^{REF}}{R_o}\right)^2 \quad \text{and} \quad E_{C.M.}^{RED} = E_{C.M.}\left(\frac{V_o^{REF}}{V_o}\right)$$

where R_o and V_o are the s-wave barrier heights and radii respectively. It is evident from these curves, that the only non-trivial isotopic dependence of the sub-barrier fusion of these systems results from the coupling of the fusion channel to the strongly varying one-neutron transfer channel. These results are in fair agreement with the available data [11] for low-energy fusion of these systems although a more detailed experimental study of the sub-barrier fusion is now clearly merited.

The results for ^{58}Ni + Sm are less easily understood. Table II lists some of the properties of the isotopes studied which, in contrast to Sn, show a transition from spherical (^{144}Sm) to well-deformed (^{154}Sm). The cross-sections for one-neutron pickup are shown plotted versus center-of-mass bombarding energy in Fig. 7. As in the case of Sn there is a strong isotopic dependence of the absolute magnitudes but in

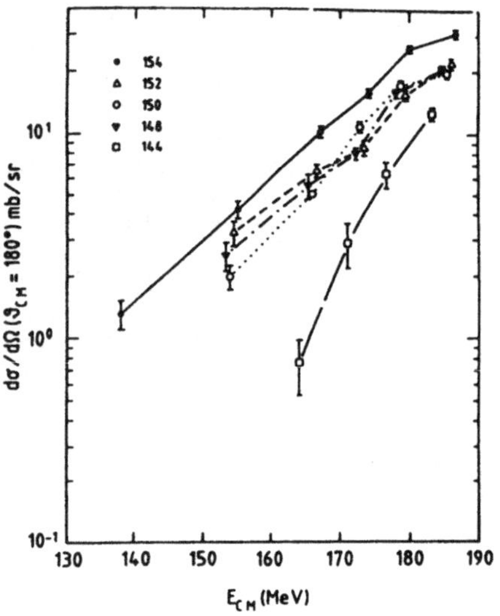

Fig. 7. Cross-sections for one-nucleon pickup in Ni + Sm reactions plotted versus center-of-mass bombarding energy.

Table II

Properties of the Even Sm Isotopes

A	B_n MeV	Q_n MeV	κ_n fm^{-1}	β_2 a)
144	10.55	-1.56	0.715	0.
148	8.14	0.86	0.628	0.11
150	7.99	1.01	0.622	0.19
152	8.27	0.73	0.633	0.25
154	7.98	1.02	0.622	0.27

a)Static ground-state deformation.

addition there now appears a dramatic variation in slope between the spherical ^{144}Sm target and the deformed targets. This variation in slope is in excess of that expected on the basis of the change of 2.5 MeV in binding energy between ^{144}Sm and the deformed targets as is illustrated in Fig. 8 where the data are

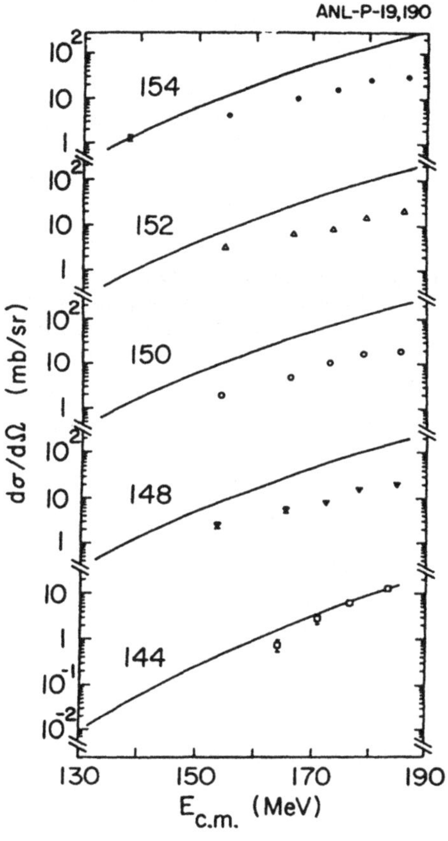

Fig. 8. Comparison between measured cross-sections for ^{58}Ni + Sm transfer and cross-sections calculated using the semi-classical model. The calculations are normalized to the highest energy point for ^{144}Sm.

compared to the predictions of the semi-classical model. For ^{144}Sm the calculation is in good agreement with the observed slope but for the heavier targets the observed slopes are much less than the semi-classical prediction. In fact, to reproduce the observed slope for ^{154}Sm, an effective binding energy of 3.6 MeV is required - some 4.4 MeV less than the ground-state value. A similar behavior is observed in the comparison between the data and the results of DWBA calculations shown in Fig. 9. As for Sn, the spectroscopic information

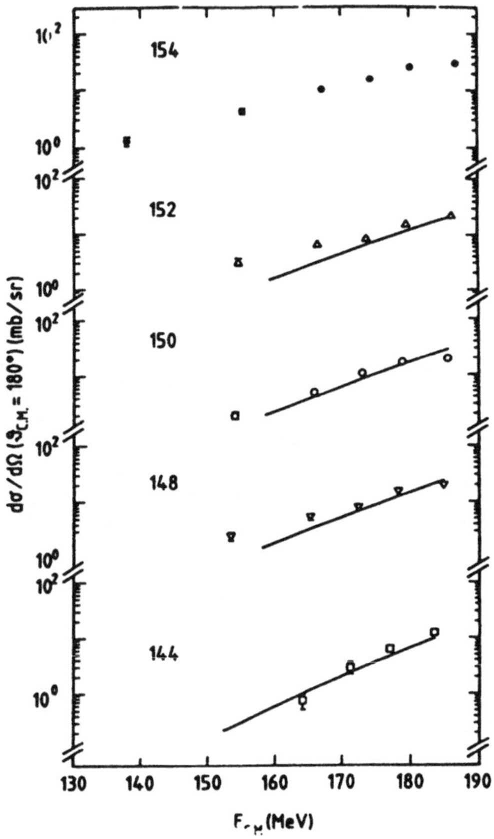

Fig. 9. Comparison between measured cross-sections for ^{58}Ni + Sm and the results of DWBA calculations.

was taken from light ion work. [12] The DWBA predictions are in good agreement with the observed magnitudes at the highest energies but fail to predict the fall in cross-section with decreasing energy for the deformed targets, the calculated cross-sections falling too steeply. We note that the ^{144}Sm data are well accounted for over the entire energy range studied.

Within the framework of the semi-classical model, it might be suspected that the observed changes in slope result simply from the additional quadrupole term in the Coulomb potential which arises from the deformed target charge distributions. This effect is readily included in the semi-classical formalism [13,14], but for randomly oriented target nuclei leads to an effect which is both small and in the wrong direction, the calculated curves now becoming slightly steeper than for a spherical target. The assumption of prolate target nuclei all aligned with their symmetry axis along the beam direction does produce a decrease in the slope over the spherical case but, for reasonable deformations, is still somewhat smaller than that observed. [14]

It is interesting to note the similarity of the present observations to those of Wirth et al., for ^{238}U + ^{238}U at sub-barrier energies. [15] The present data for ^{154}Sm are compared with the θ_{cm} = 180° data points for ^{238}U + ^{238}U in Fig. 10. The two lines are, in each case, a line drawn through the data

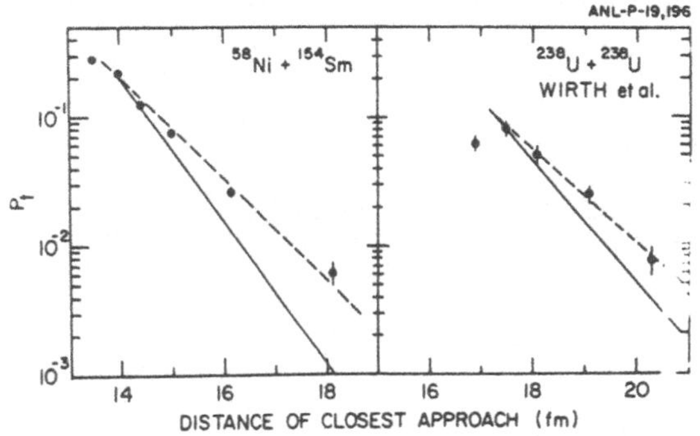

Fig. 10. Comparison between transfer probabilities for ^{58}Ni + ^{154}Sm and ^{238}U + ^{238}U and the predictions of the semi-classical model using ground-state binding energies (solid curve). The dashed curve through the points is to guide the eye.

to guide the eye and the semi-classical expectation using ground state binding energies. In both cases the enhancement of the data over the semi-classical expectation is similar. The ^{238}U + ^{238}U data are of particular interest in that it has been found for angles corresponding to more glancing collisions the data were well accounted for by the semi-classical calculation which indicates that the enhancement of transfer is connected to head-on rather than larger impact parameter collisions. It appears therefore that the enhancement of transfer at large distances is a quite general effect associated with the interactions between heavy ions in which at least one of the reaction partners is deformed. It does not seem that the magnitude of the deformation has any significant effect as the enhancement appears similar between the more weakly deformed and strongly deformed Sm isotopes and between all of these and the well deformed-deformed ^{238}U + ^{238}U case.

Similar "slope anomalies" have been reported by Juutinen et al., [16] for several different projectiles incident on Dy target nuclei all at bombarding energies above the barrier. In these cases, in which gamma ray energy and multiplicity information were also obtained, it was concluded that the occurrence of the slope anomaly was correlated with small angular momentum transfer. In the present data it is true that the low bombarding energy and detection angle also favor low angular momentum transfer but this evidence is only circumstantial. It seems nevertheless that such anomalies maybe a general phenomenon and that further study will give us new insight into the interplay between structure and dynamics in heavy ion reactions.

This work was performed in collaboration with C.N. Pass, P.M. Evans, A.E. Smith (Oxford); J.S. Lilley, J. Simpson, J.R. Smith (Daresbury); A.N. James (Liverpool); and B.R. Fulton (Birmingham). Support from NATO Collaborative Research Grant 0076/87 is gratefully acknowledged. This research was supported by the U.S. Department of Energy, Nuclear Physics Division, under contract W-31-109-ENG-38.

References

[1] Proceedings of the International Conference on Fusion Reactions below the Coulomb Barrier, Cambridge, Mass. June 1984 ed. S.G. Steadman (Springer Verlag, Berlin 1985).

[2] A.N. James, et al., Nucl. Inst. and Meth. A267, 144 (1988).

[3] A.N. James, these proceedings.

[4] R.R. Betts, et al., Phys. Rev. Lett. 59, 978 (1987).

[5] C.N. Pass et al., submitted to Nucl. Phys.

[6] C.N. Pass et al; these proceedings.

[7] E.J. Schneid et al., Phys. Rev. 156, 1316 (1967).

[8] P.E. Cavanagh et al., AERE Harwell Report No. AERE-R5901 (unpublished).

[9] S.C. Pieper et al., ANL Report No. ANL-76-11 (unpublished).

[10] C.H. Dasso, S. Landowne and A. Winther, Nucl. Phys. A407, 221 (1983).

[11] K.T. Lesko et al., Phys. Rev. C34, 2155 (1986).

[12] S. Gales et al; Nucl. Phys. A398, 19 (1983).

[13] R.K. Gupta et al., Phys. Rev. C37, 1502 (1988).

[14] R.R. Betts et al., to be published.

[15] G. Wirth et al., Phys. Lett. B177, 282 (1986).

[16] S. Juutinen et al., Phys. Lett. B192, 307 (1987).

HOW WELL DO WE UNDERSTAND QUASI-ELASTIC REACTIONS AT ENERGIES CLOSE TO THE BARRIER

K. E. Rehm
Argonne National Laboratory
Argonne, IL 60439

In collisions between two heavy nuclei a wide spectrum of different reaction modes is observed covering the range from simple processes like elastic scattering to complicated multistep transfers and fusion. During the past few years it has become clear that these different reaction types are not separate processes but are strongly related. Because the strength of the various reaction modes varies with the mass of the projectile (e.g. for Coulomb excitation and for transfer reactions) experiments with medium mass projectiles allows us to investigate the interrelations of the various reaction modes and their dependence on bombarding energy, neutron and proton number in more detail. One of the most well known cases for an interference between two reaction types is the influence of inelastic scattering and transfer reactions on the strength of the fusion process at subbarrier energies[1]. There are also several examples showing how inelastic scattering can influence the transfer reactions[2].

On the theoretical side heavy ion reactions are usually analyzed using models that were developed first for light ion induced reactions: the optical model for elastic scattering and the DWBA for more inelastic processes like transfer and inelastic scattering[3]. Some of the assumptions going into these approximations, however, are not valid for heavy ion induced reactions. Inelastic excitation, in particular, is sometimes so strong that it has to be included explicitly in a full coupled channel (CC) calculation with all its complexity. While this is rather straightforward for the case of inelastic scattering[4], a similar calculation including transfer reactions is considerably more complicated. Generally it can be stated, however, that well established and successful models exist for reactions which get their main contributions from large partial waves and several examples will be given below. For reactions involving low partial waves various statistical model codes have been developed which usually give reliable predictions for the strength of compound nucleus formation at least at energies above the Coulomb barrier. The region between fusion and quasi-elastic reactions, is not well understood theoretically. This region is associated with deep inelastic collisions, which are complex multiparticle reactions involving transfer of several protons and neutrons. Some models in this region are extensions of a statistical treatment (diffusion type models)[5], while other approaches start from a DWBA treatment[6].

In this contribution I want to discuss to what extent experiments in the field of quasi-elastic scattering are understood within the framework of various theoretical models and in what areas more work is needed. The experiments I want to discuss were performed with the Argonne Superconducting accelerator (ATLAS) and an Enge split pole magnetic spectrograph with its associated focal plane detectors for particle identification[7]. This setup provides good mass and Z information (up to the mass 100 region) at reasonable solid angles and, if one is willing to run with thin targets good energy resolution can also be obtained. An example for the energy resolution from the system ^{37}Cl + ^{208}Pb is given in Fig. 1a while data for the Z-separation

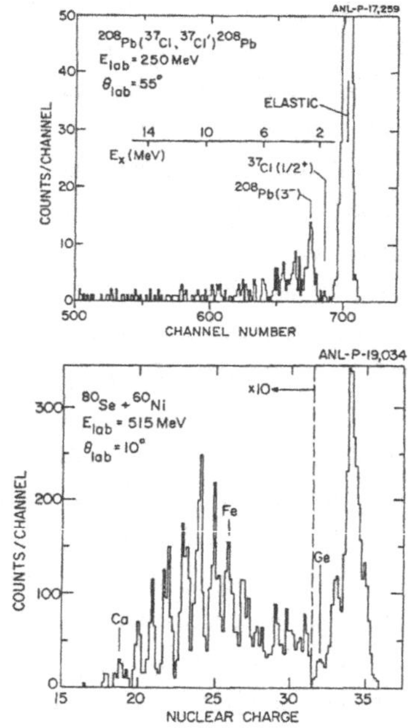

Fig. 1. Energy spectrum (top) for the reaction ^{37}Cl+^{208}Pb and nuclear charge spectrum (bottom) for the system ^{80}Se+^{60}Ni, as measured with a split pole magnetic spectrograph and a position sensitive ionization chamber in the focal plane.

are shown in Fig. 1b obtained from the scattering of ^{80}Se ions on ^{60}Ni.

The area of elastic scattering has been discussed in several other contributions to this symposium[8]. It should be mentioned, however, that an extension of the work of Ref. 8 towards heavier masses is extremely difficult. The decreasing energy resolution for even heavier beams caused mainly by target effects puts a strong limitation on the choice of projectile and target nuclei. In particular at low energies, where most of the interesting effects occur, target effects become more and more severe. The heaviest system where pure elastic scattering has been measured so far is still the ^{86}Kr + ^{208}Pb system, measured about 10 years ago at GSI[9].

For heavy ion induced inelastic scattering, the next more complex process, several studies have been performed with projectiles up to ^{86}Kr. As mentioned above the main difference to inelastic scattering induced by light ions is that in many cases several channels have to be treated coherently in a coupled channel calculation. An example of this effect is shown in Fig. 2 for the two systems

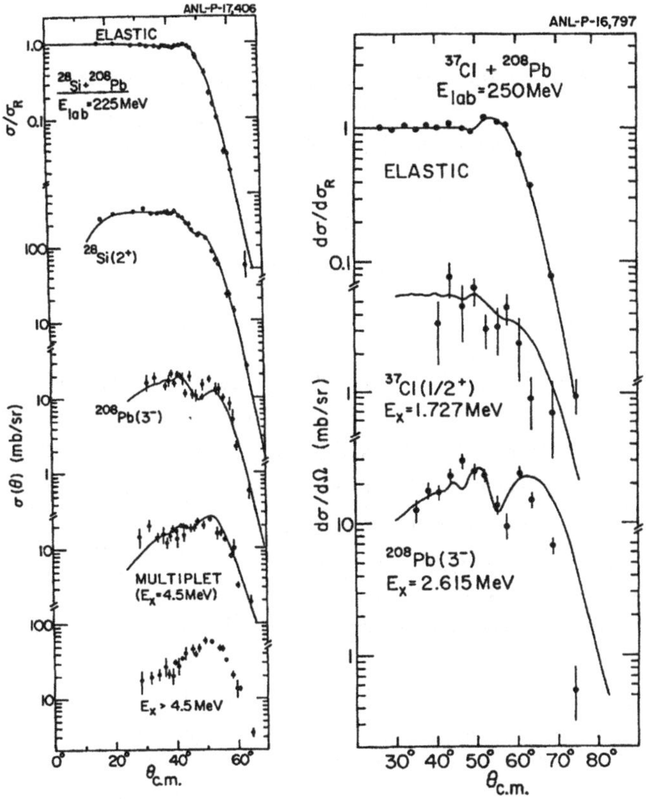

Fig. 2. Angular distributions for elastic and inelastic scattering for the systems ^{28}Si and ^{37}Cl on ^{208}Pb. The solid lines are the results of CC (^{28}Si) and DWBA (^{37}Cl) calculations.

^{28}Si,^{37}Cl + ^{208}Pb. The ^{28}Si + ^{208}Pb system requires a full coupled channel treatment[10] because the 2^+ state in ^{28}Si is strongly excited, while the ^{37}Cl + ^{208}Pb case can be calculated within the simple one-step DWBA[11]. It should be mentioned that except for the optical potential and the size of the reorientation matrix element all other coupling parameters were take from measured B(eλ)-values and no adjustment to the absolute cross sections was made. With this in mind the agreement is rather impressive both for the strongly excited collective 2^+ state, but also for the single-particle transition in ^{37}Cl. Similar observations have been made for other systems and one can state that for low-lying states CC and DWBA calculations give reliable absolute cross sections. It should be mentioned, however, that not everything is completely understood in this field. As can be seen from Fig. 2 for the ^{37}Cl + ^{208}Pb system the 3^- state in ^{208}Pb is poorly described in particular at larger angles. Another example is given in Fig. 3 where the angular distributions for the 3⁻ state in ^{208}Pb are given for various

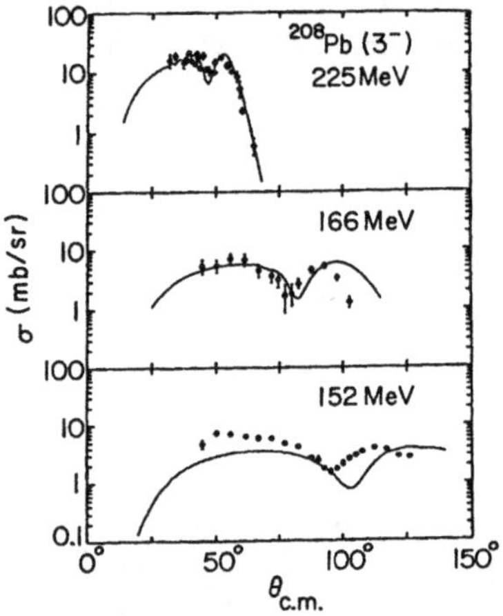

Fig. 3. Angular distributions for the excitation of the 3⁻ states in ^{208}Pb by ^{28}Si at different bombarding energies. The sold lines are the result of CC calculations.

bombarding energies[12]. The solid lines are the result of CC calculations using the same potential which gives excellent agreement for elastic scattering and 2^+ excitation. There is a considerable overprediction of the cross section in the backward hemisphere for this presumably strongly collective 3⁻ state at the lowest bombarding energies. The reason for this discrepancy is not completely understood at the moment. Recently it was observed that for the excitation of the 3⁻ state in ^{208}Pb by ^{16}O at low bombarding energies an energy-dependent interaction potential had to be used[13]. It would be interesting to see if similar effects could explain the ^{28}Si + ^{208}Pb system.

While a reasonable amount of experimental data is available for inelastic scattering one-nucleon transfer reactions to resolved states involving medium mass projectiles (A>20) have not been studied very extensively. As for inelastic scattering nuclei around closed shells had to be used for projectile and/or target in most cases. An example of an energy spectrum[14] for the $^{40}Ca(^{28}Si,^{29}Si)^{39}Ca$ reaction is shown in Fig. 4.

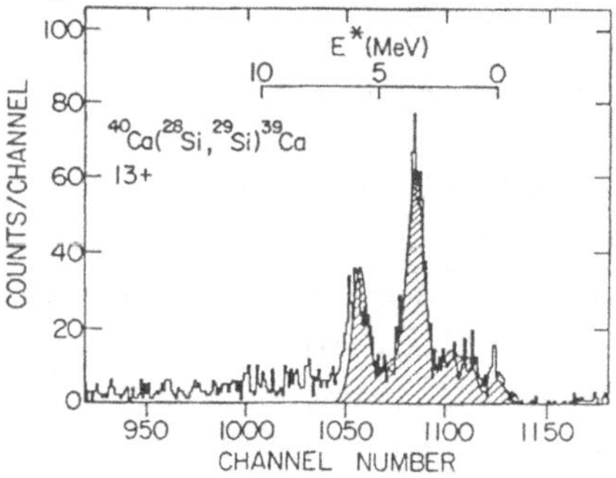

Fig. 4. Energy spectrum for the one-neutron transfer reaction ^{40}Ca (^{28}Si,^{29}Si)^{39}Ca at E_{lab}=225 MeV. The cross hatched area is a theoretical spectrum calculated within DWBA and spectroscopic factor from light ion reactions.

Good agreement with the low-lying energy spectrum is obtained with a DWBA treatment, indicated by the cross-hatched area in Fig. 4 using spectroscopic factors taken from light ion reactions. Deviations are generally less than 50% which is of the same order of magnitude as the uncertainty of spectroscopic factors. For the higher excitation energy region spectroscopic factors are in many cases not available. Caused by the larger level density in heavier nuclei transitions to individual states can not be resolved experimentally. Therefore only the energy integrated strength in the low excitation energy range can be analyzed. Two examples for the reactions ^{28}Si + ^{208}Pb (Ref. 10) and ^{58}Ni +^{208}Pb (Ref. 15), are shown in Fig. 5.

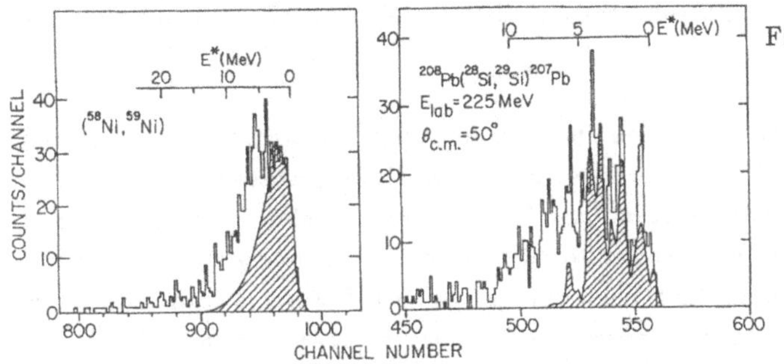

Fig. 5. Energy spectra for the one-neutron transfer reactions (^{58}Ni, ^{59}Ni) (left) and (^{28}Si, ^{29}Si) (right) on ^{208}Pb, measured at E_{lab}=375 MeV and E_{lab}=225 MeV, respectively. The cross hatched area is explained in Fig. 4.

For the one-particle transfer reactions one can therefore summarize that deviations between theory and experiment by about factors of two are not uncommon.

Two-nucleon transfers are the next more complicated processes. While two-particle transfers played an important role in the study of pairing effects in light ion reactions similar experiments with projectiles heavier than mass 20 populating individual states have been carried out only recently[16]. These experiments are very difficult because the transfer strength to low-lying states is usually very small. In order to get a general idea about the problems which have been encountered I want to mention only a few results involving ^{16}O projectiles. In Fig. 6 the ratio between the measured cross section and the DWBA prediction (assuming a cluster transfer)

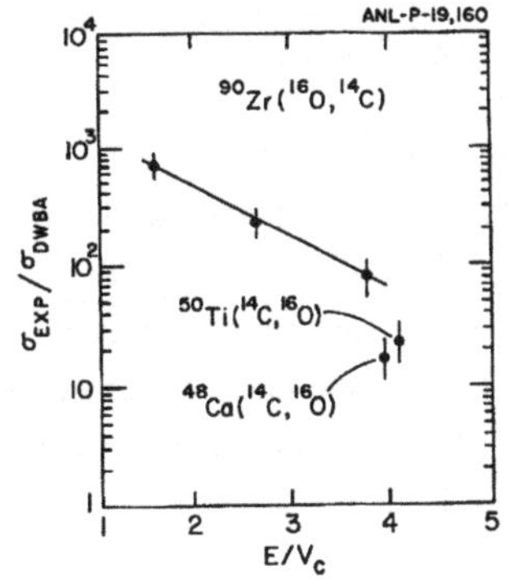

Fig. 6. Ratio of the experimental cross sections and the DWBA-predictions (assuming a cluster transfer) for the two-proton transfer reaction ^{90}Zr(^{16}O,^{14}C)^{92}Mo measured at different bombarding energies. The (^{14}C,^{16}O) data are from Ref. 18.

is plotted for the two-proton transfer reaction $^{90}Zr(^{16}O,^{14}C)^{92}Mo(0^+)$ as function of the bombarding energy[17]. Some data for the $(^{14}C,^{16}O)$ reaction on lighter targets are also included[18]. The discrepancy between theory and experiment is about 700 for low energies and decreases at higher energies. The inclusion of 2-step processes[19] reduces the discrepancy by about a factor of 4-5 and is therefore not able to explain the large experimental cross sections at low bombarding energies. Compared with the good agreement observed for simpler processes like inelastic scattering and one-nucleon transfer it surprising to find that similar DWBA or CC calculations suddenly underpredict measured cross section by several orders of magnitude. This shows that our theoretical understanding of nucleon transfer changes dramatically between one and two-particle transfer reactions.

Similarly more complicated heavy ion induced multiparticle transfer reactions can not be described within present models. The theory for these processes needs considerable improvements, in particular at low bombarding energies, where one would expect that the processes are usually easier to understand. Unfortunately not much development has been going on in that field during the last few years although there is a real need for new ideas. At this point I will leave the reactions involving transitions to individual nuclear states and discuss the behavior of more global quantities like energy-integrated transfer cross sections.

In the beginning I will discuss the one-neutron transfer reactions which for most cases show the largest energy-integrated cross sections. Fig. 7 shows

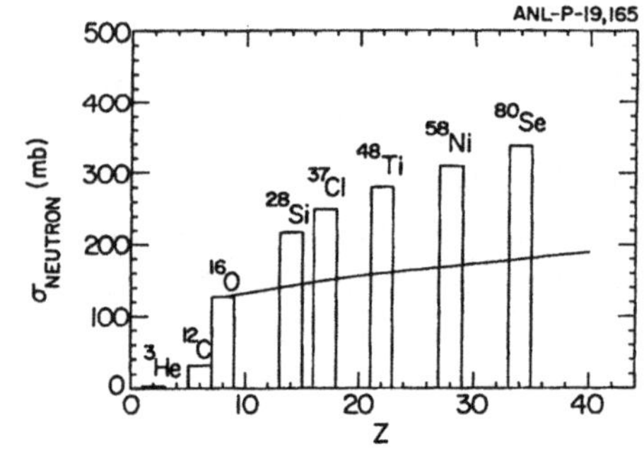

Fig. 7. Neutron transfer cross sections for reactions induced by various projectiles on ^{208}Pb at energies about 25% above the barrier. The solid line is the expected mass dependence calculated within a single geometrical model and normalized to the case of ^{16}O.

energy and angle-integrated cross sections for various projectiles bombarding a ^{208}Pb target at energies about 25% above the Coulomb barrier. It is clear that the absolute strength of the cross sections increases dramatically going from the light particles like ^{3}He (Ref. 20), ^{12}C (Ref. 21) and ^{16}O (Ref. 22) to the heavier projectiles. In addition to the Q-values, which influence the transfer strength for these reactions, the number of single particle states which can be populated in the transfer is another factor responsible for the strong increase in the energy and angle integrated transfer cross sections. With only these simple ingredients (phase space and Q-value) one already gets a quite good understanding of the absolute strength of the transfer cross sections. From simple matching conditions one expects that for neutron transfer processes states at excitation energies corresponding to Q-values around Q=0 should be strongly populated. If one considers a reaction with a positive Q_{gg}-value then a large number of excited states can be populated whereas a smaller integrated cross section is to be expected for a reaction with negative Q_{gg}-value. One therefore expects a cross section behavior as sketched schematically in Fig. 8. The experimental data for one-neutron transfer reactions induced by

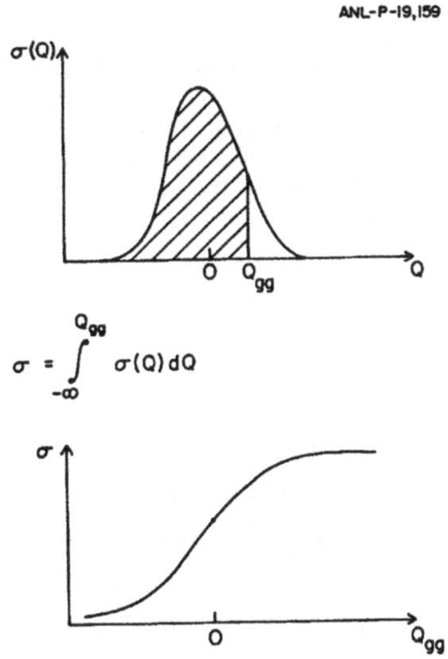

Fig. 8. Schematic for the expected Q_{gg} dependence of energy-integrated neutron-transfer cross sections. See text for details.

projectiles from ^{28}Si to ^{86}Kr and targets from ^{58}Ni to ^{208}Pb are shown in Fig. 9 (top). One clearly observes the exponential increase as function of Q_{gg} and if,

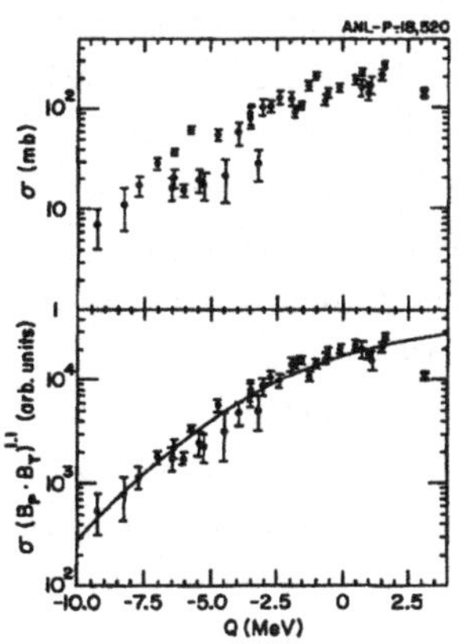

Fig. 9. top: One neutron transfer cross sections for various reactions induced by medium-mass projectiles on heavy targets (A_t>120) plotted as function of the ground-state Q-value Q_{gg}. bottom: same as top, but corrected for binding energy effects (see Ref. 23).

in addition, effects of the binding energy are taken into account all data points follow a smooth behavior[23], as illustrated in the bottom part of Fig. 9.

The next question is the extent to which charged particle and multi-particle transfer reactions can be described within the same framework. Several theoretical calculations have shown that multi-particle transfer processes proceed mainly via sequential transfer routes[2,24]. In Fig. 10 reduced (i.e. binding energy corrected)

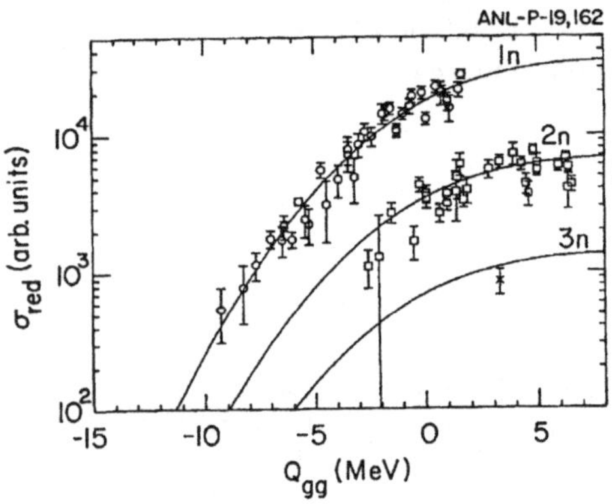

Fig. 10. Binding energy corrected cross sections for one, two and three-neutron transfer reactions induced by medium mass projectiles on heavy targets. See text for details.

energy and angle-integrated two-neutron transfer cross sections are plotted as function of the groundstate Q-value for various systems. It is clear that their dependence on Q_{gg} is similar to the one observed for one-neutron transfer but the absolute strength is smaller by about a factor of 5. The only case of a three neutron transfer reaction is smaller yet by an additional factor of 5. The solid curves are obtained from the one-neutron transfer reactions scaled down by the appropriate factors of 5 and 25, respectively. The agreement is generally quite good.

Contrary to the observations for neutron transfer reactions, the optimum Q-value for proton transfer depends on the bombarding energy and the charge of projectile and target. Cross sections for charged particle transfer reactions are therefore expected to be energy, Q-value and projectile dependent. For these reasons a comparison of different systems is not feasible without introducing additional parameters. In order to compensate for the energy and charge dependence, Q-value corrected (along similar lines as in Ref. 23) cross section ratios for 2-neutron/1-neutron and 2-proton/1-proton transfer reactions are plotted in Fig. 11 as function of the charge of the projectile. The ratio for the neutron transfers is about 0.2, but about twice that value is observed for proton transfer reactions. It should be noted that proton pairing effects (i.e. populating states in even and odd-Z ejectiles) can not be solely responsible for this behavior, because the cross section

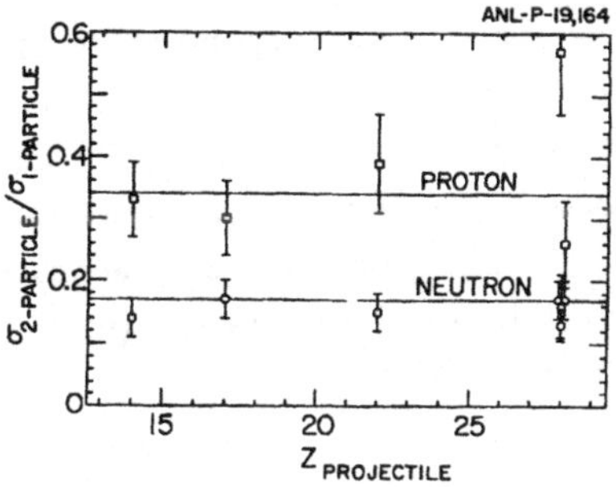

Fig. 11. Ratio of two-particle and one-particle - transfer cross sections for proton and neutron transfer, respectively. The solid lines serve to guide the eye.

ratio does not change in going from even Z projectiles (^{28}Si or ^{48}Ti) to an odd Z projectile ^{37}Cl. For the cases investigated so far it therefore seems that the 2-proton transfer reactions show more enhancement if compared to the corresponding neutron transfer reactions. Similar observations have been made recently for proton transfer studies in the system ^{88}Sr + ^{144}Sm (Ref. 25).

Let me mention two other topics in quasi-elastic reactions which have been discussed in the recent literature. One is the so-called 'slope anomaly' of neutron transfer reactions. The slope of the angular distribution for two-neutron transfer reactions at forward angles shows, if plotted vs. the distance of closest approach a slope similar to the one obtained from the angular distribution for one-neutron transfer. From binding energy arguments one would expect, however a difference by a factor of two. Heavier deformed nuclei, in particular, show large deviations from the expected behavior[26]. An example of a case where the theoretically expected slopes are observed is the system ^{58}Ni + ^{208}Pb as shown in Fig. 12.

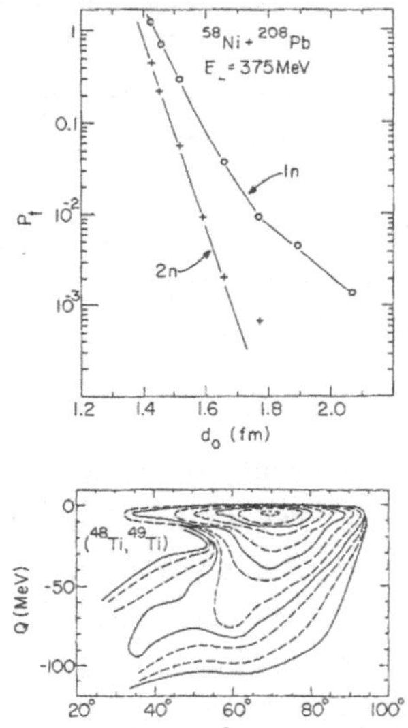

Fig. 12. top: Transfer probability for the one and two-neutron transfer reactions (^{58}Ni,^{59}Ni) and ^{58}Ni,^{60}Ni) on ^{208}Pb, plotted as function of the distance of closest approach.
bottom: Wilczynski - plot for the one-neutron transfer reaction ^{208}Pb(^{48}Ti,^{49}Ti)^{207}Pb.

The deviations from a straight line at the largest distances corresponding to the most forward angles is caused by the inclusion of deep inelastic processes as can be seen from a Wilczynski plot for the (^{48}Ti,^{49}Ti) reaction shown in Fig. 12b. We have analyzed the slopes for one- and two-neutron transfer reactions for a large number of systems involving projectiles from ^{48}Ti up to ^{80}Se. The slopes of the one-neutron transfer reactions (open circles) together with the theoretical expectation (solid line) are plotted in Fig. 13 as function of the mass number of

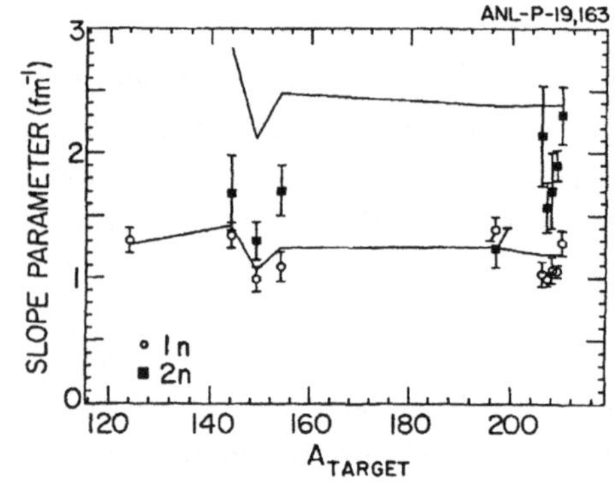

Fig. 13. Slope parameter for 1n and 2n-transfer reactions induced by medium weight projectiles for various target nuclei. See text for details.

the target and good agreement is generally observed. A different picture evolves, however, for the two-neutron transfers (full squares in Fig 13). Only very few nuclei (in particular ^{58}Ni + ^{208}Pb) are in agreement with the slope expected for a two-neutron reaction. Other projectiles (e.g. ^{48}Ti, ^{64}Ni) on the same target (^{208}Pb) show large deviations from the expected behavior. Reactions with ^{58}Ni on spherical and deformed Sm isotopes (mass 144, 149 and 154, respectively) show all similar deviations from the solid line. The largest discrepancy is found for the reaction ^{57}Fe + ^{197}Au where the two-neutron slope is the same as the slope for the one-neutron transfer reaction. From the data which were analyzed so far we can therefore conclude that the slopes of the angular distributions for a two-neutron transfer reaction induced by projectiles up to the mass 80 region is not particularly different for a spherical or for a deformed target. There seems to be, however, a strong nuclear structure dependence as seen e.g. from the results with the ^{208}Pb target and different projectiles.

The last question I want to address has to do with the influence of quasi-elastic transfer reactions on the strength of the fusion process. It has been shown that a coupling to inelastic scattering in many cases can explain the large subbarrier fusion enhancement observed particularly in systems involving medium mass projectiles[1]. The variation of the enhancement within a given projectile-target combination involving different isotopes, however, is not easily understood by including inelastic excitation alone. One example is the much discussed ^{58}Ni+ ^{58,64}Ni system[27], but many others have been measured. The variations within such systems have been attributed generally to the influence of transfer reactions. Experimentally it has been observed that for heavier projectiles (Ti, Ni) neutron transfers are the most important channels whereas for lighter systems proton transfers can also be strong[28]. Generally the strength of the quasi-elastic transfers scales with the observed fusion enhancement. One of the largest fusion enhancements measured so far involves the system ^{64}Ni + ^{100}Mo (Ref. 29). If the enhancement is measured by the energy shift necessary to bring the experimental excitation function in agreement with a one-dimensional barrier penetration calculation, then ^{64}Ni + ^{100}Mo requires a shift of about 8 MeV. Calculations for this system involving higher order couplings[27] do not converge including first and second order effects[30]. Coupling to inelastic channels requires β-values which are about 50% larger than the measured deformation parameters[29]. From the systematics of transfer cross sections involving Si....Se projectiles on heavy targets (Sn...Pb) we would expect[23] neutron transfer cross sections which are of the order of 120 mb. We have recently measured transfer cross sections for the systems ^{58,64}Ni + ^{100}Mo at energies slightly below and above the Coulomb barrier by bombarding Ni targets with a ^{100}Mo beam. Surprisingly the measured (preliminary) cross sections are smaller than the values quoted above by about a factor of 5. Two-neutron transfer cross sections are even smaller by another factor of 4. In ^{64}Ni + ^{100}Mo we have therefore a system where so far all our theoretical attempts of coupled channel treatments at least up to second order are unsuccessful. The nuclei around mass 100 therefore seem to be a challenge, both in terms of their small transfer cross sections and in terms of their large fusion enhancement.

From these few examples it is clear that our understanding of quasi-elastic reactions, i.e. processes where relatively small amounts of energy, mass and charge are exchanged during the nuclear interaction, is limited to the most simple processes, like elastic and inelastic scattering and one-nucleon transfer reactions. More complicated processes like two-particle transfers, which are important doorway states towards more complex reactions (deep inelastic scattering and fusion) are not well understood. Furthermore several nuclear structure effects are observed in these reactions which still have to be explained.

I would like to thank all my colleagues who participated in the experiments, in particular A. van den Berg, B. Glagola, J. Kolata, D. Kovar, W. Kutschera, S. Sanders, M. Vineyard, R. Vojtech, F. Wolfs, and T. F. Wang. This research was supported by the U.S. Department of Energy, Nuclear Physics Division, under contract W-31-109-ENG-38.

References:

1. Proc. of the Symp. "The Many Facets of Heavy Ion Fusion Reactions", Argonne (1986) ANL-PHY-86-1 and references therein.
2. R.J. Asciutto and E.A. Seglie, in "Treatise on Heavy Ion Science" ed. by D.A. Bromley Plenum Press New York, 1985, Vol. 1.
3. S. Landowne and A. Vitturi, in "Treatise on Heavy Ion Science" ed. by D.A. Bromley Plenum Press New York, 1985, Vol. 1.
4. M.H. Macfarlane and S.C. Pieper, ANL report 76-11 (rev. 1), (1978).
5. H.A. Weidenmüller, in "Prog. in Nuclear and Particle Phys., Pergamon Press (1980).
6. H.H. Wolter, in "Windsurfing the Fermi sea", Vol. II, Elsevier 1987 and references therein.
7. J.R. Erskine et al., Nucl. Inst. Meth. 135, 67 (1976), K.E. Rehm and F.L.H. Wolfs, subm. to Nucl. Instr. Meth.
8. J.S. Lilley, contribution to this symposium, M.A. Nagarajan, contribution to this symposium, A.M. Stefanini, contribution to this symposium.
9. J.C. Lie et al., Phys. Rev. Lett. 47, 1039 (1981).
10. J.J. Kolata et al., Phys. Rev. 30C, 125 (1984).
11. K.E. Rehm et al., to be published.
12. R.J. Vojtech, Thesis, Univ. of Notre Dame, (1987), unpublished.
13. C.S. Lim et al., Phys. Lett. 200B, 263 (1988).
14. M.F. Vineyard et al., Phys. Rev. 33C, 1325 (1986).
15. K.E. Rehm et al., Phys. Rev. Lett. 51, 1426 (1983).
16. C.Y. Wu et al., submitted for publication.
17. K.E. Rehm et al., to be published.
18. F. Videbaek et al., Nucl. Phys. A451, 131 (1986), and private communication.
19. P.P. Tung et al., Phys. Rev. C18, 1663 (1978).
20. W.P. Alford and D.G. Burke, Phys. Rev. 185, 1560 (1969).
21. K.S. Toth et al., Phys. Rev. C14, 1471 (1976).
22. F. Videbaek et al., Phys. Rev C15, 954 (1977).
23. A.M. van den Berg et al., Phys. Lett. 194B, 334 (1987).
24. T. Kammuri, Nucl. Phys. A259, 343 (1975).
25. R. Künkel et al., GSI scientific report 1986, GSI 87-1, p.41.
26. S. Juutinen et al., Phys. Lett. 192B, 307 (1987).
27. H. Esbensen and S. Landowne, Phys. Rev. 185B, 15 (1987).
28. A.M. Stefanini et al., Phys. Lett 185B, 15 (1987).
29. M.L. Halbert et al., Proc. 6th Adriatic Intern. Conf. on Nucl. Phys. 1987.
30. S. Landowne, private communication.

TRANSFER REACTIONS FOR THE ^{50}Ti + ^{90}Zr SYSTEM BELOW THE COULOMB BARRIER

H. J. Kim
Oak Ridge National Laboratory
Oak Ridge, Tennessee 37831, U.S.A.

Experimentally observed enhancements of fusion cross sections are often attributed to result from coupling interactions between fusion and transfer channels. Existing experimental results [1] indicate that large fusion enhancement generally goes with large neutron-transfer cross section and thus supports this attribute, but observations of more specific relations between these quasielastic channels have not yet been reported. Here we present some experimental results that pertain to fusion-transfer coupling interactions.

Transfer reactions between heavy nuclei at near-barrier energies are well described by semiclassical theories in which the reactions are assumed to occur on classically prescribed orbits; examples can be found in Refs. 2-6. A simple and clear picture emerges from these examples: at energies below and near the barrier, nucleon-transfer reactions proceed via the tunneling of valence nucleons between the interacting nuclei on the classical orbit, but the tunneling evolves into a grazing, direct reaction as the energy exceeds the interaction barrier.

With analysis of the results in semiclassical approaches in mind, cross sections for a number of element- (Z) and mass- (M) identified ejectiles resulting from the ^{90}Zr-beam bombardment of ^{50}Ti were measured for four bombarding energies in the 276- to 305-MeV range and lab angles ranging from 8 to 23 degrees with 2- and 3-degree steps. The traditional role of target and projectile was reversed in this study in order to measure large c.m. angle differential cross sections. Target and target-like particles ejected from thin (30 and 70 $\mu g/cm^2$) targets were detected by a position-sensitive gas-filled detector system [7] placed at the focal plane of the ORNL split-pole spectrometer. The Z of ejectiles was identified from the energy (E) and energy-loss (ΔE) data; the M was identified from the E and magnetic rigidity ($B\rho$) data. Figure 1 illustrates typical Z and M resolutions attained. ^{50}Ti counts from an Si(Au) detector placed at a fixed angle (42°) provided normalization factors for absolute cross sections. (Elastic scattering was assumed to be pure Rutherford at this angle.) A single peak with a low-energy tail dominates all observed target-ejectile (^{50}Ti) focal-plane spectra. The "elastic" peak broadens, and its tail becomes more extensive with increasing angle (at given energy) and with increasing energy (at given angle), indicating the contribution to the peak from numerous but weak inelastic excitations. The resolution (1.8 MeV) was not sufficient to resolve individual peaks. A single broad (> energy resolution) peak or bump also dominates the transfer-ejectile (^{49}Ti and ^{51}V) spectra. Energy-integrated differential cross sections for ^{49}Ti, ^{50}Ti, and ^{51}V ejectiles were obtained from the focal-plane spectra for the dominant charge states q = 18, 19, and 20.

Among the many versions of semiclassical theory in the literature [4,8-10], that described by von Oertzen et al. [4] is followed in this report. Briefly, a quasielastic reaction occurs on the classically prescribed orbit with a probability P that is particular to the reaction. Thus, the differential cross section for a given channel i is

$$(d\sigma)_i = P_i(d\sigma)_{qe} = P_i(1 - p_a)(d\sigma)_{sc} , \quad (1)$$

where $\Sigma P_i = 1$, $(d\sigma)_{qe}$ and $(d\sigma)_{sc}$ are quasielastic and theoretical (e.g., Rutherford) scattering cross sections, and p_a is the absorption probability into such nonquasielastic channels as fusion, deep-inelastic, etc. The absorption in semiclassical theory is given by the mean-free-path attenuation of incident flux [9]. Experimentally, the quasielastic cross section $(d\sigma)_{qe}$ is obtained by adding contributions from all observed channels, and the ratio of transfer to quasielastic cross section $(d\sigma)_i/(d\sigma)_{qe}$ is the probability P_i. The probability can be factored as P = tSF for one-nucleon transfer reactions. Here t is the intrinsic probability (e.g., transmission coefficient in the tunneling model); S is the factor to account for nuclear structure (spectroscopic factors); and F is the reduction factor to account for the orbit mismatch or perturbation (caused by Q-value, changes in angular momentum L, M, Z, etc.). For the whole orbit, t can be approximated by its value t(D) at the perihelion, where the nuclei are separated by the apsidal distance D, and with this approximation the transfer probability becomes

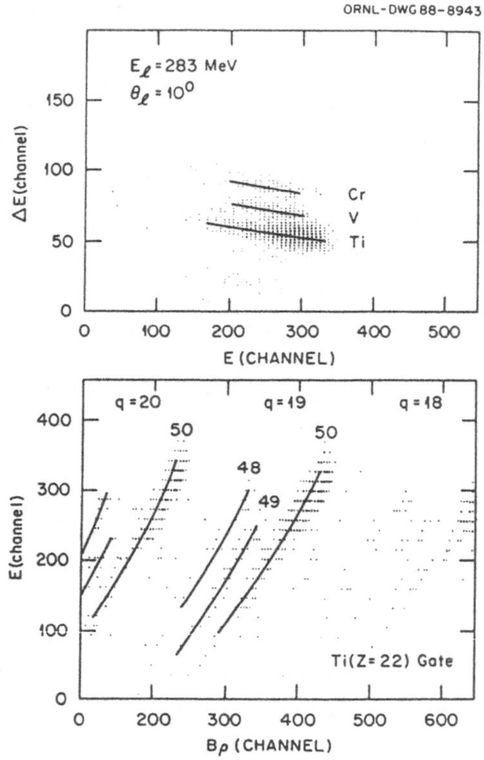

Fig. 1

$$P_i[D(E,\theta)] = Ct_i(D) , \quad (2)$$

where C stands for the product SF. An implicit dependence of D on orbit parameters E and θ are shown. The factor C generally depends on collision dynamics as well as on nuclear structure [2,4,5]; however, previous studies [4-6] have shown that for such cases as this, where many individual transitions are included in the energy-integrated cross section, it is insensitive to or even independent of E and θ.

The ratios of differential cross sections measured at four different energies were combined into common plots using the relation for Coulomb orbits, $D_c = [(Z_p Z_t e^2)/(2E)][1 + \csc(\theta/2)]$, and are shown in Fig. 2. If the short-range attractive nuclear force could be ignored, the probability should follow a straight line on a semi-log plot; i.e., $P(D_c) \simeq \exp(-2kD_c)$, where k is a constant. But, as illustrated in Fig. 2, this is not the case. The slopes of the solid lines, which were drawn to represent large D_c results, were calculated in the manner described below. As was pointed out by Körner et al. [11] and Christensen et al. [2], the deviation of the kind seen in Fig. 2 can result from the distortion of Coulomb orbits by the tail of the attractive nuclear potential. The tail causes P to rise faster by pulling, or refracting, the Coulomb orbit inward, forcing the actual apsidal distance to be smaller than the corresponding D_c. The rise ceases when the strong-absorption regime is reached, and the probability turns over, producing a shape similar to those seen in Fig. 2.

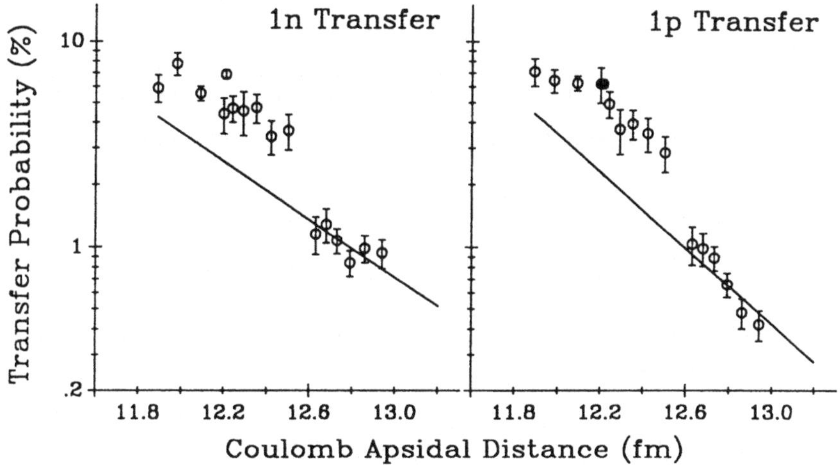

Fig. 2

To find a more realistic relation between the actual apsidal distance D and angle θ, a nucleus-nucleus potential was added to the Coulomb deflection field and then the orbit integral,

$$\theta = \pi - 2L(E,D)\int_D^\infty [E - U_N - U_c - U_L]^{1/2} \frac{dr}{r^2} , \qquad (3)$$

was numerically evaluated. In this integral L(E,D) is the classical angular momentum; and U_N, U_c, and U_L are the nucleus-nucleus, Coulomb, and centrifugal potentials, respectively. A Woods-Saxon equivalent of Lozano and Madurga's exponential potential [12], which represents the tail region especially well, was adopted for this calculation. The left panel of Fig. 3 shows the interaction potential $U_N + U_c$ as a function of internuclear distance. The top of the interaction barrier, which is

105.6-MeV high, is reached at 11.30 fermi. The θ vs. D relation for the c.m. energies corresponding to the four bombarding energies is shown in the right panel for back angles. (Since D is directly related to the impact parameter or L, the lower figure shows, in effect, classical deflection functions of the system.) Also included in the same figure is D_c (solid curves) for reference. As expected, the

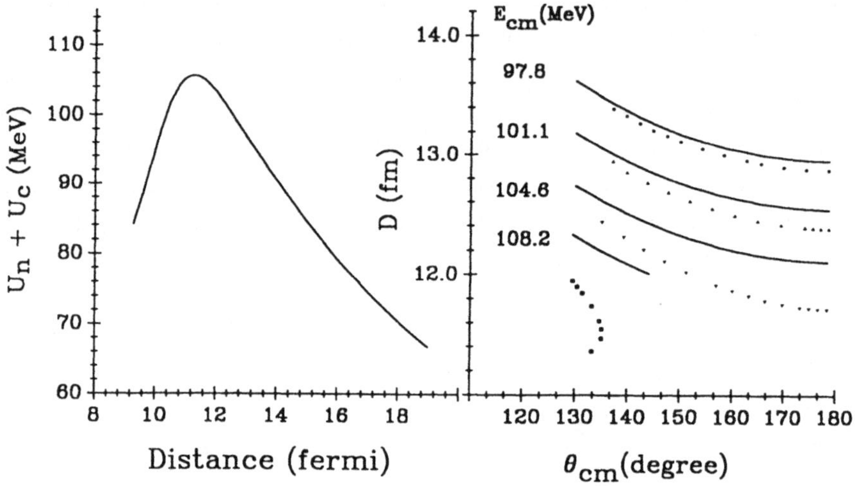

Fig. 3

added attractive potential does reduce apsidal distances. The orbits are so modified that the relation between θ and D is still unique for three lower energies, but not for the highest. The interaction barrier ($U_N + U_C$) is exceeded at the highest energy (108.2 MeV, see left portion of Fig. 3), and the relation is multi-valued (in the angular range shown) because the paths of severely refracted (pulled inward) trajectories, which turn around at inner turning points, can merge with the paths followed by more distant trajectories, which turn around at outer turning points. (See, for example, Ref. 9 for these and more complicated trajectories.) Small changes of the values of diffuseness or strength parameters of the nucleus-nucleus interaction potential do not alter the θ vs. D relation materially.

The transfer probabilities replotted vs. new apsidal distance are presented in Fig. 4. Although the theory is not applicable, the cross-section ratios of those close collisions for which the outer turning points do not exist are included in the figure, for sake of completeness. They are shown in the hatched areas. The solid lines are from the tunneling model of Brosa and Gross [13]. A valence nucleon which is initially bound in the potential well of the core nucleus eventually acquires a sufficient probability to be found in the potential well of the other nuclei if the distance separating the nuclei becomes close enough. In this model the nucleon-transfer probability is the transmission coefficient of the nucleon to tunnel through the barrier that is created by the overlapping tails of the two potential wells.

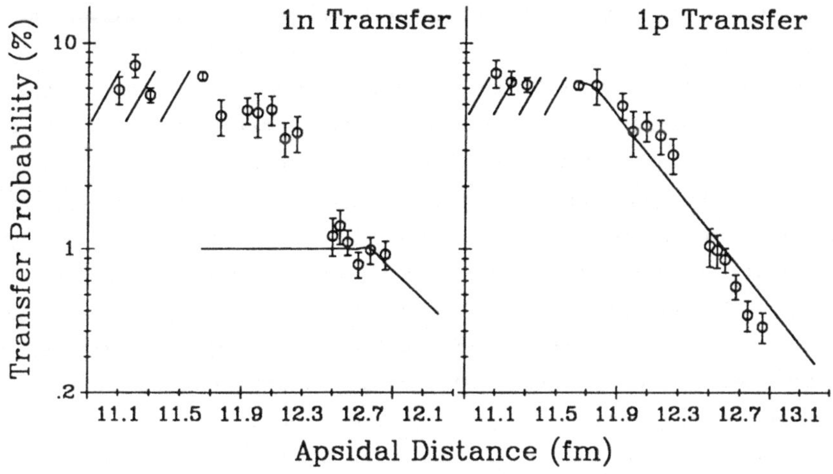

Fig. 4

Figure 5 shows profiles of the combined potential wells (in a plane perpendicular to the axis joining the nuclei) for the proton, $^{50}Ti + (^{89}Y + p)$, and for the neutron, $^{90}Zr + (^{49}Ti + n)$, cases at two representative internuclear distances. The standard potential wells [14] and the known nucleon binding energies, $B_N = 10.9$ and $B_\rho = 8.4$ MeV, were used for these figures and for the calculation of nucleon transmission t(D). The WBK- and parabolic-methods were used for the t(D) calculation with the frozen potential assumption. Both methods gave results that are practically identical: t(D) is of $\exp[k(D_o - D)]$ form, where k and D_o are constants. This is the form expected from the DWBA [8] and other theories [10,15] at large internuclear distances. The value of k for the neutron is primarily determined by the binding energy, but the combination of binding plus the Coulomb

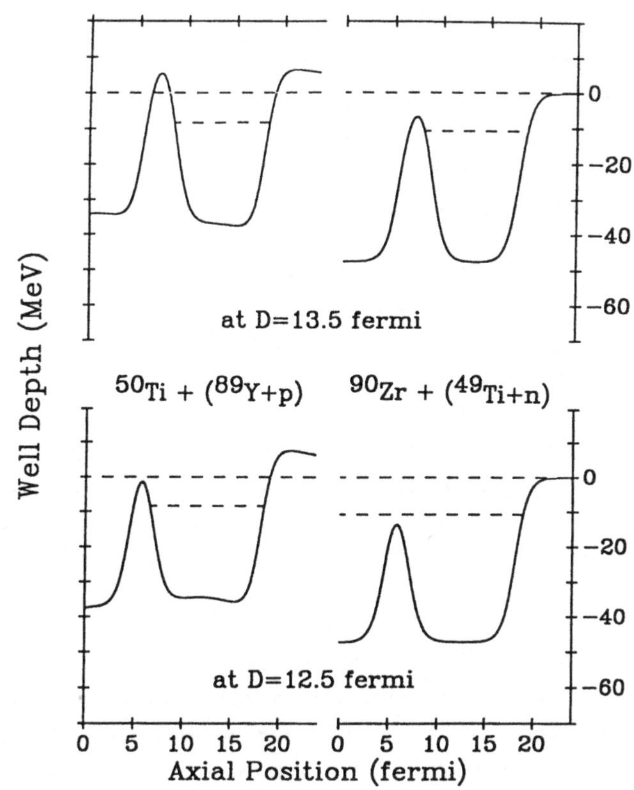

Fig. 5

repulsion energy is the dominant factor determining the proton k value. As shown in Fig. 5, the barrier for the proton is more substantial than the barrier for the neutron even though the neutron binding energy is larger. Accordingly, the proton k value is about 30% larger. For a given binding energy (for neutron), or binding plus Coulomb energy (for proton), the well radii, i.e., $r_0A_p^{1/3}$ and $r_0A_t^{1/3}$, are the critical factors that determine the value of D_0. The Coulomb repulsion causes the proton D_0 value to be significantly smaller (by ~1 fermi in this case) than the neutron D_0 value. Diffuseness and depth of the wells influence the values of k and D_0 only mildly. The nucleon transmission coefficient t(D) rises linearly on a semi-log plot as the barrier becomes lower with decreasing D, but saturates once the barrier height becomes equal to or lower than the binding energy (i.e., t(D) = 1 for $D < D_0$). (For example, t(D) = 1 for the case depicted in the lower right-hand corner of Fig. 5.) This feature of the model is illustrated by the theoretical results (solid lines) shown in Fig. 4, after scaling to represent the data. The theory reproduces the observed 1p-transfer feature very well, including the saturation at correct D. The same theory, on the other hand, gives much smaller 1n-transfer probability. This follows because the theory predicts an early onset of the saturation. A substantial (0.3 f or more) shift of this onset can be obtained if a correspondingly large change of the well radii ($\Delta D_0/D_0 = \Delta r_0/r_0$) is made. But such a change results in "wrong" nucleon wells and hence is not a viable procedure. Reasonable changes (±5%) of the value of the diffuseness or depth do not yield comparable shifts of D_0.

The saturation of the transfer probability is based on the one-dimensional tunneling model, in which the transfer probability is assumed to be the transmission coefficient of the barrier at the saddle point. This is a reasonable assumption for distant collisions since the contribution away from the saddle point is insignificant due to the rapidly increasing barrier, but needs closer scrutiny for $D = D_0$ and smaller. For example, if the tunneling becomes multidimensional [16] inside D_0, or if, equivalently, a neck [17,18] is formed, then the neutron-transfer probability continues to increase through this newly acquired mechanism. Whatever its nature, Fig. 4 shows that the mechanism affects the neutron transfer but not the proton transfer.

Stelson [19] found the principal coupling mechanism that enhances fusion of heavy nuclei at subbarier energies to be the free flow of neutrons between the interacting nuclei. He fitted an extensive body of fusion cross sections with his distributed-barrier model and found that the threshold of barrier distribution systematically occurs near the apsidal distance where the tunneling probability of the least-bound neutron of the system first becomes 100% (or free flow). The threshold energy for the present system was determined from the experimental data [20] following the distributed-barrier model. It is 100.8 MeV. Figure 6 illustrates how the threshold energy was obtained from the data. The head-on collision apsidal distance at the threshold is 12.4 fermi (see Fig. 3). This distance is very close to D ~ 12.5 f where the 1n-transfer data begin to deviate from the tunneling theory prediction. The similarity of these two distances suggests the free flow of neutron is responsible for the large neutron-transfer probability as well as the enhancement of the fusion cross sections. It is very tempting to associate the neutron flow with the initial stage of the neck formation [17,18] since the neutron flow is the earliest process that connects the nuclei during the collision as their tails begin to overlap. An immediate consequence of the neck formation would be the enhancement of fusion cross section [17]. If the above association is valid, the neck should also enhance the neutron-transfer cross section. This expectation contrasts with the mechanism considered in Ref. 21 where the fusion cross section is enhanced at the expense of the neutron-transfer cross section.

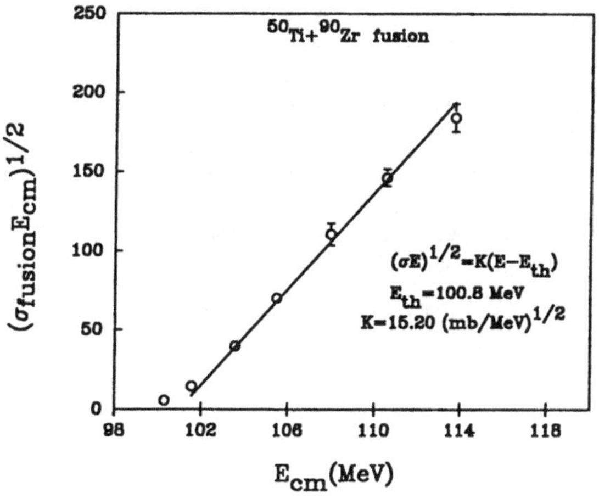

Fig. 6

This investigation is being extended to include similar studies of neighboring systems with the hope of gaining further insight into the quasielastic reactions between heavy nuclei near the barrier.

The work presented here is a collaborative effort from many colleagues. I want to thank particularly and acknowledge contributions from J. Gomez del Campo, P. H. Stelson, and D. Shapira. This research was sponsored by the Division of High Energy and Nuclear Physics, U.S. Department of Energy, under contract DE-AC05-84OR21400 with Martin Marietta Energy Systems, Inc.

REFERENCES

1. See, for example, J. Wiggins et al., Phys. Rev. C **3**, 1315 (1985); A. M. Stefanini et al., Phys. Rev. C **30**, 2088 (1984); R. Pengo et al., Nucl. Phys. **A411**, 255 (1983).
2. P. R. Christensen et al., Nucl. Phys. **A207**, 33 (1973).
3. K. Alder et al., Nucl. Phys. **A191**, 399 (1972).
4. W. von Oertzen et al., Z. Phys. A **326**, 463 (1987).
5. S. Siekmann et al., Z. Phys. A **307**, 113 (1982).
6. G. Wirth et al., Phys. Letts. **177B**, 282 (1986); G. Himmele et al., Nucl. Phys. **A404**, 401 (1983).
7. D. Shapira et al., Nucl. Instrum. Methods **169**, 77 (1980).
8. L.J.B. Goldfarb and W. von Oertzen, in Heavy-Ion Collisions, edited by R. Bock (North-Holland, Amsterdam, 1979), Vol. 1, Chapter 2.
9. R. A. Broglia and A. Winther, Heavy-Ion Reactions Lecture Notes (The Benjamin/Cummings, Reading, Massachusetts, 1981), Vol. 1, Chapter 3.
10. W. Norenberg and H. A. Weidenmuller, Lecture Notes in Physics: Introduction to the Theory of Heavy-Ion Collisions, edited by J. Ehlers, K. Hepp, H. A. Weidemuller, and J. Zittartz (Springer-Verlag, Berlin, 1976), Chapters 5 and 7.
11. H. J. Körner et al., Phys. Rev. C **7**, 107 (1973).
12. M. Lozano and G. Madurga, Nucl. Phys. **A334**, 349 (1980).
13. U. Brosa and D. H. E. Gross, Z. Phys. A **298**, 91 (1980).
14. A. Bohr and B. R. Mottelson, Nuclear Structure, Vol.1, p. 236 ff (Benjamin, New York, 1969).
15. A. Trabert, D. Trautmann, and K. Alder, Z. Phys. A **281**, 391 (1977); A. Trabert, D. Trautmann, and K. Alder, Z. Phys. A **274**, 213 (1975).
16. H. J. Krappe et al., Z. Phys. A **314**, 23 (1983).
17. C. E. Aguiar et al., Nucl. Phys. **A472**, 571 (1987).
18. A. Iwamoto and K. Harada, Z. Phys. A **326**, 201 (1987).
19. P. H. Stelson, Phys. Lett. **205B**, 190 (1988).
20. P. H. Stelson et al., Phys. Div. Prog. Rep., Sept. 30, 1985, ORNL-6233 (April 1986), p. 122.
21. R. Broglia et al., Phys. Rev. C **27**, 2433 (1983).

Correlation between Transfer and Fusion in Heavy Ion Reactions at the Coulomb Barrier

R. Bellwied, J.V. Kratz
Institut für Kernchemie, Universität Mainz

W. Reisdorf, D. Schüll
GSI Darmstadt
B. Kohlmeyer
Universität Marburg
R. Künkel
HMI Berlin

The fusion of massive heavy nuclei ($Z_1Z_2 > 1600$) is characterized by interesting phenomena. In very heavy, fissile systems ($Z_1+Z_2 > 100$) with dynamically suppressed fusion ('extra-push' /1/) the missing fusion cross section shows up in a process, termed quasi-fission, that forms fission-like fragments /2,3/ without ever forming a compound nucleus. Such a process has also been reported /4/ in fission studies on intermediate mass nuclei with fission barriers vanishing at high angular momenta. It is interesting to investigate in what reaction channels the missing cross section shows up when compound nuclei with a still sizeable macroscopic fission barrier are prevented from being formed by dynamical entrance channel effects /5/. In the macroscopic extra-push picture the missing fusion cross section is expected to reappear in massive transfer channels, because the nuclear complex which is then being formed is expected to have a considerable life-time. However, systems at the extra-push threshold also show a nuclear structure dependent 'microscopic' correlation between the sub-barrier behavior of fusion and the number of valence particles or holes. Keller et al. /6/ show that for Zr+Zr-reactions the system with the closed neutron shell, $^{90}Zr+^{90}Zr$, has a much steeper excitation function below the Coulomb barrier than the systems $^{90}Zr+^{92,96}Zr$. In recent years strong theoretical efforts have been made to explain these effects in the framework of quantum mechanical coupled channel calculations /7,8/. Also, only in the closed-shell system the experimental barrier agrees with the nominal fusion barrier, while there is an increase in the barrier height in reactions with $^{92,96}Zr$ due to the presence of extra valence neutrons. Similar observations were made for other isotopic series, e.g. for $^{124}Sn+^{90,92,94,96}Zr$ /5/ and for a series of ^{100}Mo induced reactions /9/. First measurements of the transfer cross sections have indicated that more copious transfer is going on in systems with the larger number of valence nucleons. It is therefore very tempting to associate the isotopic trends in sub-barrier fusion with quasi-elastic reactions (excess energy used <u>before</u> reaching the barrier). The 'macroscopic' extra-push (excess energy needed <u>after</u> passing the barrier) would then be added on top of the former.

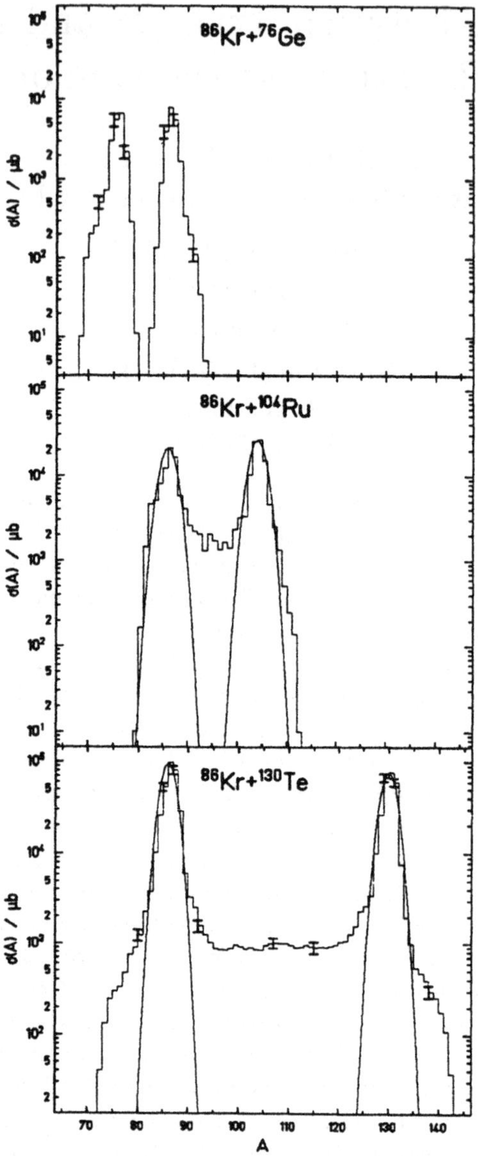

Figure 1: Mass distributions for ^{86}Kr with ^{76}Ge, ^{104}Ru and, ^{130}Te at 3.22, 3.54, 3.96 MeV/u in a histogram plot. Typical error bars are shown. The quasi-elastic peaks are overlaid by the shape of the peaks in the Kr+Ge-reaction.

We have performed a series of experiments in order to further assess the influence of both 'microscopic' and 'macroscopic' effects on fusion and the competing direct reaction channels. In a first experiment we determined complete mass distributions for collisions of ^{86}Kr with ^{76}Ge, ^{104}Ru and, ^{130}Te at the Bass-model fusion barrier with a radiochemical technique. These systems span the region from unhindered fusion (^{76}Ge) to a massive dynamical hindrance of fusion (^{130}Te) as suggested by the extra-push concept /1/ and experimental evidence /5/. The mass distributions show striking differences, see Fig.1: for Kr+Ge it consists of just two components, quasi-elastic transfer and complete fusion. For the heavier systems the compound-nucleus formation is increasingly suppressed and the missing fusion cross section seems to reappear in binary reaction channels with large mass flow towards symmetry. For Kr+Te we observe a broad symmetric distribution spanning 70 mass units with an integral cross section of 30 mb. This value is only a part of the deficit in the fusion cross section observed in ref./5/, however. This component is expected to be a mixture of deep inelastic transfer, quasi-fission and complete fusion-fission reactions. The large mass flow comes about because these dinuclear systems pass over the barrier and have time to grow a large neck /2/ before they reseparate into two fission-like fragments.

At the same time the quasi-elastic reactions show an interesting feature. The transfer peaks in all three reactions have similar shapes and can be approximated by one gaussian curve with the same width. But in the absolute magnitude of the cross section there is an increase of about a factor seven from Kr+Ge to Kr+Te (40 mb to 280 mb). This factor is not due to different Q_{gg}-values because all three systems have nearly the same Q-value matrix.

In order to investigate whether this strong increase in the quasi-elastic cross section is a general phenomenon, we have evaluated the systematics for a large number of sub- and near-barrier transfer data. The data were converted to an uniform Q_{gg}-value and to the same incident energy. The Q-value reduction was carried out with the DWBA-code PTOLEMY /10/. The energy conversion in the sub-barrier region was based on the semiclassical theory. The results are summarized in Fig.2. Coming from the heavy side (U+U) one observes a slow, rather smooth decrease of the transfer cross section until one reaches Z_1Z_2=2000, at which point the smooth behavior is disrupted by a strong fall-off. A detailed study of the overlap of the mass densities of the colliding nuclei at the fusion barrier distance as a function of Z_1Z_2 by Berdichevsky and Reisdorf /11/, predicts just a steady linear decrease. We suggest two possible explanations for this interesting phenomenon: 1) since it cannot be excluded that the nucleus-nucleus potential of the most heavy systems no longer has a pocket, it is possible that the smooth trend above Z_1Z_2=2000 is characteristic of systems without a pocket; the sudden fall-off would then coincide with the development (as Z_1Z_2 gets smaller) of a sizeable pocket which by its presence could cause diversion from the QE channels into more complex (deep-inelastic or quasi-fission) channels. 2) alternatively, coming from the low Z_1Z_2 side, the sudden rise near Z_1Z_2=1600 could coincide and correlate with the onset of the extra-push phenomenon for fusion /12/.

Clearly, this observation called for more detailed data around the 'critical' Z_1Z_2. In order to obtain complete kinematic information we have measured Z,A,Θ, and Q-value distributions for transfer reactions by using the GSI magnetic spectrometer /13/ for two systems close to the critical Z_1Z_2=1600. The spectrometer consists of six ion optical elements, one 45 deg. bending magnet, three quadrupoles, and two sextupoles, to focus the beam and obtain the focal plane tilt angle and

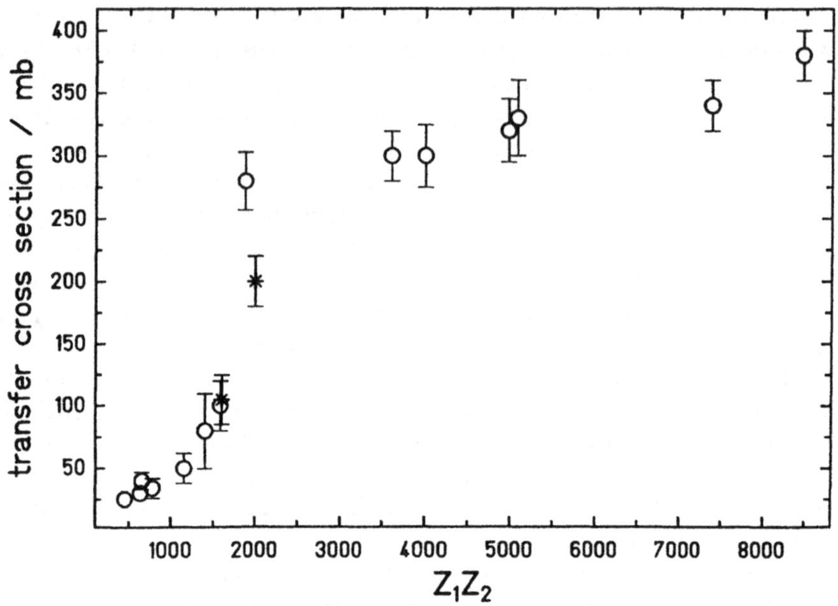

Figure 2: Integral quasi-elastic transfer cross section as a function of Z_1Z_2 for the same incident energy (E/B=1.02) and the same Q_{gg}-value (-3.5 MeV). The crosses are preliminary results (see text).

the momentum resolution. The elements are mounted on a turntable which can be rotated around the target. The detector system consists of a hybrid ionisation chamber with a position sensitive proportional counter as focal plane detector, and a time of flight counter with a secondary emission channel plate detector with a carbon foil as start detector at the entrance of the spectrometer, and a szintillation foil as stop detector. In the ionisation chamber we measure the energy loss and the rest energy. The catcher anode of the start detector is subdivided into 2mm wide strips, allowing an accurate entrance angle determination. With these values it is possible to evaluate the mass number A, the nuclear charge Z, the electric charge state q, and the kinetic energy E. As systems we choose the $^{96}Zr+^{90}Zr$ system, where fusion is obviously correlated with the microscopic action of valence neutrons, and $^{124}Sn+^{96}Zr$ where, in addition, the barrier is drastically shifted upwards by about 28 MeV /5/. We choose inverse kinematics and detected, in the heavier system the target-like recoil nuclei, because under these conditions they have the higher velocity and give a better chance for identification. Mass and charge resolution are demonstrated for the heavier system in Fig.3. These preliminary results show again, that far more transfer channels are open in the heavier of the two systems, see Fig.3b. The width of the Gaussian fitted to the measured mass distribution, see Fig.3a, is comparable to the value extracted from the Kr-induced reactions. An analysis of the energy loss spectra (TKEL=-Q) of the transfer products reveals only a small energy damping. The bulk of the distributions is peaked below 30 MeV. In the framework of a Viola-systematic the TKEL for fission-like events is about 60 MeV. Therefore, the measured transfer events can be attributed to the quasi-elastic regime.

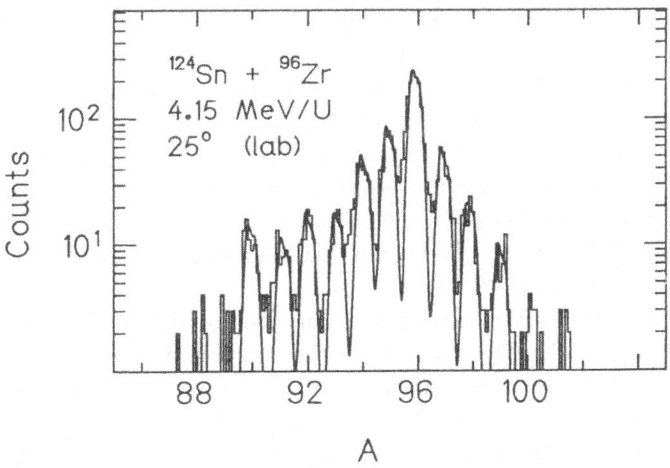

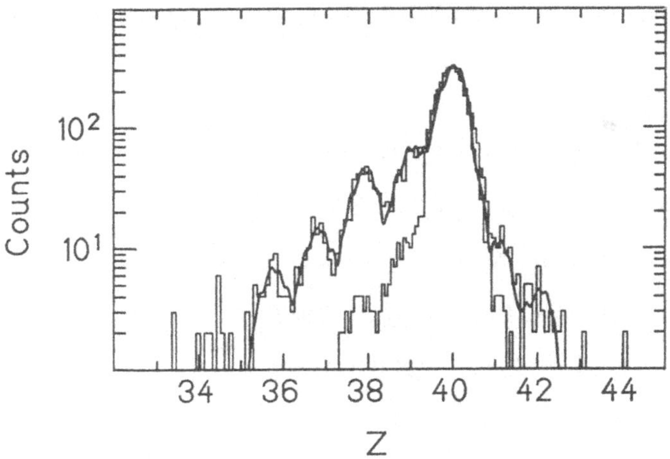

Figure 3: a.) mass distribution for the target-like recoil products in the system ^{124}Sn+^{96}Zr at 4.15 MeV/u for θ_{lab}=25^0, b.) Charge distributions for the ^{124}Sn+^{96}Zr at 4.15 MeV/u for θ_{lab}=25^0 (main histogram), and the ^{96}Zr+^{90}Zr at 4.15 MeV/u for θ_{lab}=35^0 (inside histogram). At these angles the nucleon exchange is dominant. The Zr+Zr curve is normalized to the Sn+Zr curve.

The integral transfer cross sections corroborate the sudden rise of the quasielastic cross sections in the vicinity of the critical value for extra-push. These preliminary data (crosses) are included in Fig.2.

Our data should allow us to obtain a rather complete balance of the various reaction channels. We have also observed fission-like events. These events, most clearly separated from the quasi-elastic transfer channels for the ^{124}Sn+^{96}Zr reaction at Z=45, have fully relaxed kinetic energies (TKEL=60 MeV) compatible with the Viola systematics. From fission systematics we estimate the width of the mass distribution and the form of the angular distribution. Under these assumptions we obtain a preliminary estimate for the fission cross section of about 30 mb. From calculations of the fusion cross section with the unshifted Bass-model barrier we expect for Sn+Zr a fusion cross section of about 200 mb. Thus again, as for Kr+Te, it seems obvious that the quasi-elastic transfer took over a large part of the missing fusion cross section.

One possible explanation might be given in a theory by Nörenberg et al. /14/ where, due to strong diabatic particle-hole excitations, the barrier is increased in the entrance channel. This diabatic rise is reversible. This has consequences for systems with a very shallow pocket at incident energies close to the classical barrier. In an extreme case trajectories that would normally pass over the barrier are then reflected into quasi-elastic channels before the nuclei can make deep contact. The 30 mb cross section for fission-like events observed both, in the $^{86}Kr+^{130}Te$ reaction and in the $^{124}Sn+^{96}Zr$ reaction, indicates that deep contact is being made as well, however, with reduced probability.

In summary, there is evidence that the disappearance of complete fusion around $Z_1Z_2=1600$, and in systems where the compound nucleus has still a well defined liquid drop barrier, is correlated with a steep rise in the quasi-elastic transfer cross section. This would indicate that a large part of the dynamical hindrance of fusion occurs <u>before</u> the colliding nuclei reach the barrier, as suggested by both, the concept of dissipative diabatic dynamics /14/ and the surface friction model /15/.

References

[1] W.J. Swiatecki, Nucl.Phys. A376,471(1982)

[2] J. Toke et al., Nucl.Phys. A440,327(1985)

[3] K. Lützenkirchen et al., Nucl.Phys. A452,351(1986)

[4] C. Gregoire et al., Nucl.Phys. A387,37c(1982)

[5] C.C. Sahm et al., Nucl.Phys. A441,316(1985)

[6] J.G. Keller et al., Nucl.Phys. A452,173(1986)

[7] S.G. Steadman, M.J.Rhoades-Brown, Ann.Rev.Nucl.Part.Sci.36,649(1986)

[8] H. Esbensen, S. Landowne, Phys.Rev.C 35,2090(1987)
S. Landowne, contribution to this conference

[9] B. Quint et al., see W. Reisdorf, contribution to this conference

[10] M.H. Macfarlane, S.C. Pieper, PTOLEMY-code, ANL-76-11

[11] W. Reisdorf, contribution to this conference

[12] A. Berdichevsky et al., GSI-preprint-87-12

[13] D. Schüll, Lecture Notes in Physics, Springer Verlag Berlin, Vol.178,(1983) p.80

[14] A. Lukasiak et al., Nucl.Phys. A426,181(1984), GSI-preprint-86-36

[15] P. Fröbrich, contribution to this conference

Fusion and Transfer around the Coulomb-Barrier of the Systems $^{33}S + ^{90,91,92}Zr$

L. Corradi, U. Lenz, K.E.G. Löbner, P.R. Pascholati, U. Quade, K. Rudolph, W. Schomburg, S.J. Skorka, H.G. Thies
Sektion Physik, D-8046 Garching

G. Montagnoli, D.R. Napoli, R. Pengo, F. Scarlassara, A.M. Stefanini, A. Tivelli
INFN, Laboratori Nazionali di Legnaro, Padova

S. Beghini, C. Signorini
Dipartimento di Fisica dell'Università and INFN, Padova

Abstract: Excitation functions of fusion below and around the Coulomb-barrier as well as near-barrier cross sections of all important transfer channels have been measured in the systems $^{33}S+^{90,91,92}Zr$. Taking into account in an approximative way the strong coupling effects between transfer and fusion and using a semi-classical approach transfer form factors were extracted and successfully employed to describe the isotopic effects in fusion enhancement. The coupling of the two-neutron pick-up channel to fusion turns out to be surprisingly strong.

Introduction

The importance of transfer channels to account for the pronounced isotopic dependence of fusion enhancement observed in many systems is well established. Still only in few cases both fusion *and* transfer cross sections have been measured[1)-3)], which allow a coupled channels analysis based not only on Q-values, but also on known transfer strengths.

A detailed calculation employing a direct reaction coupled channels code for the silicon-nickel fusion reaction has been carried out by Landowne et al.[4)]. In this analysis it was shown that the main effect of transfer on fusion originates from the bulk of the two-neutron pick-up reaction to excited states, even though the one-neutron transfer cross sections are larger than those for two-neutron transfer.

In the present contribution a rather crude analysis of the coupling effects between fusion and transfer in the systems ^{33}S on $^{90,91,92}Zr$ is applied with the advantage of simplicity and insight into the redistribution of cross sections due to the action of coupling effects. Again the two-neutron pick-up plays a dominant role. Its strikingly small observed cross section (see fig. 4) can be looked at as the *result* of its exceptionally strong coupling to fusion: A large part of the reflected flux is lost in favour of the absorbed flux.

From the measured transfer angular distribution (at small angles, where absorption is small) an estimate of the transfer form factors is obtained within a simplified semi-classical approach. With the help of a modified version[5)] of the finite range code CCFUS[6)] the transfer form factors and transfer Q-values are then employed to calculate the fusion enhancements. Good agreement to the experimental enhancements is achieved.

Experiments

The fusion excitation functions have been measured at the XTU-tandem accelerator of the L.N.Legnaro using the electrostatic separator[7] of this laboratory. A plot of the cross sections is displayed in fig. 1. The extracted relative asymptotic barrier shifts $\Delta D_1 = D(^{91}Zr) - D(^{90}Zr)$ and $\Delta D_2 = D(^{92}Zr) - D(^{90}Zr)$ are largely free of uncertainties from the choice of the nuclear potential and experimental errors in the absolute cross sections. From fig.1 we deduce $\Delta D_1 = (0.50 \pm 0.15)$ MeV and $\Delta D_2 = (1.28 \pm 0.20)$ MeV. The absolute asymptotic barrier shifts with respect to a one-dimensional tunnelling calculation amount to 2.4±0.6, 2.9±0.6 and 3.7±0.6 MeV, respectively. In our analysis given below we concentrate on the *relative* barrier shifts. Since the low-lying collective states of the Zr-isotopes have similar properties these relative shifts reflect the isotopic differences of the transfer reactions.

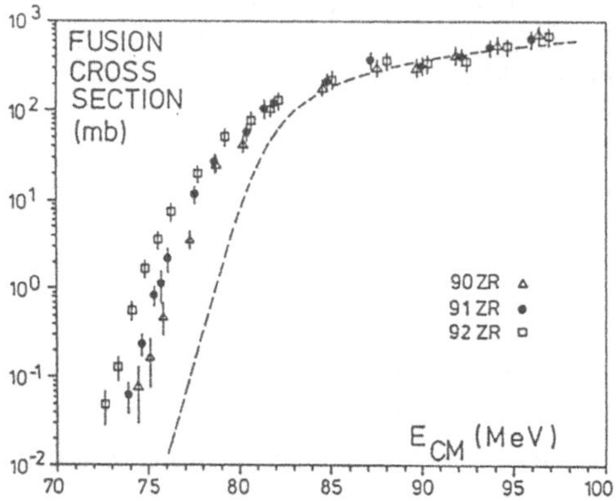

Fig.1: Fusion cross sections. Dashed: 1-dimensional tunnelling of ^{90}Zr.

Double-differential transfer cross sections $d^2\sigma/d\Omega dQ$ have been obtained at E_{cm}=82 and 88 MeV (112 and 120 MeV Lab.) employing a time-of-flight telescope in connection with a ΔE-E_{rest} ionization chamber[8]. The timing was performed with the help of two mirror-type

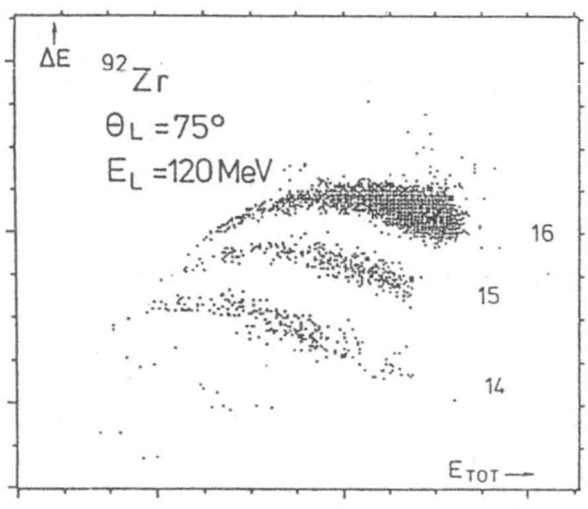

Fig.2: ΔE-E matrix of projectile-like transfer products.

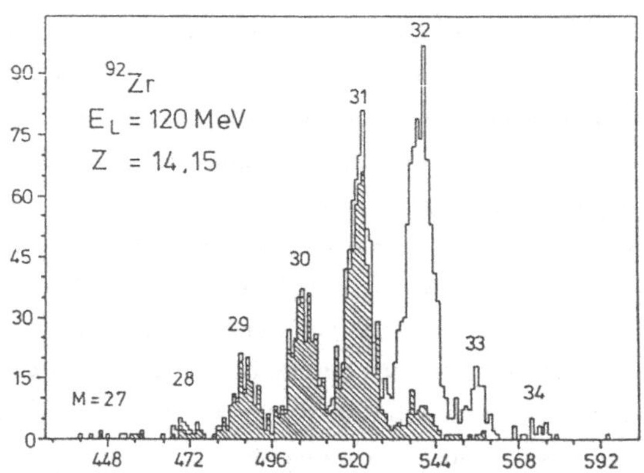

Fig.3: Mass spectrum of transfer products with Z=14 and Z=15 (hatched).

channel plate time detectors with a total time resolution of typically 0.7 % at a flight path of 1.5 m. Position determination in the ionization chamber which had a solid angle of 0.7 msr allowed a correction for kinematic energy shifts. The good resolution obtained in Z and mass can be seen in fig's 2 and 3. The angle- and Q-integrated cross sections are shown in fig.4. As usual the one-nucleon transfer cross sections dominate. The small two-neutron pick-up cross section is worth noting.

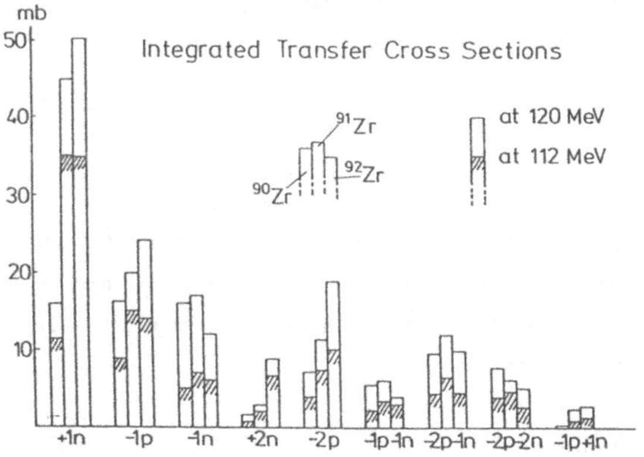

Fig. 4: The measured angle- and Q-integrated transfer cross sections. The symbol +1n denotes the one-neutron pick-up channel, etc.. The experimental errors amount to at least 15%, in cases of very small cross sections up to 30%.

Extraction of transfer form factors

As a first step of the analysis first order corrections to the scattering angles and distances of closest approaches were applied[9], which amounted to up to 10 degrees and a few percent, respectively. In the spirit of the semi-classical formalism the transfer cross sections of the transfer channel i were assumed to be

$$(d\sigma^{(i)}/d\Omega)_{tr} = P^{(i)}(\Theta)\,(1 - P_a^{(i)}(\Theta))\,(d\sigma/d\Omega)_{Ruth.} \qquad (1)$$

with the transfer function

$$P^{(i)}(\Theta) = N_i \sin\Theta/2 \exp\{-2\alpha_i(D(\Theta) - D_c)\} \quad \text{for } D \geq D_c\text{, else } P^{(i)}(\Theta) = N_i \sin\Theta/2 \qquad (2)$$

At energies well below the barrier this expression would describe the transfer probability. The parameter D_c was fixed at $1.40\,(A_t^{1/3} + A_p^{1/3})$ fm and is - within limits - not critical. It is about equal to the quarter point distance. The wave numbers α_i were deduced from the average binding energies. In the case of two-proton stripping consistent fits to the angular distributions were only possible with a somewhat reduced α_i of 1.1/fm in place of the calculated 1.25/fm.

In eq.1 the quantities $P_a^{(i)}(\Theta)$ describe the absorption into more complex transfer channels or fusion. They are assumed to depend somewhat on the transfer channel. The *average* absorption, as derived from the elastic (plus inelastic) cross sections and given by

$$1 - \overline{P}_a = (d\sigma/d\Omega)_{el} / (d\sigma/d\Omega)_{Ruth.}\,, \qquad (3)$$

was parametrized by the expression

$$1 - \overline{P}_a = \exp\{-(\Theta_{lab} - \Theta_c)/\Theta_a\} \quad \text{for } \Theta_{lab} \geq \Theta_c\text{, else } \overline{P}_a = 0 \qquad (4)$$

with Θ_c and Θ_a fitted to the observed elastic cross sections.

The normalization factors N_i were then adjusted such that the angular distributions at *small* angles, where the absorption $P_a^{(i)} = 0$, were described adequately, with the constraint, that the same N_i must result for both projectile energies. Particularly the angular distributions of the +2n and also to lesser extend the +1n and -2p channels, which are expected to be most strongly coupled to fusion due to their favourable Q-values, showed deviations at *large* angles indicating additional absorption in excess to what is described by $\overline{P}_a$. This extra absorption is interpreted - as explained in the introduction - as an additional loss of transfer cross section due to a strong tendency of the +2n-process to proceed into the fusion channel.

In an approximative manner this channel coupling effect was described by a correction $-\Delta\Theta_c^{(i)}$ and $-\Delta\Theta_a^{(i)}$ to Θ_c and Θ_a respectively in eq. 4. The cross sections $\Delta\sigma_i$ lost due to this extra absorption are compensated by corresponding gains in other reaction channels, e.g. fusion, such that $\sum \Delta\sigma_i = 0$ and, according to eq. 3

$$\sigma_{reaction} = \int \overline{P}_a (d\sigma/d\Omega)_{Ruth.} d\Omega \, . \tag{5}$$

In fig. 5 a comparison is made between $\sigma_{reaction}$ and the sum of the experimental fusion and all observed transfer cross sections, showing a satisfactory consistency. It is interesting to note that the extension of the absorption to smaller angles in the +2n-channel reflects itself in an apparent reduction of the slope factors α_{2n} (typically about 0.6/fm instead of 1.3/fm), if extracted from cross sections including the region between $\Theta_c - \Delta\Theta_c$ and Θ_c. The $\Delta\Theta_c$'s amount to only a few degrees for +1n, but to around 20 degrees for +2n. In fig. 6 the angular distribution fits are illustrated by two examples.

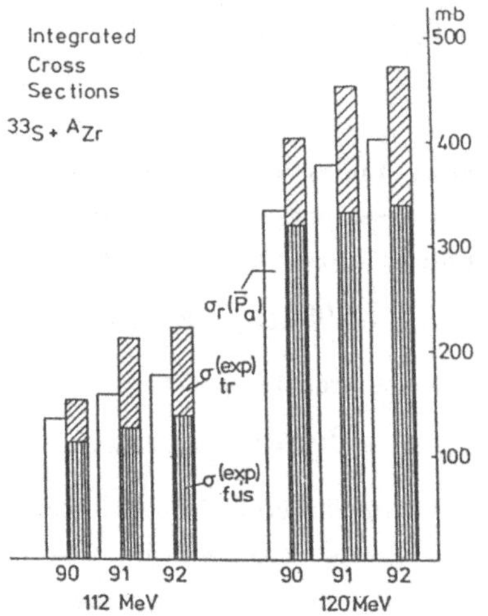

Fig. 5: Comparison of the total experimental cross section to the one calculated from the absorption according to eq. 5.

Finally the form factors were written as

$$F_i(r) = F_0^{(i)} \frac{R_1}{r} \exp\{-\alpha_i (r - R_1)\} \tag{6}$$

with $R_1 = 1.2\,(A_t^{1/3} + A_p^{1/3})$ and the strength parameter $F_0^{(i)}$ extracted from the transfer probabilites (eq. 2) using the simplified relation

$$P^{(i)}(\Theta) = \left[1/\hbar \int F_i(r(t))\, dt\right]^2 \tag{7}$$

Essentially it is $F_0^{(i)} \sim \sqrt{N_i \alpha_i}$, independent of the absorption and the projectile energy. The Q-dependence which should be included into the integral was taken into account by assuming a proper Q-distribution of the cross sections (see below).

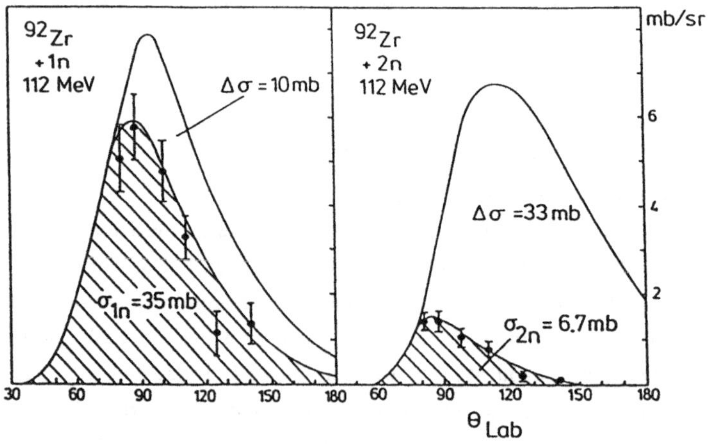

Fig. 6: Two examples of the evaluation of the angular distributions. Hatched: Experimental cross section fitted using parameters N_i, $\Delta\Theta_c^i$ and $\Delta\Theta_a^i$, (s. Text). Due to the exponential dependences involved, the $\Delta\sigma$'s are subject to relatively large uncertainties.

Coupled channels calculations of fusion

The evaluation described in the preceding section was, in fact, performed as a function of the excitation energy, assuming a Gaussian Q-window positioned at the optimum Q-value with a reasonable width. For the coupled channels calculation two energy channels per transfer reaction with excitation energy dependent α's and F_o's were defined: one covering the first MeV above ground state with $Q_{tr} = \tilde{Q}_{gg} - 0.5$ MeV and the corresponding part of F_o and one covering the bulk of the transfer events with an average $Q_{tr} = \tilde{Q}_{gg} - 3.5$ MeV. In addition one representative E2-mode of the projectile as well as one E2- and one E3-mode for each target nucleus were taken into account in order to allow for inelastic excitations. Their total effect on ΔD_1 and ΔD_2 is negligible.

		F_o(MeV)	Q (MeV)	α/fm	$F(r_b)$(MeV)
^{90}Zr	+1n	0.56	-0.90	0.75	0.16
		1.54	-3.90	0.68	0.50
	+2n	2.4	-3.05	1.38	0.28
		5.8	-6.05	1.32	0.76
	-2p	0.6	+0.45	1.10	0.12
		4.1	-2.55	1.00	1.02
^{91}Zr	+1n	0.25	+3.88	0.67	0.08
		2.20	+0.88	0.59	0.81
	+2n	2.6	-0.94	1.35	0.32
		8.8	-3.94	1.28	1.19
	-2p	0.65	+1.30	1.10	0.14
		5.15	-2.70	1.00	1.22
^{92}Zr	+1n	0.43	+2.45	0.70	0.13
		2.57	-0.55	0.63	0.89
	+2n	2.4	+2.40	1.29	0.32
		14.3	-0.60	1.21	2.13
	-2p	0.7	+2.30	1.10	0.16
		7.0	-0.70	1.00	1.84

Table 1: Transfer coupling parameters. $F(r_b)$ denotes the form factor at the barrier radius r_b.

As mentioned above only the +1n, +2n and -2p transfer channels were included, in view of their trends of the transfer Q-values and form factors The other channels are not expected to contribute much to the differences of the barrier shifts. Thus altogether 9 channels entered into the coupled channels calculation. The most important parameters are listed in table 1. Some results in terms of the absolute and relative asymptotic barrier shifts can be seen in table 2. The transfer form factors at the barrier radius turn out to lie in the reasonable range of 0.1 to 2 MeV.

The comparison of the experimental ΔD's with those calculated including all relevant transfer channels is very satisfactory, in particular in view of the very approximative character of the calculations. Disregarding the three ground state transfers reduces the relative barrier shifts by only around 10%. This is in accordance with the finding for the systems Si+Ni (Ref.4). Table 2 shows also that ΔD_1, the relative barrier shift of ^{91}Zr, is essentially determined by the *one*-neutron pick-up, while ΔD_2 receives its main contribution from the *two*-neutron pick-up. Both facts mainly reflect the different ground state Q-values and thus the Q-values assigned to the bulk of the transfer strength at about 3.5 MeV excitation.

The absolute barrier shifts resulting from the calculation turn out to be smaller than the experimental ones. It should be noted, however, that many more transfer channels, in particular the -1p channel and the more complex ones, exist, which are expected to contribute to the enhancements, although not significantly to the isotopic differences.

Transfer channels coupled to fusion*)	absolute barrier shifts (MeV) D(^{90}Zr)	D(^{91}Zr)	D(^{92}Zr)	relative shifts (MeV) ΔD_1	ΔD_2
+1n, +2n and -2p	1.77	2.37	2.87	0.60	1.10
+1n only	1.68	2.16	1.92	0.48	0.24
+2n only	1.67	1.81	2.35	0.14	0.68
none (inelastic only)	1.65	1.74	1.65	0.09	0.01
Experimental shifts	2.4±0.6	2.9±0.6	3.7±0.6	0.50 ±0.15	1.28 ±0.20

*)in addition to the inelastic channels mentioned in the text.

Table 2: Results of the coupled channels calculation in comparison to the experimental shifts.

In summary, we have measured both the sub-barrier fusion excitation functions and the cross sections of all important transfer channels for the reactions $^{33}S+^{90,91,92}Zr$. Neutron and proton transfer products could be observed separately. The analysis outlined above cannot replace a full coupled channels calculation, but shows that transfer cross sections and fusion enhancement are found in accordance, if - and only if - the strong coupling situation is taken into account. In particular the two-neutron pick-up reaction seems to support fusion very effectively, which explains its dominant influence observed in many other sub-barrier fusion experiments.

Stimulating discussions with C.H. Dasso and H. Wolter are greatly acknowledged. This work was supported by the Bundesministerium für Forschung und Technologie of the FRG.

References:

1) K.E. Rehm et al., Phys.Rev.Lett. 55 (1985) 280
M. Beckerman et al., Phys.Rev.Lett. 45 (1980) 1472 and Phys.Rev. C25 (1980) 2427
S. Landowne and S.C. Pieper, Phys.Rev. C29 (1984) 1352
M. Beckerman, Phys.Rep. 129 (1985) 145
H.A. Aljuwair et al., Phys.Rev. C30 (1984) 1223

2) W. Henning et al., Phys.Rev.Lett. 58 (1987) 318

3) A.M. Stefanini et al., Nucl.Phys. A456 (1986) 509
Y. Sugiyama et al., Phys.Lett. B176 (1986) 302
A.M. Stefanini et al., Phys.Lett. B185 (1987) 15

4) S. Landowne et al., Phys.Rev. C35 (1987) 597

5) Code CCFUSB, S.J. Skorka, unpublished

6) C.H. Dasso and S. Landowne, Phys.Lett. B183 (1987) 141

7) S. Beghini et al., Nucl.Instr.Meth. A239 (1985) 585

8) K. Rudolph et al., Nucl.Instr.Meth. A204 (1983) 407

9) R.A. Broglia and A. Winther in Heavy Ion Reactions, Vol.I, p.181 ff, Benjamin/Cunnings Publ. Comp., Inc. 1981.

MEASUREMENTS OF 180° SUB-BARRIER TRANSFER REACTION CROSS SECTIONS IN S + Mo, Nb SYSTEMS

M.G. Herman, L.L. Lee, Jr., R.J. Vojtech
Department of Physics, SUNY at Stony Brook
S. B. Gazes, M. Satteson, J. Boyle
NSRL University of Rochester

The enhancement of sub-barrier fusion cross sections over those predicted by one-dimensional barrier penetration remains one of the intriguing problems of low energy heavy-ion physics. It is now generally accepted that there is a "softening" of the barrier caused by the presence of quasi-elastic reaction channels which can serve as doorways to the fusion process. The details of the channel coupling and the relative importance of the various possible reactions remain subjects for speculation. It is then clearly necessary to have good experimental measurements of the various reaction strengths contributing to the total cross section as the incident energy is varied through the barrier region.

There now exist a number of reliable measurements of fusion excitation functions. Measurements of corresponding quasi-elastic reaction cross sections exist in only a few cases. They are made difficult by the fact that the yields of projectile-like reaction products are backward peaked at energies near the barrier. In reactions such as the S + Mo systems studied here this has the result that the reaction products of interest have such low laboratory kinetic energies that detection and identification are serious experimental problems. However, there is a forward target-like recoil associated with each backward reaction product and detection of these recoils provides a corresponding measurement of the reaction cross section. This has been demonstrated by Betts et. al. [1] for ^{58}Ni + ^{A}Sn reactions for which the Recoil Mass Separator at the Daresbury Nuclear Structure Facility was used to detect 0° recoils corresponding to far backward angle light reaction products. Measurement of the mass spectrum of the recoils provides a measurement of the total differential cross section for transfer of one or more nucleons between the target and the beam. There is no determination of whether the transferred nucleons are neutrons or protons, although Q-value considerations may provide guidance in many cases. The method has one convenience in that, at energies near and below the barrier, the total yield for masses near the target mass can be assumed to correspond to the Rutherford elastic scattering cross section. This provides an internal normalization for what would otherwise be a quite difficult measurement.

This paper reports use of this method to measure 180° transfer reaction cross sections for the reaction of ^{32}S with targets of ^{92,98}Mo and ^{93}Nb. The corresponding fusion excitation functions for ^{32}S + ^{92,98}Mo have been measured by Pengo et. al. [2]. Their measurements indicate little subbarrier enhancement for the ^{92}Mo target but very large enhancements for the ^{98}Mo target. As far as we know there are no fusion measurements for ^{32}S + ^{93}Nb. Rough measurements made in conjunction with the present experiment suggest ^{93}Nb may behave similarly to ^{98}Mo.

The present experiment made use of the University of Rochester Recoil Mass Spectrometer (RMS) [3] to detect the zero-degree recoils in a position-sensitive silicon detector in the focal plane of the RMS. A then proportional counter was used to provide an energy loss measurement which was needed to discriminate between the target-like recoils and beam remnants scattered within the RMS. A recoil mass spectrum for the ^{92}Mo target at an incident laboratory energy of 114 MeV is shown in Figure 1. The central mass 92 corresponds to elastic (plus inelastic) scattering near 180° and the peak identified as 91(93) represent single nucleon pickup (stripping) in the projectile. Measurable yields corresponding to single-nucleon (and some two-nucleon) transfer have been obtained for incident energies as low as 20 MeV (lab) below the nominal Coulomb barrier. In this paper we will confine our discussion to the single-nucleon transfer results. Some measurements of two-nucleon transfer, both pickup and stripping, have been obtained but the results are not yet ready for presentation.

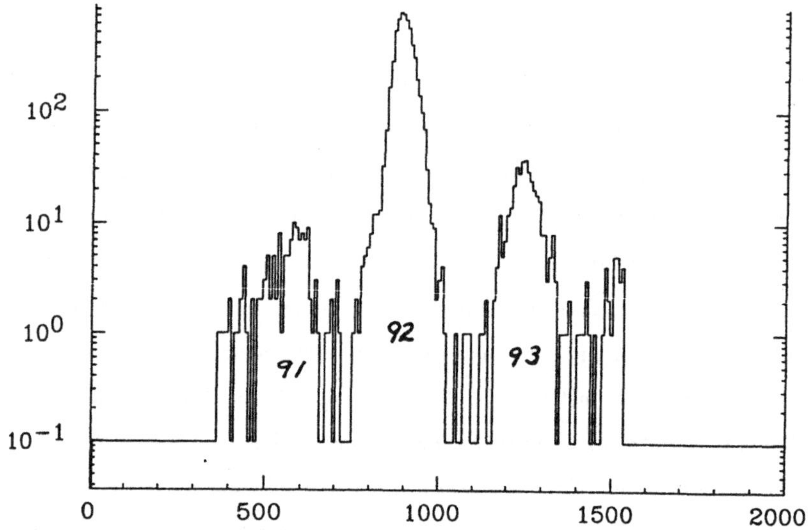

Fig. 1

Results for ^{92}Mo and ^{98}Mo targets are shown in Figure 2. Several features are notable. While all of the cross sections decrease approximately exponentially with decreasing energy it is evident that the transfer cross sections for the ^{98}Mo target are much larger than those for the ^{92}Mo target. We also note that for the ^{92}Mo target stripping is more likely than pickup while pickup is stronger than stripping for the ^{98}Mo target. While, as noted above, these experiments are unable to measure the charge of the transferred nucleon, calculations suggest that the results correspond to single-proton stripping and single neutron pickup, largely from Q-value considerations. The data for the ^{93}Nb target are quite similar to those for ^{98}Mo; large cross sections are observed for both stripping and pickup. The ground-state Q-values Q_{gg} for the reactions under consideration here are calculated in Table I (-indicates pickup, + indicates stripping).

Table I

Target	Q_{gg}			
	-1n	-1p	+1n	+1p
^{92}Mo	-4.04	-4.61	-6.97	-4.78
^{98}Mo	0	-7.52	-9.11	-2.36
^{93}Nb	-0.09	-3.67	-7.72	-0.28

Below the Coulomb barrier the 180° transfer cross sections can be expected to decrease exponentially with the distance of closest approach d for zero impact parameter, the slope being determined by the wave number of the transferred nucleon. In figure 3 the present cross section measurements are plotted as a function of d; the exponential fall off is clearly evident. The lines accompanying the data represent the results of DWBA calculations performed with the code PTOLEMY[4]. In performing these calculations it was assumed that we were observing only proton stripping and neutron pickup. Spec-

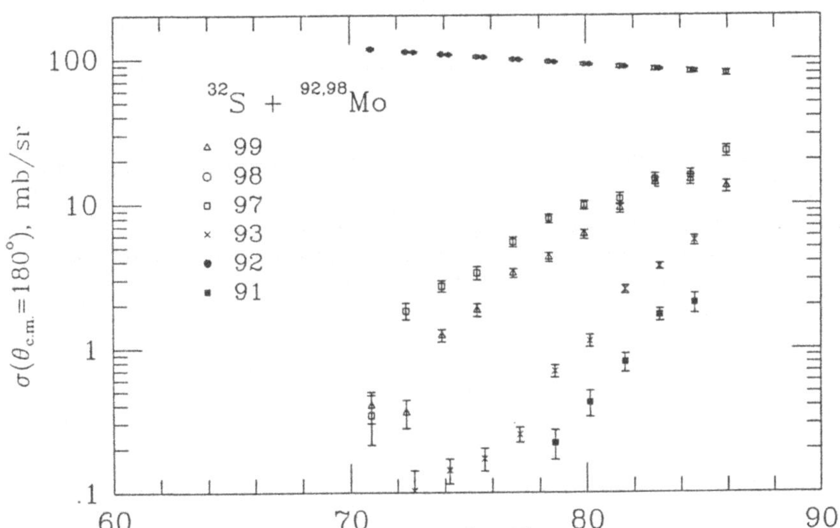

Fig. 2

troscopic factors were obtained from the literature when available and from shell-model considerations; the solid lines represent the sum of all contributions. As would be expected the calculations reproduce the slope rather well with the possible exception of the (weak) stripping on the ^{92}Mo target. In most cases the magnitude of the cross section is also predicted remarkably well. An exception is for the ^{98}Mo target where the calculations overpredict the pickup strength and underpredict the stripping. Note that ^{98}Mo is a vibrational nucleus; inclusion of coupling to inelastic channels may alter the results of the calculations.

In summary, the detection of zero-degree target-like recoils has enabled the measurement of total 180° nuclear transfer strength for the ^{32}S + $^{92,98}Mo$, ^{93}Nb systems. More than an order of magnitude more transfer is measured for the ^{98}Mo target than for the ^{92}Mo target, suggesting a strong correlation between transfer strength and the large sub-barrier fusion enhancement. All of the present measurements exhibit the expected exponential decrease with increasing distance of closest approach. The cross section for the ^{92}Mo and ^{93}Nb targets are consistent with the predictions of DWBA calculations, while, as cited earlier for the ^{98}Mo target the calculations overpredict the pickup and underpredict the stripping.

It is evident that the transfer channels should be incorporated into any complete description of the fusion process in these systems. The details of the appropriate calculations remain uncertain. The growing availability of transition reaction strengths, such as those measured here, now provides material for the calculations. Note that far below the barrier transfer is by far the major contributor to the total reaction cross section.

References

[1] R.R. Betts et.al., Phys Rev. Letters **59**, (1987) 978 and R.R. Betts, invited paper this symposium.

[2] R. Pengo, et. al., Nuclear Physics, **A411**, 1983) 255.

[3] T.M. Cormier and P.M. Stwertka, Nuclear Instruments and Methods, **184**, (1981) 423.

[4] M.H. Macfarlane and S.C. Pieper, Argonne National Laboratory Report No. ANL 76-11 (1978) (unpublished).

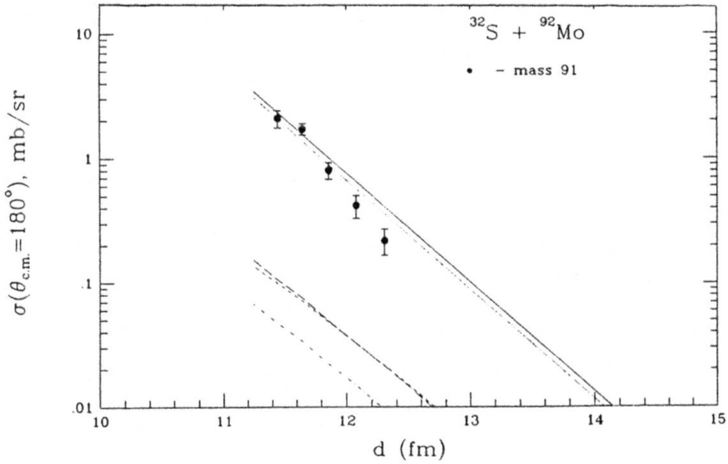

$^{32}S + ^{92}Mo$
• – mass 91
$\sigma(\theta_{c.m.}=180^o)$, mb/sr
d (fm)
10
1
.1
.01
10
11
12
13
14
15

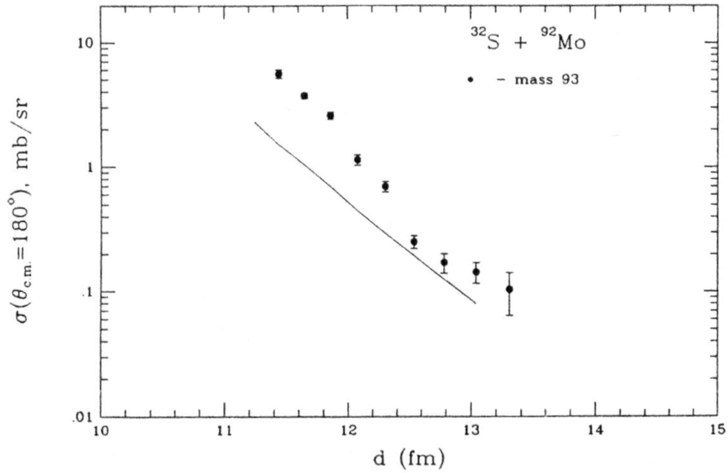

$^{32}S + ^{92}Mo$
• – mass 93
$\sigma(\theta_{c.m.}=180^o)$, mb/sr
d (fm)
10
1
.1
.01
10
11
12
13
14
15

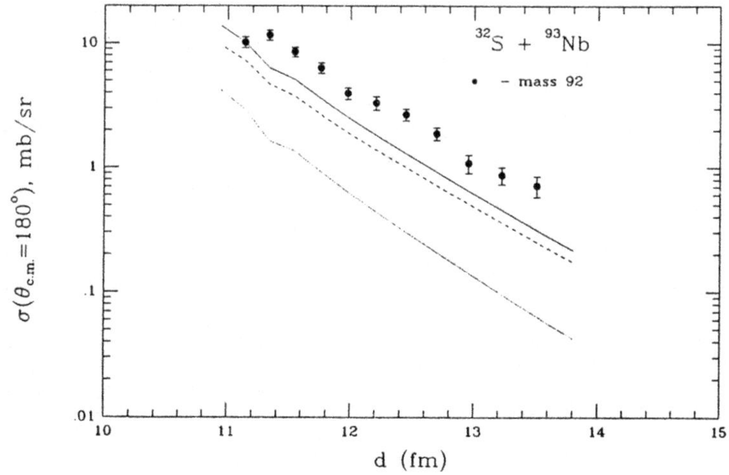

$^{32}S + ^{93}Nb$
• – mass 92
$\sigma(\theta_{c.m.}=180^o)$, mb/sr
d (fm)
10
1
.1
.01
10
11
12
13
14
15

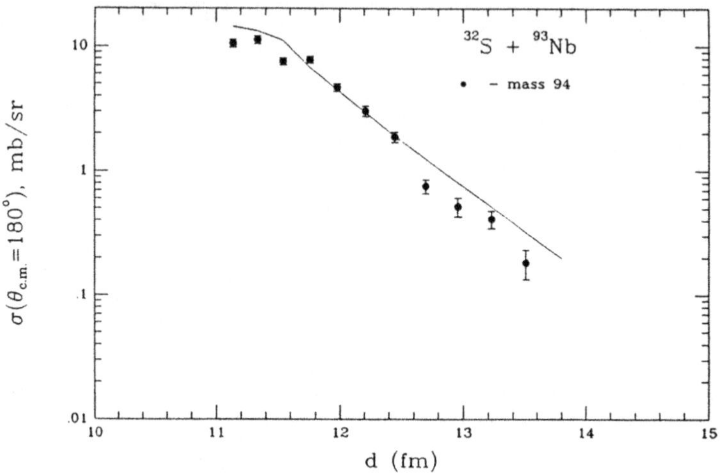
$^{32}S + ^{93}Nb$
• – mass 94
$\sigma(\theta_{c.m.}=180^{\circ})$, mb/sr
d (fm)

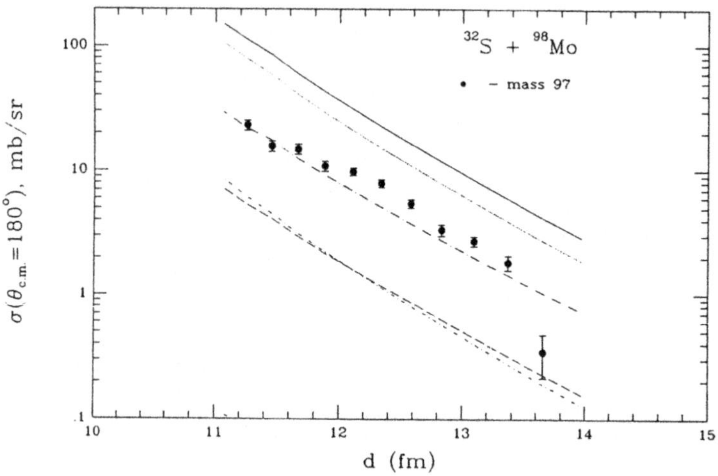
$^{32}S + ^{98}Mo$
• – mass 97
$\sigma(\theta_{c.m.}=180^{\circ})$, mb/sr
d (fm)

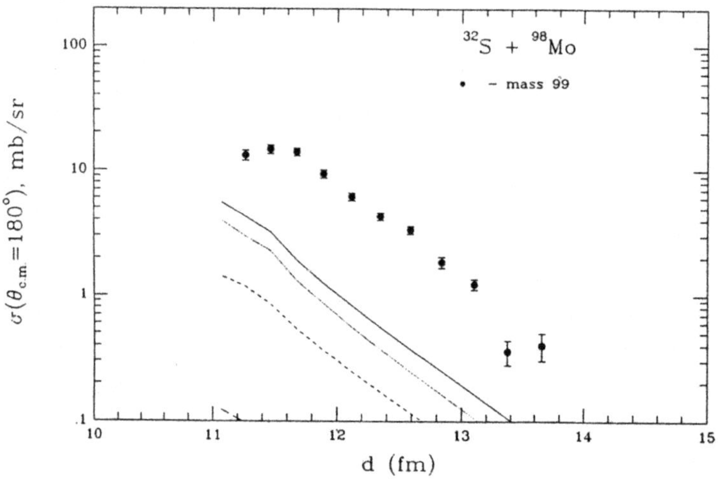
$^{32}S + ^{98}Mo$
• – mass 99
$\sigma(\theta_{c.m.}=180^{\circ})$, mb/sr
d (fm)

SEMICLASSICAL DESCRIPTION OF MULTIPAIR TRANSFER PROCESSES IN HEAVY-ION REACTIONS

P. Lotti
Dipartimento di Fisica "Galileo Galilei"
Università di Padova and INFN, Padova, Italy

Abstract

Pair transfer processes involving superfluids system are studied in terms of phase space distributions constructed in a product representation which blends both ordinary- and gauge-space degrees of freedom. The time evolution of these distributions is followed by solving a collection of classical equations of motion. The excitation of the pairing rotational degree of freedom - corresponding to a variation in the number of particles - is induced by the ion-ion potential, which the superfluid character of the target makes to be deformed. Different orientations leading to the same final mass transfer may produce (in the case of large pairing deformation) interference effects in the final population of the members of the pairing rotational band.

In this contribution it is illustrated the application of a semiclassical treatment to the multipair transfer process between heavy ions, in analogy with the description of the Coulomb excitation of a rotor [1]. In the formulation we will consider the case in which the target nucleus displays a superfluid phase. The most natural framework for the description of the nuclear structure of the target is the BCS formalism in which the number of particles operator $\hat{N}$ is associated with rotations around a symmetry axis in an abstract space, the so-called *gauge space*. Particle–particle correlations are responsible of a static deformation of the superfluid nucleus in this abstract space and it is possible to establish a complete analogy with respect to the particle–hole correlations which lead to a static or dynamic deformation of the nuclear shape in the ordinary space [2]. For instance one can introduce an intrinsic hamiltonian for the superfluid system which, expressed in terms of the gauge coordinates, has the same form as the intrinsic hamiltonian of a deformed system with a symmetry axis.

An essential element in this analysis of multipair transfer processes is the characterization of the interplay between the ordinary reaction variables and those in the abstract space; the link between the two set of variables is established assuming a deformed density in the *gauge space* [3–4]. This deformation describes the finite content in number of particles which is present in the intrinsic state.

The deformed nuclear density contains a dependence of the gauge degree of freedom which shall be referred as the *gauge angle* [2]:

$$\rho(r,\phi) = \rho(r - R(\phi))$$

where [3]

$$R(\phi) = R_0 \left(1 + \frac{2\beta_p}{3A}\cos(2\phi)\right)$$

The parameterization of the nuclear radius as a function of ϕ leads to an expression for the ion-ion potential which shall represents an interaction term coupling the ordinary spatial variables with the gauge coordinates. The total hamiltonian which describes the reaction process may be written in the following form:

$$H = H_{rel} + H_{int} + V_{coup}$$

where the different terms on the right-hand-side,

$$\begin{aligned} H_{rel} &= \frac{p_r^2}{2\mu} + \frac{p_\varphi^2}{2\mu r^2} \\ H_{int} &= \frac{p_\phi^2}{2\Im} \\ V_{coup} &= \frac{V_0}{1+\exp\left(\frac{r-R(\phi)}{a}\right)} + \frac{Z_1 Z_2 e^2}{r} \end{aligned}$$

describe the relative motion, intrinsic motion and coupling interaction respectively. Note that the conjugate momentum to the gauge angle, p_ϕ, corresponds to the variable ΔN representing variations in the number of particles in the system.

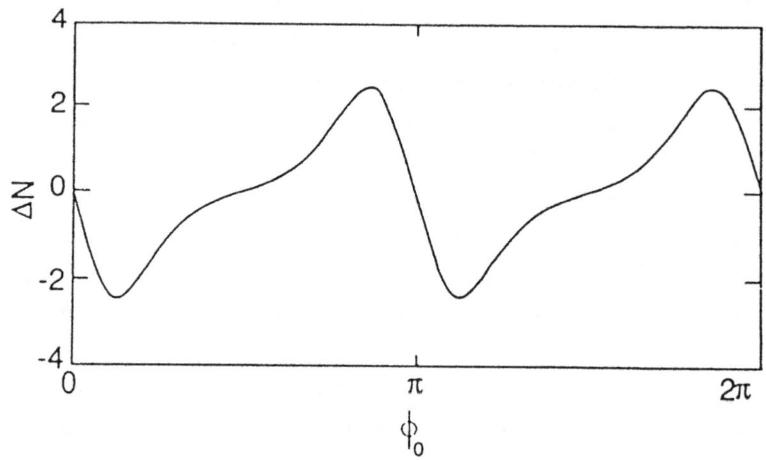

Fig.1. Final values of p_ϕ (i.e. the final mass transfer ΔN) obtained in the semiclassical approach as a function of the angle ϕ_0 characterizing the initial orientation of the rotor in *gauge space*. The calculation refers to the case of the reaction ^{40}Ca+^{116}Sn (for details see the text).

To provide a concrete illustration it has been considered the reaction ^{40}Ca+^{116}Sn;

from the binding energies of the even tin isotopes a moment of inertia $\Im = 5\hbar^2$ MeV^{-1} is extracted. A bombarding energy E_{CM} of 120 MeV and an impact parameter $\rho = 3$ fm were chosen, conditions which lead to a near-grazing collision. A characteristic value of $\beta_p = 18$ was assumed to define the strength of the coupling.

In fig.1 it is shown the final value of p_ϕ (i.e.,ΔN) as a function of the initial orientation in the *gauge space* ϕ_0. The existence of two values of the initial orientation of the rotor in the abstract space which lead to the same value of the final mass transfer is certainly in analogy with the Coulomb excitation process: our expectation is then to observe oscillations in the population probability of a given state of the pairing rotational band of the superfluid nucleus, i.e., in the probability of transfer a given number of particles. The classical probability for the different multi-particle transfer processes is obtained by converting the uniform distribution in ϕ_0 according to

$$P(\Delta N) = P(\phi_0)\frac{d\phi_0}{d(\Delta N)} = \frac{1}{2\pi}\frac{d\phi_0}{d(\Delta N)}$$

Given the multi-valued character of the function $\phi_0(\Delta N)$, this expression is understood to include a summation over the separate fragments of the curve which lead to the same value of ΔN.

Note the characteristic rainbows that appear at the edges of the classically allowed values of ΔN. As noted before, the classically allowed values of a given ΔN receive contributions from different initial orientation angles, eventually merging at the maximum allowed value ΔN_{max}. From this observation an improved prescription for projecting into number-of-particles eigenstates can be implemented. To this end, one can associate to each contribution an amplitude [5–6] with a complex phase $e^{i\Phi/\hbar}$, obtained by evaluating along the trajectory the classical action integral

$$\Phi = -\int_{-\infty}^{+\infty}(r\dot{p}_r + \phi\dot{p}_\phi)dt$$

Now, making use of the symmetries of the function $\Delta N(\phi_0)$, it is straightforward to derived the following property concerning the action integral Φ:

$$\Phi(\phi_0 + n\pi) = \Phi(\phi_0) - n\pi\Delta N \qquad n = 0, \pm 1, \pm 2, ... \tag{1}$$

The semiclassical probability $P(\Delta N)$ is then constructed as

$$P(\Delta N) = \mid -i\sqrt{P_1}e^{i\Phi_1} + \sqrt{P_2}e^{i\Phi_2} \mid^2$$

(P_k and Φ_k, k=1,2 are respectively the modulus of the classical probability and the action integral corresponding to the two values of ϕ_0 in the interval $[0, \frac{\pi}{2}]$ which lead to the same value of ΔN). If one counts all the contributions in the range $-\infty \leq \phi_0 \leq +\infty$ it turns out that only the ones corresponding to $\Delta N=$ even give a probability $P(\Delta N) \neq 0$. In fact all the contributions to $P(\Delta N)$ for $\Delta N \neq$even coming from the full range of variability of ϕ_0 are out of phase as shown in eq. (1). Furthermore, the values of the probability $P(\Delta N)$ for $\Delta N=$ even obtained restricting the calculations keeping ϕ_0 in the interval $[0, \frac{\pi}{2}]$ coincide with those we would obtain performing the full calculation, while the values corresponding to $\Delta N \neq$even are meaningless being dependent on the choice of the interval of ϕ_0.

The probability distribution is compared in fig. 2 with the result obtained in the pure classical approach (namely without interference). Since one of the functions is continuous while the other is discrete, what should be comparable is the sum of the ordinates corresponding to the quantal results with the area under the classical curve.

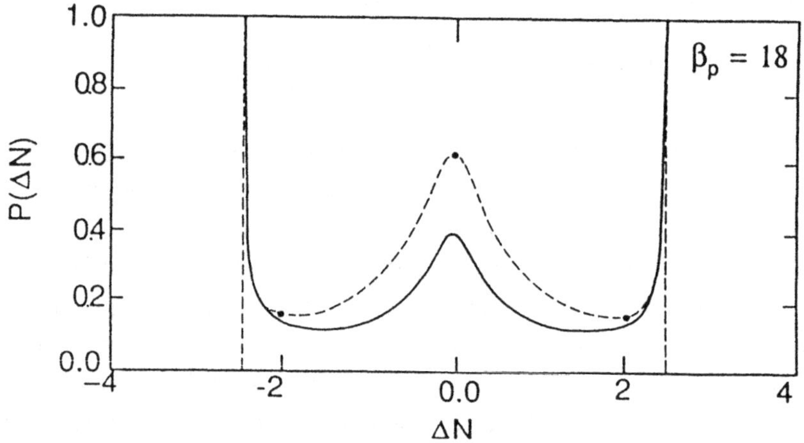

Fig.2. Semiclassical probability distribution $P(\Delta N)$ (full dots) associated with the different mass partitions. The dashed line has been drawn to guide the eye. In both cases the continuous curves refer to the prediction of the pure classical picture, namely without any interference.

In order to investigate the features of the distribution $P(\Delta N)$ in a case in which many pairs of particles are exchanged we have arbitrarily increased the pairing deformation parameter β_p by a factor of 10, leaving unaffected the trajectory of relative motion. This leads to a simple scaling of the values of ΔN_{max} and of the gauge action integral, which now becomes the dominant ingredient in determining the phase differences. The associated probabilities for different final channels are shown in fig. 3, again compared with those obtained ignoring the interference. The distribution has now developed marked oscillations around the pure classical results. This pattern is quite similar to the ones displayed by the final spin population in the inelastic excitation of a rotor and in [7] within this context. Note that we have constructed our probability distributions $P(\Delta N)$ from the information provided by only one impact parameter. A more correct treatment should sample all partial waves.

The action integrals Φ defined in the previous section are pure real numbers. Consequently, the phases $e^{i\Phi/\hbar}$ calculated in the applications - although essential to determine the interference patterns - did not introduce any attenuating effects. The action integrals provide, however, a way to extend the formalism into the classically forbidden region. Along the lines of the present contribution, this could be achieved by extending the integration of the classical equations of motion into a more general regime in which the dynamical variables are allowed to assume complex values. The imaginary parts of the action integrals Φ then provide the exponential damping factors in the probabilities for transfer processes with $\Delta N > \Delta N_{max}$.

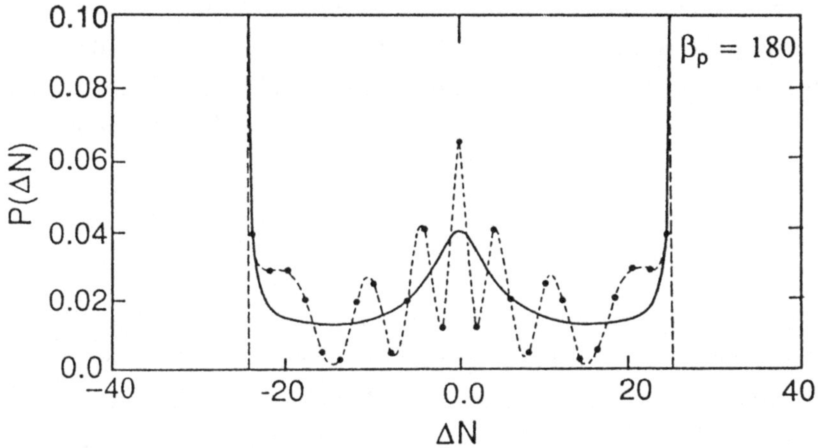

Fig. 3. As in figure 2, but with a larger value of the parameter β_p.

One can avoid the introduction of complex equations of motion and implement, instead, a simpler prescription. Indeed, an accurate estimation of the complex action integrals can also be made by extrapolating into the forbidden region the results obtained with the real trajectories. We refer to [4] for a more detailed explanation of this point.

Conclusions.

We may summarize the results of this analysis saying that in the multipair transfer processes the characteristic range of the coupling makes the nature of the problem significantly different from the semiclassical treatment of Coulomb inelastic excitation. Oscillations in the population probability of neighboring systems as a function of number of particles can hardly be expected for typical values of the pairing deformation parameters.

On the other hand, from the experimental point of view there is now evidence that, under special conditions, it is possible to observe the transfer of one, two and even more pairs of identical nucleons between the reaction partners [8].

This fact encourages to extend the theoretical investigation of the multipair transfer process within the semiclassical approach to reactions for which experimental data are available. One case of particular interest is the case of the reaction ^{208}Pb+^{152}Sm [9] at bombarding energies close to the Coulomb barrier; considering the heavy charge of the reaction partners and the superfluid character of ^{152}Sm, it shall be particularly interesting to analyse the excitation of the rotational band in ^{152}Sm associated with the transfer of neutron pairs. In fact, the excitation of the member of the rotational band in the A+2 system can proceed via different reaction mechanism. One possibility is that nucleus ^{152}Sm is first excited to a state of its rotational band and later receives one pair or more of particles from the projectile. A second way of ending in the same final state is that first the transfer occurs and then isotope $^{152+2n}$Sm is excited in its rotational band. These two processes may give rise to interesting interference effects, which could be useful in detecting evidence of the occurrence of Berry phase phenomena [10]. The use of the semiclassical treatment of this problem could lead to a simple identification of the situations for which this h occurs.

References

[1] C.H. Dasso, in Proceedings of the La Rabida International Summer School on Nuclear Structure and Reactions, Eds. G. Madurga and M. Lozano (World Scientific, 1986);

[2] A. Bohr, B. Mottelson, Nuclear Structure, Vol.II,W.A. Benjamin Press, New York 1975; P. Ring and P. Schuck, The Nuclear Many-Body Problem, 1980, Springer-Verlag;

[3] C.H. Dasso and A. Vitturi, Phys. Rev. Letters 59 (1987) 634;

[4] P. Lotti, A. Vitturi and C.H. Dasso, Nucl. Phys. 474 (1987) 274;

[5] W.H. Miller, J. Chem. Phys. 53(1970)1949;

[6] C.H. Dasso, S. Landowne. R.A. Broglia and A. Winther, Z. Phys. A 317(1984)187; C.H. Dasso, Proceedings of XXI International Winter Meeting on Nuclear Physics, Bormio,1983, p. 722; S. Landowne, in Nuclear Physics, Proceedings of the Nuclear Physics Workshop, Eds. C.H. Dasso, R.A. Broglia and A. Winther (North Holland, 1982), p. 421;

[7] A. Winther, in Proceedings of the International Workshop on Semiclassical and Phase Space Approaches to the Dynamics of the Nucleus, Aussois, 1987, to be published.

[8] W. von Oertzen, in *Frontiers in Nuclear Dynamics*, edited by R.A. Broglia and C.H. Dasso, Plenum, New York (1985) p.241
W. von Oertzen, H.G. Bohlen, B. Gebauer, R. Kunkel, F. Pulhofer and D. Schull, Z. Physik **A326** (1987) 465

[9] C. Lauterbach, P. Butler and H.J. Wollersheim, in *Nuclear Structure 1985*, edited by R.A. Broglia, G.B. Hagemann and B. Herskind, Elsevier Science Publishers B.V., (1985) p. 405

[10] R.S. Nikam, P. Ring and L.F. Canto, Z. Physik **A324** (1986) 241

TRANSFER AND INELASTIC CHANNELS AROUND THE COULOMB BARRIER

A. Rapisarda and M. Baldo
INFN, Corso Italia 57, 95129 Catania, Italy

Abstract: *The role of transfer and inelastic channels around the Coulomb barrier is investigated within a microscopic model for heavy ion reactions. In particular fusion and transfer reactions for the systems* $^{40}Ca + ^{40}Ca$ *and* $^{58}Ni + ^{58}Ni$ *are considered. The theoretical predictions compare quite well with the experimental data.*

In the last few years, it has become apparent that the coupling among different channels is essential in the dynamics of heavy ion reactions at energies around the Coulomb barrier. The increasing of fusion cross section below the barrier [1], the polarization potential in elastic scattering [2] are typical phenomena in which inelastic and transfer processes play a crucial role. These findings indicate that a realistic description of heavy ion reactions at low energies must correctly take into account all the degrees of freedom of the dinuclear system.

A model has been developed in which both transfer and inelastic processes are treated microscopically, considering the relative motion in a classical way [3]. The model is based on a mean field approach with a realistic description of the nuclear structure, considering a finite number of single-particle levels near the Fermi energy. Therefore the Hamiltonian for a nucleon feeling the two wells can be written as

$$H = H_o + V(t),$$

where $H_0 = H_1 + H_2$ describes the internal structure of the two nuclei, while V is the interaction

$$V(t) = \begin{pmatrix} W_1 & Z \\ Z^\dagger & W_2 \end{pmatrix},$$

which takes into account the transfer of particles (through the operators Z and $Z^\dagger$) and the single-particle inelastic excitation inside each nucleus (W_i).

Solving the equation

$$i\hbar\dot{\varrho} = [V(t), \varrho(t)] \tag{1}$$

for the density matrix ϱ defined as

$$\varrho_{bb'} = <\Psi(t)|(a_{b'})^\dagger a_b|\Psi(t)>,$$

one is able to follow the evolution of the system at each instant of time. Then the mean average of any one body observable Q is given by

$$<Q(t)> = \mathrm{tr}(\varrho(t)Q).$$

Eq. (1) must be solved together with the newtonian equations of motion [3,4] .In this way one has a complete description of the evolution of the system. At the present stage, mean field distortions as well as residual interaction are not considered, but in principle the model can be extended [4]. For the moment this picture should give a fairly good description of peripheral heavy ion reactions at low energies .

Applications

The model is very general and it has already been used to study the role of frictional forces [3,4]. This study proved to be very successful and showed the importance of both transfer and inelastic channels. An example is reported in ref.4) where the reaction $^{86}Kr + ^{154}Sm$ at 610 MeV is discussed. From that investigation one sees how only considering both transfer and inelastic processes one gets the experimental trend, at variance with the classical proximity model [4]. These discrepancies could be due to quantal and memory effects not considered in the classical picture.

Studies of this kind are still in progress to investigate if the same agreement is reached with different systems. Lately several measurements were done at the Laboratori Nazionali del Sud in Catania.

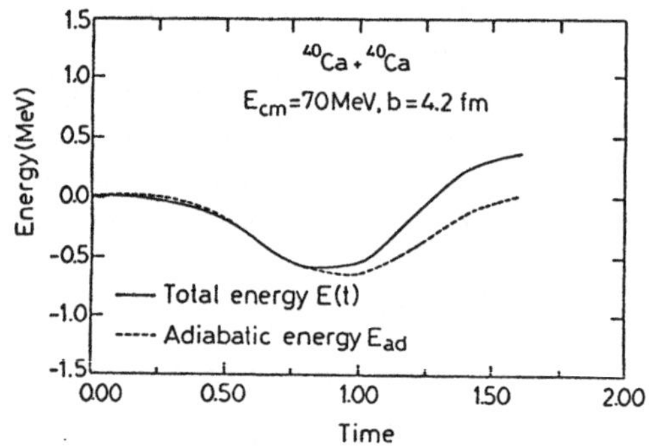

Fig.1. For the system $^{40}Ca + ^{40}Ca$ and for a trajectory above the barrier the adiabatic energy and the total one are plotted as a function of time, see text. The time units are $0.6 \cdot 10^{-21}$ sec. ($\hbar$ = c = 1).

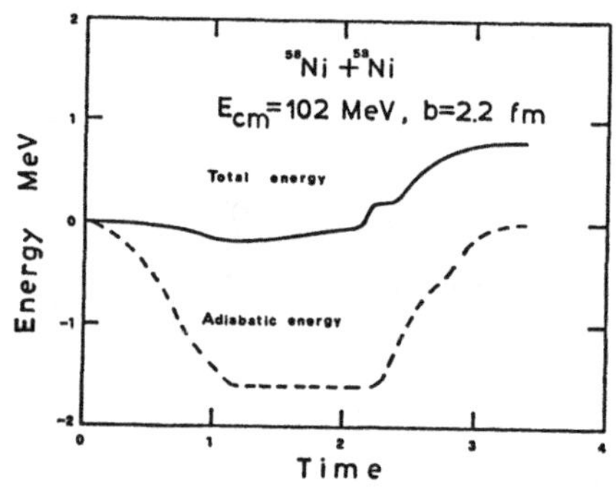

Fig.2. The same of fig.1 for the system $^{58}Ni + ^{58}Ni$.

In the present work we would like to focus the attention on the role these degrees of freedom play in fusion and transfer reactions near the Coulomb barrier. For this reason we consider the two systems $^{40}Ca + ^{40}Ca$ and $^{58}Ni + ^{58}Ni$.

a) Fusion

The problem of fusion and especially of sub-barrier fusion cross-section has been one of the most discussed during the last years. The fact that a very pronounced enhancement over the theoretical predictions has been found [1)] has been for long time an open question. Recently it has been demonstrated the fundamental role played by the inelastic and transfer channels, which coupled to the entrance ones act as doorway states increasing fusion probability [1)] . Up to now only inelastic channels have been taken into account microscopically ,while the contribution of transfer has been investigated only in a phenomenological and qualitative way[1)].

In our model it is possible to estimate the role of transfer microscopically, calculating the transmission coefficients entering into the formula for fusion cross-section

$$\sigma_F = \sum_l \pi \lambda\!\!\!^{-2}(E)(2l+1)T_l(E). \tag{3}$$

In a one-dimensional treatment the transmission factor $T(E)$ at the energy $E < V_B$, where V_B is the height of the barrier, is given by

$$T(E) = \exp\left(-2\int_{x_1}^{x_2}\sqrt{\frac{2\mu}{\hbar^2}(V(x)-E)}dx\right), \tag{4}$$

In the last equation, $V(x)$ is the potential barrier through which one has tunnelling, μ is the reduced mass and x_1,x_2 are the two turning points for which $V(x_1) = V(x_2) = E$.

In the adiabatic limit, i.e. when the internal motion of the nucleons is faster than the relative one, the problem reduces to a tunnelling through the adiabatic potential and $V(x) = V_{ad}(x)$. The latter is obtained by introducing the polarization potential generated by transfer and inelastic processes [5)].

A *dynamical* calculation can be done extending this formula to the case of a smoothly time-dependent barrier. In this case eq. (4) can be written like

$$\begin{aligned} T(E) &= \exp\left(-2\int_{t_1}^{t_2}\sqrt{\frac{2\mu}{\hbar^2}[V(x(t))-E]}vdt\right) \\ &= \exp\left(-\frac{2}{\hbar}\int_{t_1}^{t_2}\mu v^2(t)dt\right), \end{aligned} \tag{5}$$

where v is the radial velocity, while t_1 and t_2 are the instants of time corresponding to x_1 and x_2. In our approach it is possible to introduce the velocity through the barrier at each time by changing sign to the acceleration of the relative motion, once the external turning point is reached. This way the system is forced to go into the classically forbidden region. This procedure is equivalent to introducing an imaginary time, cf. ref. 5).

The study on $^{40}Ca + ^{40}Ca$ has already been presented [5] , but it is interesting to review the main results in comparison with the new calculations obtained for the system $^{58}Ni + ^{58}Ni$ which behaves differently.

In both cases one sees a big enhancement of the experimental fusion cross section due to the coupling with transfer and inelastic channels. However while in the case of $^{40}Ca + ^{40}Ca$ we are in the adiabatic limit for the system $^{58}Ni + ^{58}Ni$ this limit is reached only at energies much smaller than the barrier. In figs. 1 and 2 it is evident the different behaviour of the total energy with respect to the adiabatic one. In the case of $^{40}Ca + ^{40}Ca$ the system tends to follow the energy of the dinuclear ground state i.e. the adiabatic energy, at variance with the case of $^{58}Ni + ^{58}Ni$. This is due to the fact that the ^{40}Ca is a closed shell nucleus while ^{58}Ni is not. For the former the levels near the Fermi energy differ of several MeVs, while for the latter this difference is around one MeV - at least for neutrons . This implies that the times associated with the internal degrees of freedom are faster in the case of ^{40}Ca than in the case of ^{58}Ni. So the adiabatic regime is strictly valid only for $^{40}Ca + ^{40}Ca$. In fact for $^{40}Ca + ^{40}Ca$ the dynamical calculations and the adiabatic ones coincides both below and above the barrier [5] .

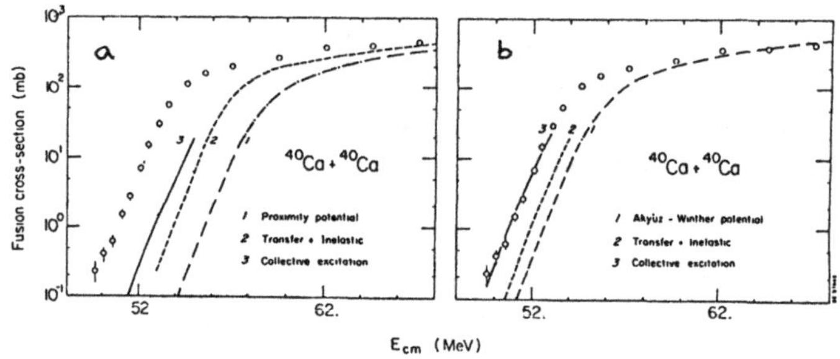

Fig.3. Comparison between the theoretical fusion cross section and the experimental data in the case of the system $^{40}Ca + ^{40}Ca$.[4] The Proximity and the Akyüz - Winther potentials were used in figs. a and b respectively. The three curves refer to different couplings, see text.

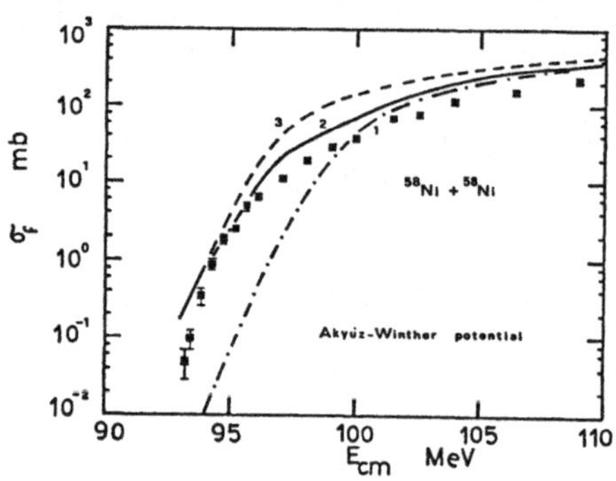

Fig.4. The experimental data for fusion cross section[4] are plotted in comparison with different theoretical predictions. The curve number 1 is the calculation without any coupling, while curve number 2 is the dynamical one including both transfer and single-particle inelastic channels. Curve number 3 is the adiabatic prediction with the same coupling of curve 2.

In fig.3 is shown the comparison between the theoretical predictions and the experimental data. The different curves refer to different couplings. Precisely curve number 1 refers to a standard calculation without any coupling, curve number 2 takes into account both transfer and single-particle inelastic excitations, while curve number 3 considers the couplings of curve number 2 plus the coupling to the 3^- and the 5^- states of ^{40}Ca. These states are the most important collective excitations for this nucleus [5]. The Proximity and the Akyüz-Winther [5] potential were used in fig.3 a and b respectively. The latter should be the most realistic being a folding-type potential which fits the elastic scattering data.

Using again the Akyüz-Winther potential one gets a good agreement also in the case of $^{58}Ni + ^{58}Ni$ as shown in fig.4. In this case, however, as already stressed, the dynamical and the adiabatic calculations coincides only at energies much below the barrier. The dynamical predictions are nearer to the experimental data. One has to notice that very recent measurements [7] confirmed Beckerman's data [1], showing a slight increase at far subbarrier energies predicted by our calculations. At variance with ^{40}Ca for the nucleus ^{58}Ni the collective excitations are not very important. The strength of the first low-lying states is only 1-2 Weisskopf units [6]. Moreover one can explain the experimental data without including these collective levels .

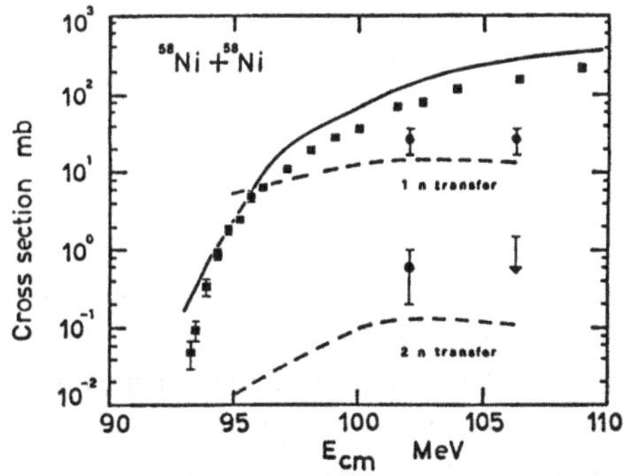

Fig.5. The dynamical predictions for the fusion cross section of fig.4 are plotted together with the theoretical cross sections for 1-neutron and 2-neutron transfer. The transfer data are from ref.8.

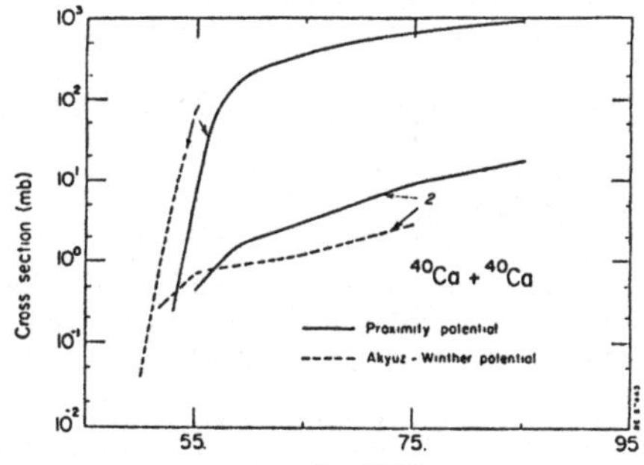

Fig.6. Fusion (curves 1) and 1-particle transfer cross sections (curves 2) for the system $^{40}Ca + ^{40}Ca$. The predictions for the Proximity (full curves) and the Akyüz - Winther potential (dashed curves) are shown. The curves for fusion are those labelled with number 2 in fig.3.

b) Transfer

As already stated in ref.8 the large value of the transfer cross section near the barrier could be responsible , even if not alone, for the enhancement in the fusion cross section. While for $^{40}Ca + ^{40}Ca$ no experimental data are available, for the system $^{58}Ni + ^{58}Ni$ there are some data, at least above the barrier [8]. In our model it is possible to extract from the final density matrix also information about the transfer cross section [5]. As it is shown in fig.5 our calculations for 1-neutron and 2-neutron transfer cross section for the reaction $^{58}Ni + ^{58}Ni$ are almost within the errors of the experimental values. For the system $^{40}Ca + ^{40}Ca$ calculations for transfer and fusion are shown in fig.6 for the two different potentials. It would be interesting to have some experimental data for transfer cross section also in this case.

In conclusion with the same model and the same potential, i.e. the Akyüz-Winther potential, one is able to get a good agreement with the experiment both for fusion and transfer. Moreover one can apply the same model both above and below the fusion barrier. One has to stress that no free parameter has been used for both transfer and fusion cross section predictions ! However it seems that the Akyüz-Winther potential could be slightly adjusted to reach a better agreement with the data.

Conclusions

The model presented seems to take in good account both the role of transfer and that of inelastic degrees of freedom. The predictions both for sub-barrier fusion and for transfer cross section are quite good and encouraging. Of course more calculations have to be done to see how far one can go. In any case this microscopic approach is very general and is able to give a realistic description of heavy ion reactions around the Coulomb barrier.

This work is part of a long term project in collaboration with R. A. Broglia and A. Winther.

References

(1) M. Beckerman, Phys. Rep. **129** (1985) 145 and refs. therein ;
S.G. Steadman and M.J. Rhoades-Brown, Ann. Rev. Nucl. Sci. **36** (1986) 649, and refs. therein.

(2) C.H. Dasso, S. Landowne, G. Pollarolo and A. Winther, Nucl. Phys. **A459** (1986) 134;
C. Mahaux, H. Ngo and G.R. Satchler, Nucl. Phys. **A449** (1986) 354;
M.V. Andrés, F. Catara, Ph. Chomaz and E.G. Lanza, Proceedings of the XXVI Winter Meeting on Nuclear Physics, Bormio (1988), Italy.

(3) M. Baldo, A. Rapisarda, R.A. Broglia and A. Winther, Nucl. Phys. **A472**(1987)333.

(4) M. Baldo and A. Rapisarda, in "The Response of Nuclei under Extreme Conditions", E. Majorana School Erice (1986),Eds. R.A.Broglia and G.Bertsch, Plenum Press 1988.

(5) M. Baldo, A. Rapisarda, R.A. Broglia and A. Winther, NBI preprint 88-05, submitted to Nucl. Phys. A;
M. Baldo and A. Rapisarda, Proceedings of the XXVI Winter Meeting on Nuclear Physics, Bormio (1988), Italy.

(6) P.M. Endt, Nuclear Data **23** (1979) 547.

(7) R. Schiker, N. Alamanos, P. Braun-Munzinger, J. Stachel and L. Waters, Phys. Lett. **B 206** (1988) 9.

(8) K.E. Rehm, F.L.H. Wolfs, A.M. van den Berg and W. Henning, Phys. Rev. Lett. **55** (1985) 280.

SPIN DISTRIBUTION

PROBING THE SPIN DISTRIBUTION IN NEAR-BARRIER FUSION REACTIONS

A. Charlop, A. Garcia, S. Gil, S. J. Luke
and R. Vandenbosch
University of Washington
Seattle, WA 98195, USA

The recognition that sub-barrier fusion cross sections are larger than expected on the basis of one-dimensional barrier penetration calculations has led to a renewed interest in fusion and competing reactions at near- and sub-barrier energies. This subject is of particular interest because of the close interplay between the nuclear structure of the participants and the dynamics of the collision. The coupling of the relative motion to both the shape and particle transfer degrees of freedom is important in enhancing the fusion cross section. The shape degree of freedom appears to be of primary importance for many systems. This is particularly born out by a recent study [1] of the fusion of ^{16}O with ^{144}Sm, which extends the earlier study of the heavier Sm isotopes by Stokstad et al.[2] The excitation function for ^{144}Sm is much steeper in the sub-barrier region than for the heavier isotopes, and can in fact be reproduced reasonably well by a one-dimensional barrier penetration calculation. It appears that transfer channels are only of major importance for systems where both target and projectile are closed shell nuclei, such as ^{16}O + ^{208}Pb. In the latter case the results of coupled channels calculations suggest that particle transfer and inelastic excitation are of comparable importance in enhancing the fusion cross sections.[3,4] Although coupled channels calculations can reproduce the fusion excitation function for ^{16}O + ^{208}Pb, there are other systems such as Ni + Ni where the most complete coupled channels calculations performed to date have been unable to reproduce the magnitude of the sub-barrier fusion cross section [5].

We have been interested in probing sub-barrier fusion in a more differential way by exploring the partial wave contributions to the total sub-barrier fusion cross section. It is helpful in thinking about what kind of information can be obtained about fusion to consider a moment expansion in the orbital angular momentum ℓ leading to fusion. The total fusion cross section σ_f is given by the zeroth moment,

$$\sigma_f = \Sigma_\ell \ell^0 \sigma_\ell$$

with $\sigma_\ell = (2\ell+1)T_\ell$. It is important to note that very different spin distributions can lead to the same fusion cross section. The first moment of the spin distribution, $\langle\ell\rangle$ is given by

$$\langle\ell\rangle = (\Sigma_\ell \ell^1 \sigma_\ell)/\Sigma_\ell \sigma_\ell$$

and can be inferred from gamma ray multiplicity measurements. As will be seen later, the second moment

$$\langle\ell^2\rangle = (\Sigma_\ell \ell^2 \sigma_\ell)/\Sigma_\ell \sigma_\ell$$

can be inferred from alpha or fission fragment angular distributions. Lesko et al. [6] have pointed out that the competition between neutron emission and fission can be sensitive to the width of the spin distribution. This approach has been pursued by Kondo et al. [7]. It is unfortunately rather sensitive to the assumed ratio of the level density parameters for the fission saddle point and the residual nucleus following neutron emission.

Even more differential information on the σ_ℓ distribution can be deduced from gamma ray multiplicity distributions measured in multiple-element crystal balls. If it were not for the dispersions in the number of gamma rays accompanying the decay of a compound state of a given J, the multiplicity distributions would enable a direct determination of the full σ_ℓ distribution. In practice there are significant corrections for the detector multiplicity resolution as well as less certain (but not too large) corrections for the dispersion in the number of gamma rays from the decay of a single J state. Particle evaporation also causes a broadening and skewing of the spin distribution prior to gamma emission. All probes of spin distributions based on gamma ray multiplicities require a knowledge of the dependence of the number of gamma rays emitted on the spin of the emitting state.

We will summarize the available data from the different types of probes at a later point in this paper. Suffice it to say at this point that relatively few of these difficult experiments have been performed, and that most of them indicate a broader spin distribution than expected theoretically. An exception to this trend is our earlier gamma ray multiplicity study of sub-barrier fusion of ^{12}C and ^{16}O with ^{154}Sm [8]. Although broader distributions than given by one-dimensional barrier penetration models were obtained, the mean spins could be accounted for by incorporating the shape degree of freedom of the deformed target.

We have recently reinvestigated the $^{16}O+^{154}Sm$ system, partly to remove a limitation of our earlier study and partly as part of a broader study of the entrance channel mass asymmetry dependence of the spin distributions. The technique used in our previous determination of M_γ relied upon tagging evaporation residues by the detection of discrete γ-rays emitted by the evaporation residues. Due to light element contamination in the target and to the complex character of the gamma decay of even-odd nuclei, we were able to obtain the contribution to M_γ from the 4n channel (^{166}Yb) only. Near and below the barrier the 3n channel begins to make a contribution, and its neglect may lead to a bias due to fractionation of the spin distribution into the different xn evaporation channels. Such a fractionation has been demonstrated [9,10] at higher spins, and it was in fact concern about possible fractionation effects that motivated us to use the more inclusive approach of exploiting fission fragment angular distribution to study the spin distribution [11,12].

We have been able to explore this fractionation effect by using a new self-supporting ^{154}Sm target which was produced by the Oak Ridge National Laboratory using a special technique that minimizes oxygen and other low-Z contaminates. We have been able to clearly identify the gamma-rays arising from the 3n channel (^{167}Yb). In particular, we were able to observe the transitions originating in the decay through the yrast band (based on the 5/2+[642] state). The E_γ= 221.7 keV ($17/2^+$ to $13/2^+$) transition of this band was used to tag the 3n channel.

The technique used for obtaining the contribution of the 3n channel selectively samples those states in the compound nucleus that feed the state $J=17/2^+$ in ^{167}Yb. Similarly the contribution from the 4n channel, tagged by the $4^+ \rightarrow 2^+$ transition, samples those states which feed this transition. In order to eliminate the bias produced by this effect, corrections were made using results from statistical decay models. The corrections were modest and fairly insensitive to the assumptions made. $\langle \ell \rangle$ was obtained using the expression

$$\langle \ell \rangle = \Sigma_x Y_{xn} \langle \ell \rangle_{xn},$$

where Y_{xn} is the relative yield for the xn channel obtained from ref. 4, and $\langle \ell \rangle_{xn}$ is the value of the first moment of the spin distribution associated with the channel xn. We can compare our new results in Fig. 1 with the older results where the contribution of the 3n channel was not included. It can be seen that the mean spin is significantly underestimated without taking into account the contribution of the 3n channel, which becomes increasingly more important as the bombarding energy is lowered. This result is consistent with statistical model expectations. We show in Fig. 2 an example of how a postulated inclusive spin distribution is fractionated into the 3n and 4n channels. We have used the PACE code of Gavron [13] to perform these calculations.

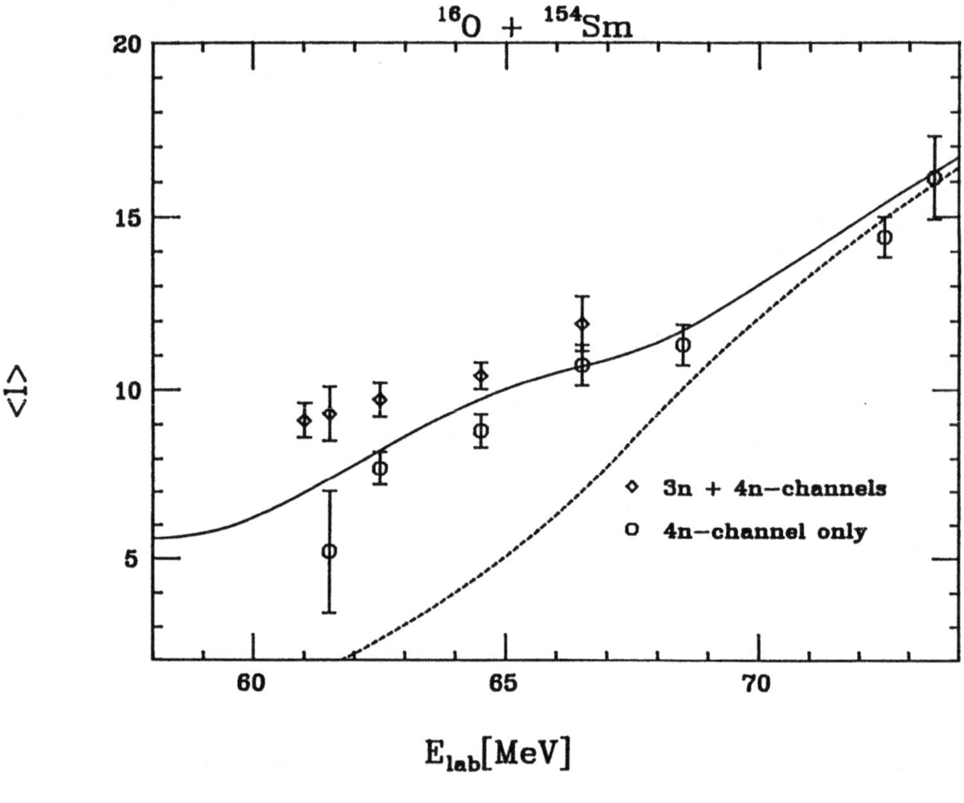

Fig. 1 Comparison of mean values of compound nuclear spin distribution leading only to the 4n channel and leading to both the 3n and 4n channel. The full curve is based on the Wong model and includes the effect of target deformation. The dashed curve is based on the sharp cutoff model.

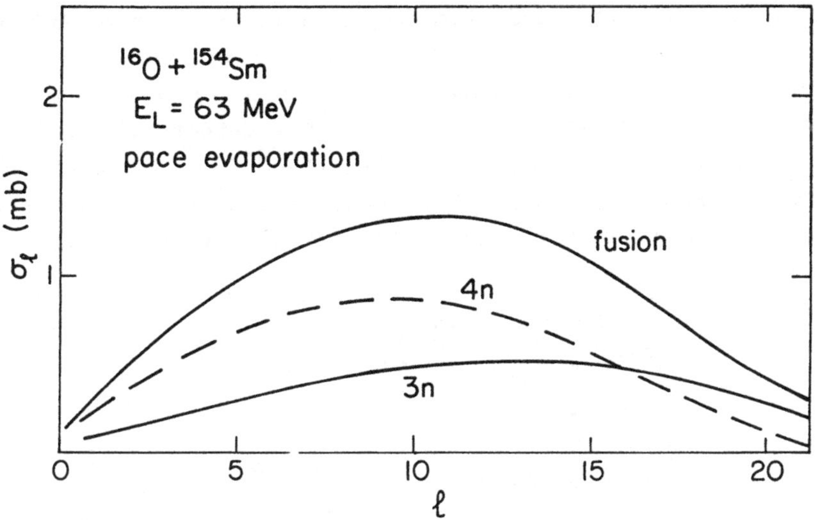

Fig. 2 Comparison of statistical model calculation of the compound nuclear spin distribution leading to the 3n and to the 4n channels.

The results of this study have led us to seek a more inclusive technique for tagging fusion reactions. We have built an electrostatic deflector to separate fusion residues from beam-like particles. The deflector and associated scattering chamber and detector box are illustrated in Fig. 3. We have also built a large-area Breskin-type gas-filled proportional counter to detect the residues. This detector

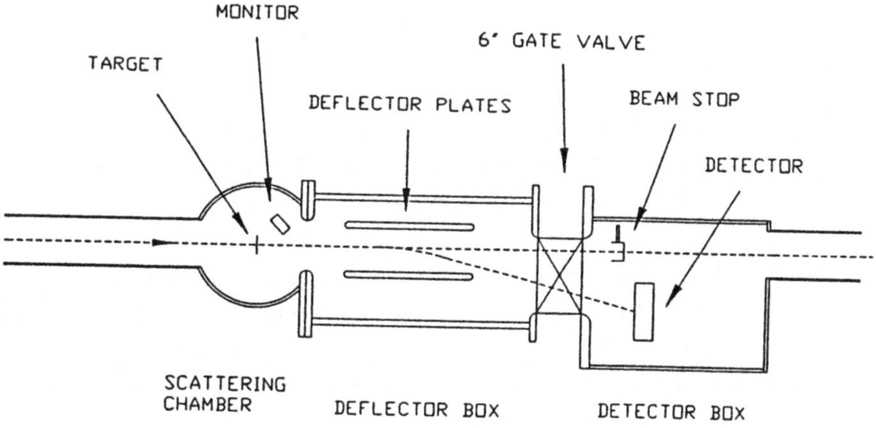

Fig. 3 Schematic view of electrostatic deflector and associated target and detector chambers.

gives both position and timing information. The latter is used together with the pulsed beam to perform time-of-flight analysis of detected particles. The gamma ray multiplicity is determined using NaI detectors in coincidence with the residue detector. As our first test of this system we have again studied the $^{16}O+^{154}Sm$ reaction. Examples of the time-of-flight singles and coincidence spectra are shown in Fig. 4. The peak at the right in the singles spectrum appears to

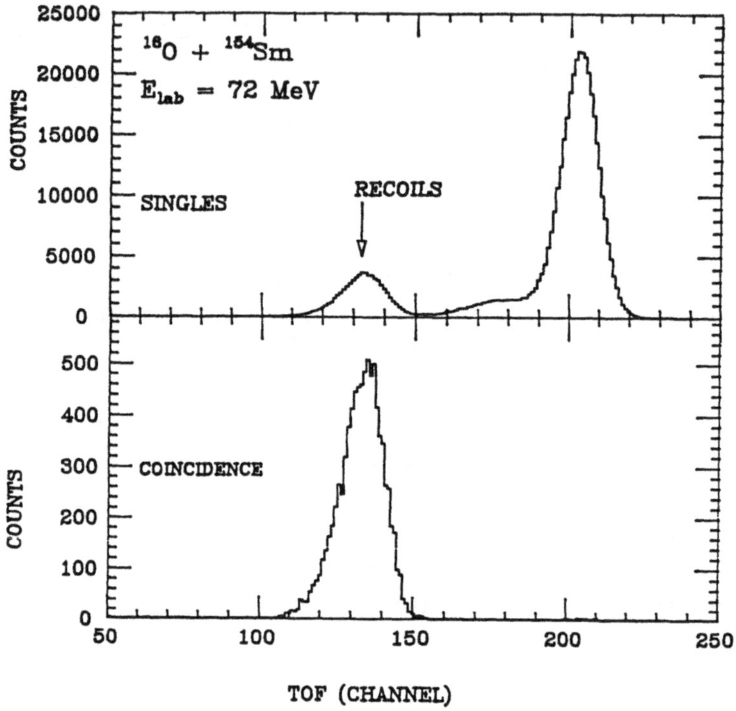

Fig. 4 The upper panel shows the singles time-of-flight spectrum and the lower panel shows the time-of-flight spectrum for events in coincidence with gamma rays.

originate mostly from slit scattered and degraded beam particles. Although we are still learning how to reduce further this background, we have already achieved higher sensitivity than we were able to with the discrete-gamma evaporation-residue tagging approach. This has enabled us to go even farther below the barrier. We show in Fig. 5 a comparison of the mean ℓ values obtained using the electrostatic deflector and the ℓ values deduced by discrete gamma ray tagging. We have recently taken data for the $^{28}Si+^{154}Sm$ and $^{28}Si+^{122}Sn$ reactions.

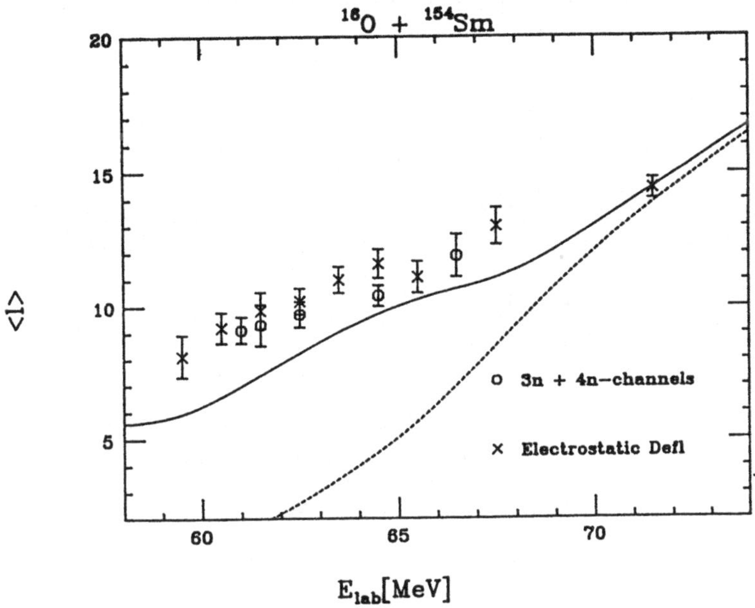

Fig. 5 Comparison of mean values of compound nuclear spin distribution as determined by the discrete gamma ray tagging technique[O] and as determined by evaporation residue tagging technique[X].

It can be seen from Figs. 1 and 5 that the mean spin inferred from these new multiplicity measurements exceeds theoretical expectations at the lower energies, similar to our findings for the $^{16}O+^{208}Pb$ system utilizing fission fragment angular distributions. We have tried to assemble most of the available data [12,14-17] on the first or second moments of the spin distribution and compare them with theoretical expectations. In Fig. 6 we plot the ratio of the experimental $\langle \ell \rangle$ value to the theoretical $\langle \ell \rangle$ value as a function of the ratio of the bombarding energy to the barrier height V_B. For those experiments which measure $\langle \ell^2 \rangle$ rather than $\langle \ell \rangle$, we have plotted the ratio of the root-mean-square values rather than the mean values. The mean and root-mean-square ratios are not very different for plausible spin distributions. Before discussing this plot we should remark that the theoretical calculations to which the experimental values are compared differ in sophistication for the different systems studied. The $^{16}O+^{208}Pb$ system has been compared with unusually complete coupled channels calculations [3,4]. The $^{16}O+^{154}Sm$ and $^{80}Se+^{80}Se$ results are compared with models [19,20] which should adequately treat effects

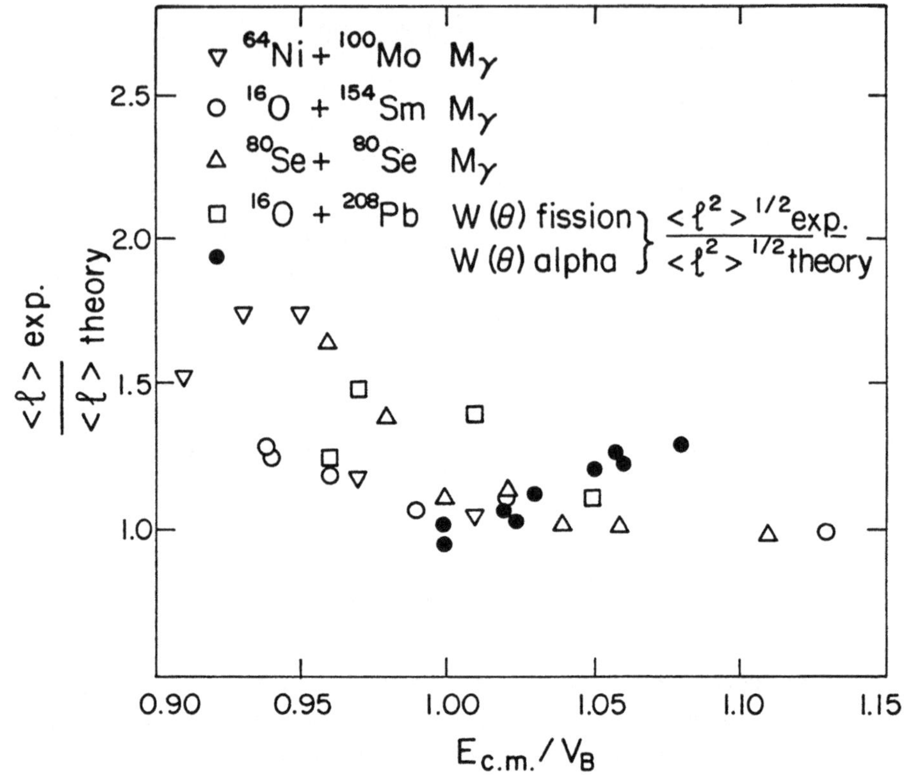

Fig. 6 Comparison of ratio of experimental to theoretical mean spin values (or ratio of rms values in csase of alpha and fission decay probes) as a function of ratio of bombarding energy to the barrier energy.

associated with the shape degree of freedom. The results associated with alpha evaporation (filled circles in Fig. 6; these points are actually for neighboring targets at a single bombarding energy and are deduced from rather small effects on recoil distributions) and for the ^{64}Ni+^{100}Mo reaction are compared with a more schematic "constant-coupling" model. In spite of these limitations, there seems to be apparent a fairly consistent pattern of a broader than expected spin distribution at sub-barrier energies. This result is not limited to observations at a single laboratory or with a single probe.

Accepting the indication in Fig. 6 that the observed mean spin is larger than expected, the question arises as to what is being missed in the theoretical models. There is really no hint in the empirical data. Nevertheless, we cannot resist speculating as to what has been left out. An assumption common to all theoretical models to date is that the inertial mass appearing in the quantum mechanical barrier penetration expression is equal to the reduced mass. Although this may be a reasonable assumption at the outer classical turning point, by the time one reaches the inner classical turning point there is appreciable density overlap and the inertial mass may differ from the reduced mass. This problem has been investigated for very light systems using the adiabatic time-dependent Hartree Fock approach [22, 23]. For $^{12}C+^{12}C$, $^{16}O+^{16}O$, and $^{40}Ca+^{40}Ca$ an appreciable increase in the inertial mass at separations close to the barrier top is obtained. It is also interesting to note that a recent treatment [24] of barrier tunneling in radioactive ^{14}C decay has employed an inertial mass that changes from the reduced mass to one estimated on the basis of shape changes of a single very deformed nucleus as the center separation decreases. Thus the effects we are probing in sub-barrier fusion reactions may have a more general context.

This work was supported in part by the United States. Department of Energy.

REFERENCES

1. D.E. DiGregorio et al., Phys. Lett. B176 (1986) 322
2. R.G. Stokstad et al., Phys. Rev. C21 (1980) 2427
3. S.C. Pieper, M.J. Rhoades-Brown, and S. Landowne, Phys. Lett. 162B (1985) 43
4. I.J. Thompson et al., Phys. Lett. B157 (1985) 250
5. H. Esbensen and S. Landowne, Phys. Rev. C35 (1987) 2090
6. K.T. Lesko et al., Phys. Rev. C34 (1986) 2155
7. Y. Kondo et al., Phys. Rev. C35 (1987) 828
8. S. Gil et al., Phys. Rev. C31 (1985) 1752
9. I.Y. Lee et al., Phys. Rev. C35 (1987) 605
10. R.D. Fischer et al., Phys. Lett. B171 (1986) 33
11. R. Vandenbosch et al., Phys. Rev. Lett. 56 (1986) 1234
12. T. Murakami et al., Phys. Rev. C34 (1986) 1353
13. A. Gavron, Phys. Rev. C21 (1980) 230
14. B. Haas et al., Phys. Rev. Lett. 54 (1985) 393
15. P.J. Noland et al., Phys. Rev. Lett. 54 (1985) 2211
16. A.M. Borges et al., in Proceedings of Symposium on the Many Facets of Heavy Ion Fusion Reactions, p. 441, Argonne National Laboratory, March 24-26, 1986
17. M.L. Halbert et al., at the 6th Adriatic International Conference on Nuclear Physics, Dubrovnik, Yugoslavia, 1987
18. R. Vandenbosch in Proceedings of Symposium on the Many Facets of Heavy Ion Fusion Reactions, p. 155, Argonne National Laboratory, March 24-26, 1986
19. C.Y. Wong, Phys. Rev. Lett. 31 (1973) 766
20. H. Esbensen, Nucl. Phys. A352 (1981) 147
21. C.H. Dasso et al., Nucl. Phys. A405 (1983) 381
22. H. Flocard et al., Nucl. Phys. A339 (1980) 336
23. P.H. Heenen et al., Nucl. Phys. A394 (1983) 525
24. F. Barranco et al., Phys. Rev. Lett. 60 (1988) 507

Intimations of Non-Compound Fission at Near-Barrier Energies

H.Rossner, J.Erxmeyer, D.Hilscher and M.Lehmann
Hahn-Meitner-Institut Berlin, 1000 Berlin 39, Federal Republic of Germany

ABSTRACT

The capture process of heavy ions has been investigated by an analysis of the fission exit channel. Mass and angular distributions were measured for fission fragments produced in the reaction $^{169}Tm(^{36}Ar, f)$ at bombarding energies of E_{lab}= 178 MeV, 189 MeV, 205 MeV and 237 MeV. At E_{lab}=189 and 205 MeV complete mass equilibration has been observed together with anisotropies of the fission fragment angular distributions consistent with predictions of the transition state model. At our lowest energy measured so far, which is roughly four MeV above the Bass barrier, we notice a small mass drift with angle and an unusually large anisotropy. This behavior is considered an indication of increased relative probability for non-compound fission at near-barrier energies.

INTRODUCTION

The investigation of fusion excitation functions in recent years has shown that fusion is a very complicated process, and many parameters have to be determined to achieve an understanding of the reaction mechanism. On the other hand, experimental excitation functions for fusion normally are very smooth and can be parameterized very well by simple models[1,2]. Obviously there is need for measurements of additional fusion properties such as the average value $\langle I \rangle$ of the spin distribution which can be deduced from the multiplicity of γ-rays emitted by evaporation residues[3], or the mean square spin $\langle I^2 \rangle$ which can be extracted from fission fragment angular distributions[4,5].

When the fusion reaction is investigated via the exit channel for fission, then a clear understanding of the fission process is necessary. Such an understanding is achieved for light ion induced reactions where the traditional transition state model (TSM)[6–8] based on the rotating liquid drop model (RLDM)[9] has proved very successful in reproducing fission fragment angular distributions. This model describes the phase space available for the fission products by properties of the saddle-point configuration of the fissioning nucleus. Within this model the anisotropy of the angular distribution $W(0^o)/W(90^0)$ is expressed by the mean square spin value $\langle I^2\rangle$, the nuclear temperature T and the moments of inertia J parallel and perpendicular to the fission axis. An approximate relation can be written as

$$\frac{W(0^o)}{W(90^o)} - 1 \approx \frac{\langle I^2\rangle}{4K_0^2} = \frac{\langle I^2\rangle}{4T * (1/J_{\parallel} - 1/J_{\perp})} .$$

For heavy-ion induced fission, however, it was shown by several groups[10−17] that the TSM model failed to reproduce measured anisotropies. This model cannot be applied when the nuclear temperature of the compound system is higher than the fission barrier, or when the trajectory of the fissioning system does not pass through the fission saddle configuration. From preliminary results of our measurements we see slight indications of non-compound fission reactions at low bombarding energies when fusion probabilities could be modified by frictional forces and neck formation, for instance.

Another point of motivation for this experiment was to extend an existing parameter set of a simple barrier-fluctuation model to heavier systems. Heavy systems have small distances s between their half-density surfaces at the Coulomb barrier, causing a stronger mutual nuclear interaction and, thus, increase the influence of competing reaction channels on the fusion process. Such a behavior has been suggested by Krappe et al[18]. They fitted fusion excitation functions for several systems using a barrier-fluctuation model based on the Krappe potential[19] by adjusting the two parameters barrier shift and fluctuation width. The global tendency of the parameters to increase with decreasing distance s has been suggested as indication for a possible formation of a neck before the Coulomb barrier is reached. Up to now our measurements for the system $^{169}Tm +^{36} Ar$ have not entered the sub-barrier energy regime to allow an accurate determination of both parameters. Nevertheless, our fusion cross sections measured so far can be reproduced by the model using parameters consistent with the existing scheme.

EXPERIMENT

We have chosen the system $^{169}Tm +^{36} Ar$ with a fissility parameter of x=0.75 and a liquid drop fission barrier height of 6.4 MeV[20] for spin zero. It is assumed that the probability for fission is still high enough[21] to neglect the yield of evaporation residues in the total fusion cross section. The mean fissility[22] x_m=0.69 is similar to the value for the system $^{208}Pb +^{24} Mg$ [13] (x_m=0.686). The compound nucleus

^{205}Fr further has a low neutron to proton ratio. Thus the emission of pre-fission neutrons can be ignored in the computation of the nuclear temperature of the transition state nucleus. Due to a small distance between the equivalent sharp sphere radii at the interaction barrier a pronounced enhancement of the sub-barrier fusion cross section is expected compared to the prediction of a one-dimensional barrier penetration model. This enhancement might be correlated with an unusually broad spin distribution of the compound nucleus.

Up to now the experiment has been performed at four bombarding energies of E_{lab} = 178 MeV, 189 MeV, 205 MeV and 237 MeV. The pulsed ^{36}Ar beam, delivered by the VICKSI accelerator in Berlin with a width of 1 nsec and a repetition rate of 1/(160 nsec), was focused on self-supporting foils of ^{169}Tm with thicknesses of 350 to 400 $\mu g/cm^2$. We used surface barrier detectors to measure energy and time-of-flight of the reaction products. Fission fragments could be well separated from projectile-like and target-like nuclei, and after an event-by-event correction of the measured energy and time-of-flight for energy loss in the target[23], and for pulse-height defect[24)] and plasma delay effects[25)] produced in the Si-detectors, the distribution of secondary masses of the fragments was obtained. The fission yield was normalized to Rutherford scattering at forward angles. The angle integrated fission cross section has been interpreted as fusion cross section and could be fitted by the barrier fluctuation model.

RESULTS AND DISCUSSION

Utilizing the spin distribution given by the barrier fluctuation model, variances K_0^2 were extracted from the anisotropies of the angular distributions and plotted in fig. 1 versus the square root of the thermal excitation energy per nucleon at the fission saddle. Using a level density parameter of a=A/10 the displayed range on the x-axis corresponds to a temperature scale of 1.3 - 2.2 MeV. The arrows indicate the shift in excitation energy when second chance fission is considered. At E_{lab}=205 MeV ($\sqrt{E^*/A}$=0.553 $(\text{MeV/nucleon})^{1/2}$) the number of neutrons emitted before scission has been determined to roughly one neutron per fission event by detecting neutrons at different angles in coincidence with fission fragments. This number overestimates the pre-fission neutron multiplicity. Based on results of an evaporation calculation we expect that half of these neutrons are emitted before the system passes through the fission saddle configuration. The prediction of the TSM model with values for the moments of inertia of the transition nucleus given by the RLDM model and with spin distributions computed with the barrier fluctuation model is shown by the solid line in fig. 1.

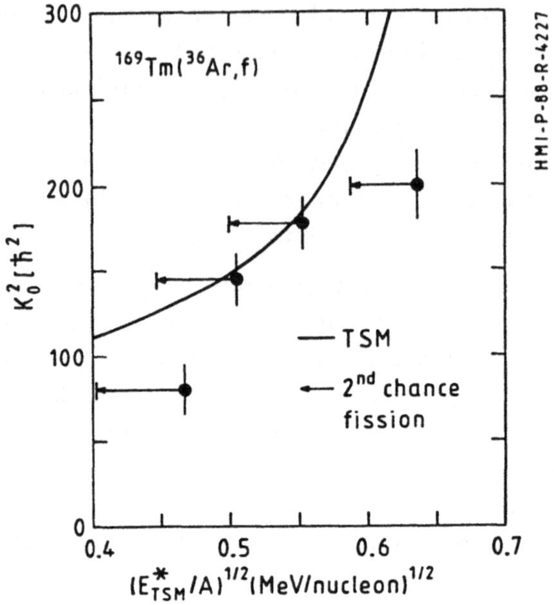

Fig. 1 : Variances K_0^2 extracted from fission fragment angular distributions are plotted versus $\sqrt{E^*/A}$ and are compared to predictions of the TSM model.

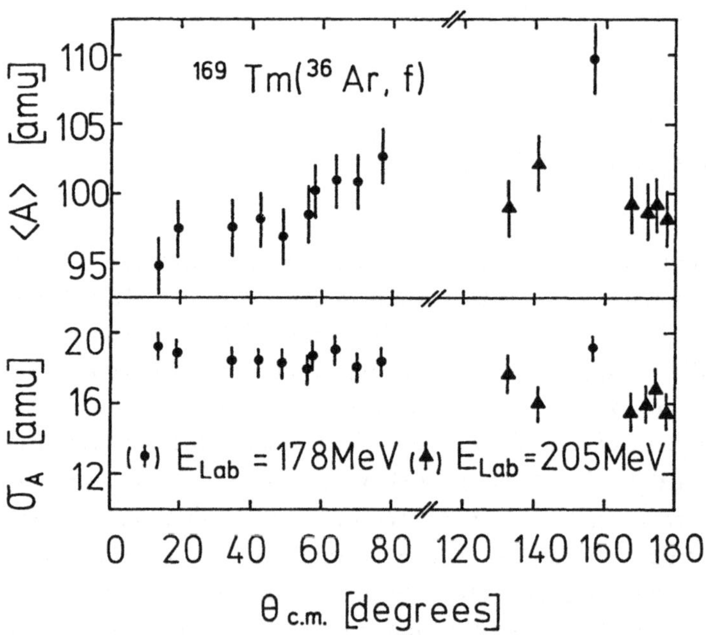

Fig.2 : First and second moments of the fission fragment mass distributions are plotted versus the center-of-mass angle at E_{lab}=178 MeV (filled circles) and E_{lab}=205 MeV (filled triangles).

Within error the variances K_0^2 extracted from the measured anisotropies are consistent with the model at E_{lab}=189 MeV and 205 MeV. At higher bombarding energy the deviation between experimental and theoretical K_0^2-values could be caused by contributions of fast fission reactions. At the lowest bombarding energy, however, the discrepancy cannot be explained by this effect. The discrepancy cannot be reduced either by adjusting the nuclear temperature, because an unrealistically high particle multiplicity before fission would have to be assumed. To reproduce the anisotropy at the energy close to the barrier within the framework of statistical models, we either have to assume a more elongated shape of the configuration that defines the phase space of the fission fragments or, as suggested by R. Vandenbosch et al.[4,5] studying more asymmetric systems, we have to assume an unusually large mean square spin value of the composite system.

At near-barrier energy we further notice a slight angle dependence of the average mass of the fission fragments (fig.2) in addition to the relatively large anisotropy of the angular distribution. The observation of this mass-angle correlation is an indication for contact times of the two nuclei shorter than the time needed for mass equilibration. If all degrees of freedom are not equilibrated at a certain configuration of the composite system, then values for K_0^2 or $\langle I^2 \rangle$ extracted from measured anisotropies using statistical models are questionable.

According to a dynamical model of Blocki et al.[22] fusion of heavy ions strongly depends on the particle velocity relative to a multi-dimensional potential energy surface and the speed related to the shift of the saddle point towards more compact shapes during mass equilibration. The formation of a neck between projectile and target before the Coulomb barrier is reached will influence the fusion path, and frictional forces may prevent trajectories of being caught behind the unconditional saddle and thus may increase the probability for non-compound fission reactions at near-barrier energies.

To summarize, we have analysed fission fragment angular distributions for the reaction $^{169}Tm(^{36}Ar, f)$ at four bombarding energies of E_{lab}=178 MeV, 189 MeV , 205 MeV and 237 MeV. Deviations from predictions of the TSM model assuming mean square spins given by a barrier fluctuation model are found at the lowest and at the highest energy measured so far, corresponding to 4 MeV and 53 MeV above the Bass barrier[26]. Both deviations are interpreted as contributions from non-compound reactions. At E_{lab}=237 MeV the fission barrier height for high spins is getting lower than the nuclear temperature. At the energy closest to the barrier we suggest that the large anisotropy of the fragment angular distribution and the observed mass-angle correlation of the fission products are due to an increased relative probability for non-compound fission.

REFERENCES

1) D.L. Hill, J.H. Wheeler, Phys. Rev 89, 1102 (1953)
2) C.Y. Wong, Phys. Rev. Lett. 31, 766 (1973)
3) R. Vandenbosch et al., Phys. Rev. C28, 1161 (1983)
4) R. Vandenbosch et al., Phys. Rev. Lett. 56, 1234 (1986)
5) T. Murakami et al., Phys. Rev. C34, 1353 (1986)
6) A. Bohr, in Proceedings of the United Nations International Conference on the Peaceful Uses of Atomic Energy (United Nations, New York, 1956), Vol. 2, p. 151
7) I. Halpern and V.M. Strutinsky, in Proceedings of the Second United Nations International Conference on the Peaceful Uses of Atomic Energy (United Nations, Geneva, Switzerland, 1958), Vol. 15, p. 408
8) R. Vandenbosch and J.R. Huizenga, Nuclear Fission (Academic Press, New York, 1973)
9) S. Cohen, F. Plasil, W.J. Swiatecki, Ann. Phys. (N.Y.) 82, 557 (1974)
10) A. Gavron et al., Phys. Rev. Lett. 52, 589 (1984)
11) B.B. Back, Phys. Rev C31, 2104 (1985)
12) B.B. Back et al., Phys. Rev C32, 195 (1985)
13) M.B. Tsang et al., Phys. Lett. 129, 18 (1983)
14) M.B. Tsang et al., Phys. Rev. C28, 747 (1983)
15) H. Rossner, J.R. Huizenga, W.U. Schröder, Phys. Rev. C33, 560 (1986)
16) V.S. Ramamurthy and S.S. Kapoor, Phys. Rev. Lett. 54, 178 (1985)
17) V.S. Ramamurthy and S.S. Kapoor, Phys. Rev. C32, 2182 (1985)
18) H.J. Krappe et al., Z. Phys. A314, 23 (1983)
19) H.J. Krappe, J.R. Nix and A.J. Sierk, Phys. Rev. C20, 992 (1979)
20) A.J. Sierk, Phys. Rev. C33, 2039 (1986)
21) L.C. Vaz and J.M. Alexander, Phys. Rep. 97, 1 (1983)
22) J.P. Blocki, H. Feldmeier, W.J. Swiatecki, Nucl. Phys. A459, 145 (1986)
23) J.F. Ziegler, J.P. Biersack, U. Littmark, The Stopping and Range of Ions in Matter, Pergamon Press, New York (1985)
24) S.B. Kaufman et al., Nucl. Instr. and Meth. 115, 47 (1974)
25) W. Bohne et al., Nucl. Instr. and Meth. A240, 145 (1985)
26) R. Bass, Phys. Rev. Lett. 39, 265 (1977)

SUBBARRIER FUSION REACTION OF $^{19}F + ^{232}Th$

Zhang Huanqiao, Lu Jun, Xu Jinchen, Liu Zuhua, Xu Kan and Ruan Ming
Institute of Atomic Energy, P.O.Box 275, Beijing, China

In recent years, there has been a lot of interest in subbarrier fusion reactions because of the significant enhancement of cross sections over the predictions from a simple model based on one dimensional barrier penetration.[1] It is generally believed that the reason for this enhancement lies in the strong coupling of elastic channel to the other quasi-elastic channels. Such coupling results in the energy-shifted barriers so as to increase the barrier penetration probabilities and broaden the spin distributions. The spin distribution broadening of compound nucleus was found for near and below the Coulomb barrier in a few cases.[2] Due to the sensitivity of the fission fragment angular distributions to the compound nucleus spin distributions, the subbarrier fusion-fission reactions are quite appealling. We report here a study of $^{19}F + ^{232}Th$ fusion-fission reaction at the bombarding energies from 85 to 105 MeV.

The experiment was performed at the HI-13 tandem Van de Graff accelerator in Institute of Atomic Energy, Beijing. A thorium target about 350 $\mu g/cm^2$ in thickness was bombarded by a collimated ^{19}F beam. The beam spot size on the target was 2-3 mm in diameter The isotopic purity of ^{232}Th target was 100%. The fission fragments were detected by mica track detectors at angles from 90° to 170°. The distance between target and mica which stuck on the inner surface of a cylinder was 4 cm. After irradiation, mica was etched by a standard procedure.[3] Two surface barrier semiconductor detectors were set at -40° and $+19.76^\circ$ directions respectively as monitors to detect the elastic scattering projectiles. The beam direction was checked by the elastic scattering measurement. The fission cross sections were normalized to the cross sections of Rutherford scattering.

The fission excitation finction of $^{19}F + ^{232}Th$ is shown in Fig. 1. The data represented by open circles are the experimental results of Leigh.[4] Our results are in good agreement with Leigh's values except for our data extending further below the Coulomb barrier.

Coupled channels theory (CCFUS code) was used to fit the fission excitation function to obtain the transmission coefficients T_l for each bombarding energy. Both the inelastic transfer channel couplings to the elastic channel included in the coupled channels calculation are listed in Table 1.

Several of the angular distributions of fission fragments together with the theoretical calculations are presented in Fig. 2. The theoretical angular distributions were calculated with the transition state theory in terms of the transmission coefficients extracted from the excitation function. This calculation requires a parameter $K_0^2 = \mathcal{J}_{eff}T/\hbar^2$. The effective moments of inertia $\mathcal{J}_{eff}$ was extracted from the systematics of Ref. (5). However, the measured values of anisotropies of fission fragments are obviously greater than the calculated ones as shown in Fig. 3.

Table 1
Inelastic transition data of the states and the parameters of transfer channels included in the coupled channels calculation

Nucleus	J^{π}	E_x (MeV)	λ	β	Transition	Q_{eff} (MeV)	F (MeV)
^{19}F	$5/2^+$	0.1976	2	0.55	-p	6.20	1.0
	$5/2^-$	1.3458	3	0.33	-2p	6.11	1.0
^{232}Th	2^+	0.0494	2	0.225	$-\alpha$	9.05	1.0
	4^+	0.1621	4	0.068	+n	-5.64	1.0
	3^-	0.7743	3	0.086			

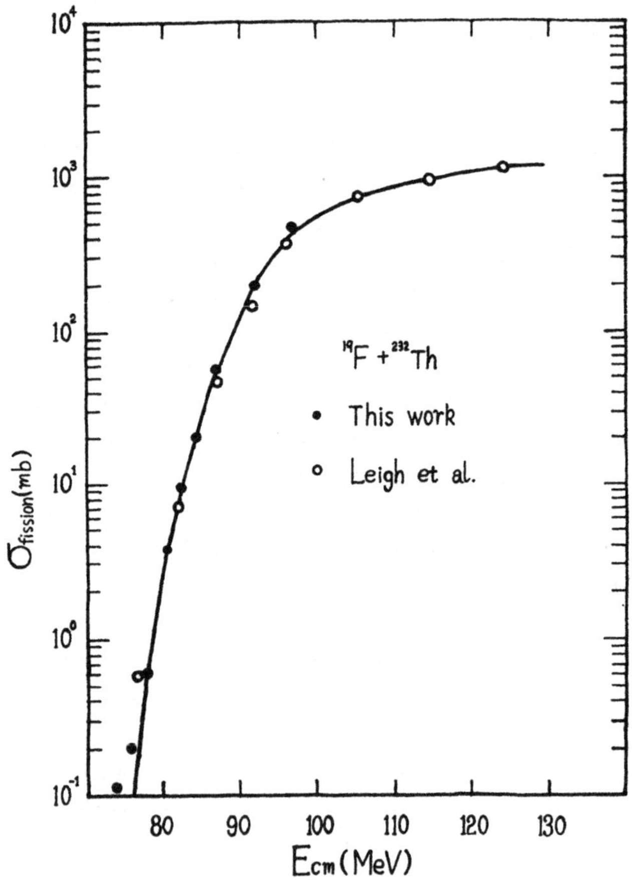

Fig. 1. Fission excitation function for the ^{19}F + ^{232}Th reaction The full line represent coupled channels calculation.

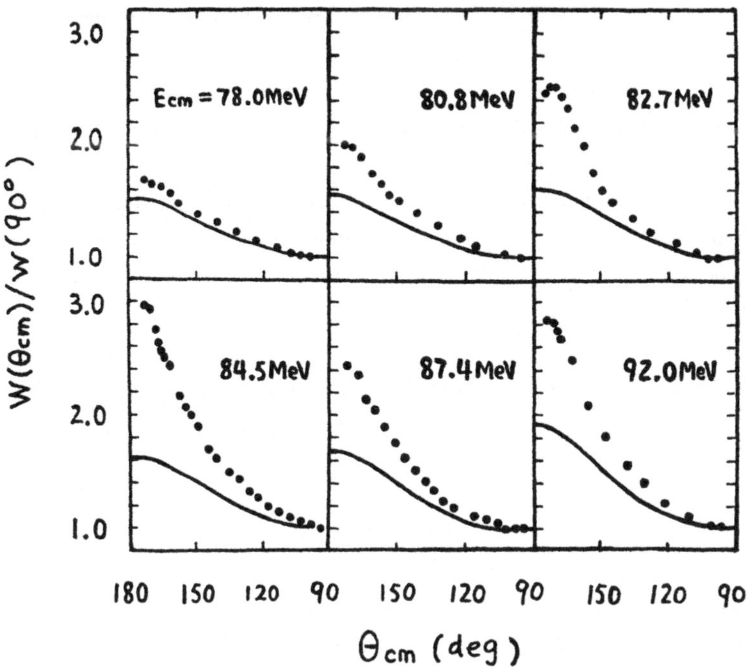

Fig. 2. Fission angular distributions for ^{19}F + ^{232}Th fusion reaction at sub-barrier energies

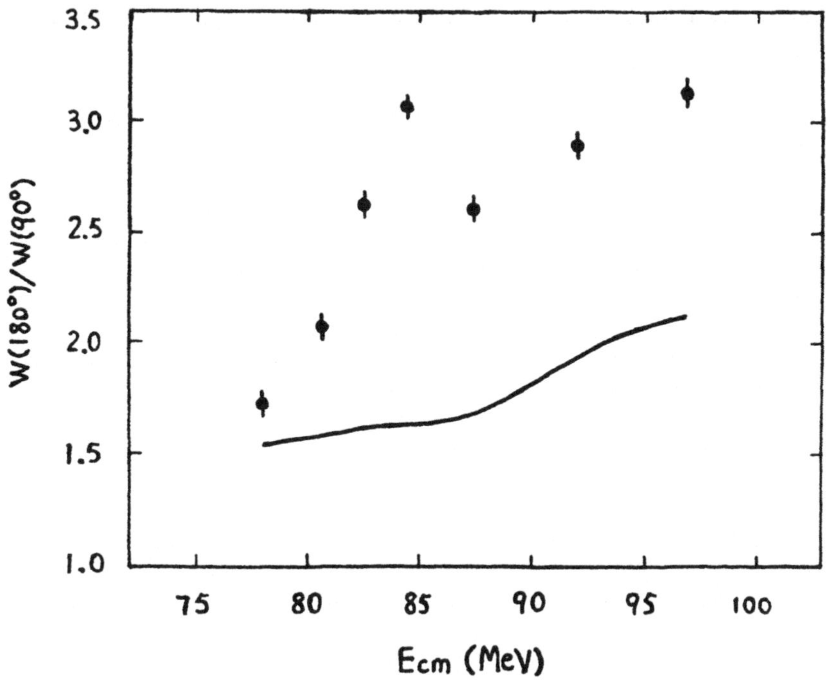

Fig. 3. Anisotropies of fission fragments for 19 + ^{232}Th subbarrier fusion fission reaction. The full line is the theoretical calculation with transition state theory.

In conclusion, we have measured fission cross sections and angular distributions to the energies considerably lower than Coulomb barrier for $^{19}F + ^{232}Th$. The excitation function is well reproduced on the basis of coupled channels theory. However, the angular distributions are much more anisotropic than expected. The preliminary impression from this study is that at the near and subbarrier energies, the present standard models are inadequate for explaining both the fusion excitation function and the angular distributions of fission fragments simultaneously. It seems to us that some kind of new mechanism may contribute to fusion-fission processes of $^{19}F + ^{232}Th$ reaction system. The data analyses are still in progress.

One of the authors (Z. H. Q.) is indebted to Dr. J. Leigh for providing us with unpublished data. This work was partly supported by National Natural Science Fundation of China

REFERENCES

1. M. Beckerman, Phys. Rep. 129 (1985) 145
2. B. Haas et al., Phys. Rev. Lett. 54 (1985) 398
   R. Vandenbosch et al., Phys. Rev. Lett. 56 (1986) 1234
3. R. L. Fleischer, P. B. Price and R. M. Walker, Nuclear Tracks in Solids : Principles and Applications (Univ. California Press, Berkeley, 1975)
4. J. Leigh, Private communication
5. R. Freifelder, M. Prakash and J. M. Alexander, Phys. Rep., 133 (1986) 315

ELASTIC SCATTERING AND THRESHOLD ANOMALY

ALGEBRAIC SCATTERING THEORY AND ITS APPLICATION TO HEAVY ION COLLISIONS

F. Iachello

Center for Theoretical Physics, Sloane Laboratory, Yale University, New Haven, Connecticut 06511

Abstract

The use of algebraic scattering theory in the study of heavy ion collisions is reviewed. It is suggested that this approach provides a convenient alternative to coupled channel calculations.

1. Introduction

Algebraic scattering theory is a convenient way to parametrize, organize and predict scattering data for complex reactions. By complex reaction is meant a situation in which the number of open channels is large enough that coupled channel calculations are needed but not so large that statistical methods are appropriate. The optimal size is ≈10-20 channels. Heavy-ion collisions at intermediate (Tandem) energies $E_{Lab} \approx 100$ MeV are in this situation. The simplification introduced by the algebraic approach gives the possibility to use intermediate energy heavy-ion collisions for extracting important nuclear structure information and for constructing a simple but complete heavy-ion reaction theory.

2. S-matrix theory

The basic assumption of algebraic scattering theory (AST) is that the fundamental quantity in scattering is the scattering matrix (S-matrix). S-matrix theory was proposed many years ago as an alternative to a theory of strong interactions[1] in hadronic physics. In heavy ion physics it has been used in a simplified form by Blair, Frahn and others[2]. For non-relativistic problems it is convenient to introduce the partial wave S-matrix, $S_\ell(k)$, where k is the center of mass momentum. In the S-matrix theory of the 60's two conditions were imposed on S, those of unitarity and analyticity. The former implies that, for single channel scattering, $|S_\ell(k)|\leq 1$, while the latter implies some particular properties of $S_\ell(k)$ when viewed as a function of the analytic continuation in the variables ℓ and k. The combination of these two properties is not sufficient to determine $S_\ell(k)$ uniquely. It provides only relations between the real and imaginary parts (dispersion relations). It has been shown recently[3] that the addition of the concept of dynamic symmetry fixes the functional form of $S_\ell(k)$.

A form particularly appropriate for heavy-ion collisions is that given by an SO(3,1) dynamic symmetry[4]. The reason is that heavy-ion collisions at intermediate energies are dominated by the Coulomb plus short range interaction, both of which have properties that can be either exactly (Coulomb) or approximately (short range) related to SO(3,1). The partial wave S-matrix arising from SO(3,1) symmetry is

$$S_\ell(k) = \frac{\Gamma(\ell+1+iv(\ell,k))}{\Gamma(\ell+1-iv(\ell,k))} \tag{2.1}$$

If v=constant, independent of ℓ and k, the S-matrix (2.1) has an exact SO(3,1) symmetry in the usual sense (i.e. the Hamiltonian has a simple functional dependence on the invariants). If $v=v(\ell,k)$ the situation is more complex, but, nonetheless the S-matrix (2.1) can be used to describe the data. If the function $v(\ell,k)$ has a simple form and it can be related to some physical quantities, then (2.1) is a convenient and useful way of parametrizing the data.

For applications to heavy-ion collisions, the quantity $v(\ell,k)$, which assumes here the role of the interaction and thus is called the "algebraic potential", must contain a part related to the Coulomb interaction and a part related to the strong interaction,

$$v(\ell,k) = v_c(\ell,k) + v_s(\ell,k) \tag{2.2}$$

The Coulomb interaction has an exact expression

$$v_c(\ell,k) = \frac{\mu\, Z_1 Z_2 e^2}{\hbar^2\, k} \tag{2.3}$$

while the strong interaction can be parametrized as a Woods-Saxon well

$$v_s(\ell,k) = \frac{v_R + i v_I}{1 + \exp[(\ell-\ell_0)/\Delta]} \tag{2.4}$$

The imaginary part in (2.4) describes the absorption.

The S-matrix (2.1) has several interesting properties. For real v it is manifestly unitary, $|S_\ell(k)|=1$. For complex v, unitarity imposes some conditions on v, which can, to a good approximation be written as Im $v \geq 0$. Because of the simple pole structure of the Γ-functions, it is also relatively straightforward to find whether there are resonances (poles of $S_\ell(k)$ for Im $k<0$) or bound states (poles of $S_\ell k$) for Re $k=0$, Im $k \geq 0$) in the problem.

The S-matrix (2.1) replaces the solution of the Schrödinger equation

$$H\,\psi = E\,\psi \quad , \; H = \frac{\vec{p}^{\,2}}{2\mu} + V(r) \tag{2.5}$$

with the optical model potential V(r),

$$V(r) = V_c(r) + V_s(r)$$

$$V_c(r) = \frac{Z_1 Z_2 e^2}{r}$$

$$V_s(r) = \frac{V_R + iV_I}{1 + \exp[(r-r_o)/\delta]} \tag{2.6}$$

It contains the same physics and produces elastic cross sections with similar properties.

3. Multichannel problems

The major advantage of AST is not so much that it can deal with elastic scattering but rather than it can be generalized to complex multichannel problems in a relatively straightforward way[*]. Consider, for example, any channel with two-bodies in the final state (Fig.1). Let n be the number of

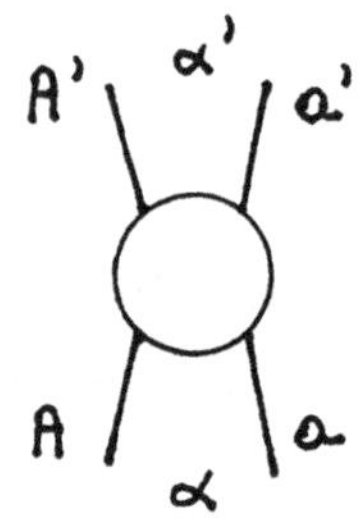

Fig.1.Schematic representation of the reaction A+a→A'+a'.The outgoing particles can be either excited states of A and a or different particles.

channels, α=1,...,n. The derivation of (2.1) from the dynamic group suggests that the generalization of (2.1) to multichannel situations is

$$S_\ell^{\alpha\alpha'}(k) = \frac{\Gamma(\ell+1+iv^{\alpha\alpha'}(\ell,k))}{\Gamma(\ell+1-iv^{\alpha\alpha'}(\ell,k))} \tag{3.1}$$

where $v^{\alpha\alpha'}(\ell,k)$ is an "algebraic potential matrix" that describes the diagonal interactions, $\alpha\neq\alpha'$, in each channel, and their coupling $\alpha\neq\alpha'$.

In heavy-ion collisions there are two types of two-body channels often considered:

(i) Inelastic channels.

The coupling potential $v^{\alpha\alpha'}$ contains here two terms, one due to the Coulomb interaction (Coulomb excitation) and the other due to the strong interaction. Coulomb excitation can be parametrized approximately in ℓ-space as ($\alpha\neq\alpha'$)

$$v_c(\ell,k) = \eta_c(\lambda)\,\frac{3}{(2\lambda+1)}\begin{cases}(\ell/\ell_{oc})^{\lambda+1}, & \ell < \ell_{oc}\\ (\ell_{oc}/\ell)^{\lambda}, & \ell > \ell_{oc}\end{cases} \quad . \tag{3.2}$$

The strong excitation is often due to a surface reaction and thus can be parametrized in the form

$$v_s(\ell,k) = -\frac{d}{d\ell}\,\frac{u_R + iu_I}{1+\exp[(\ell-\ell_0)/\Delta]} \quad . \tag{3.3}$$

If one assumes that the ratio of the real to imaginary part of the coupling potential is the same as in the diagonal potential, Eq.(3.3) can be written as

$$v_s(\ell,k) = -\,\eta_s\,\frac{d}{d\ell}\,\frac{v_R + iv_I}{1+\exp[(\ell-\ell_0)/\Delta]} \quad . \tag{3.4}$$

(ii) Transfer Channels.

For these channels Coulomb excitation is not present and one is left only with the strong excitation. For a surface reaction a convenient parametrization is again given by (3.3).

The multichannel S-matrix (3.1) replaces the solution of the coupled channel Schrödinger equations

$$(\nabla^2 + k_\alpha^2)\psi_\alpha(\vec{r}) = \sum_{\alpha'} V^{\alpha\alpha'}(\vec{r})\psi_{\alpha'}(\vec{r}) \quad , \tag{3.5}$$

with coupling potentials that contain a Coulomb excitation term

$$V_c(r) = \gamma_c(\lambda) r_c\,\frac{3Z_1Z_2e^2}{(2\lambda+1)}\begin{cases} r^{\lambda}/r_c^{\lambda+2}, & r < r_c\\ r_c^{\lambda-1}/r^{\lambda+1}, & r > r_c\end{cases} \tag{3.6}$$

and a strong excitation term

$$V_s(r) = - r_o \frac{d}{dr} \frac{U_R + iU_I}{1+\exp[(r-r_o)/\delta]} \quad ,$$

$$U_R + iU_I = \gamma \, (V_R + iv_I) \quad . \tag{3.7}$$

The advantage of the algebraic approach is in this case considerable, since the solution of the multichannel scattering problem is given in closed form (no differential equations to solve!). In practice, the way in which the cross sections for excitation of the various channels are calculated is that of diagonalizing the potential matrix $v^{\alpha\alpha'}$, constructing the eigen-S-matrix,

$$D = \begin{pmatrix} \frac{\Gamma(\lambda_1)}{\Gamma(\lambda_1')} & 0 & \cdots & 0 \\ 0 & \frac{\Gamma(\lambda_2)}{\Gamma(\lambda_2')} & \cdots & 0 \\ \cdot & \cdot & \cdot & \cdot \\ 0 & 0 & \cdot & \frac{\Gamma(\lambda_n)}{\Gamma(\lambda'_n)} \end{pmatrix} \tag{3.8}$$

and from D, obtaining

$$S = Z \, D \, Z^{-1} \tag{3.9}$$

where Z is the matrix that diagonalizes v. The matrix v is a complex, symmetric, matrix,

$$v^T = v \tag{3.10}$$

Because imaginary terms are contained in the potential matrix, v, it is important to check the unitarity bound

$$|S_{11}|^2 + |S_{12}|^2 + \ . \ . \ . + |S_{1n}|^2 \leq 1 \tag{3.11}$$

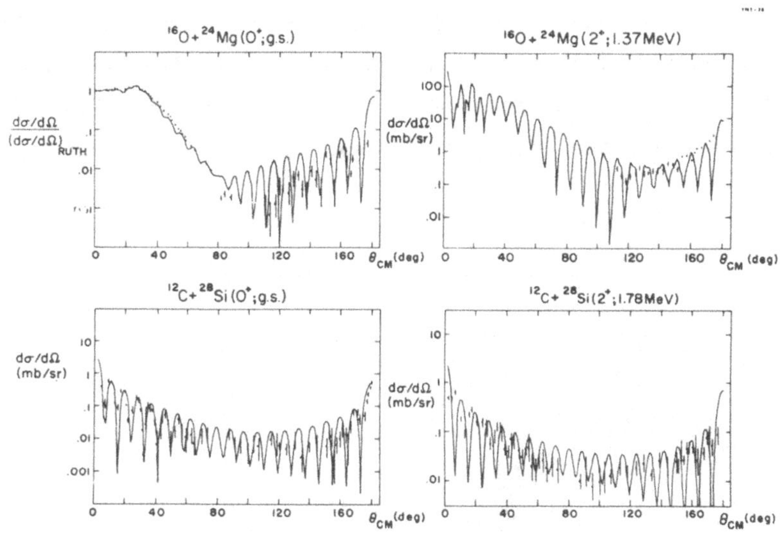

Fig.2. An example of application of algebraic scattering theory to a four channel reaction. The reaction channels are discussed in the text.

4. Applications to heavy ion collisions

An example of application of this approach to heavy ion collision is provided by the study of the four reaction channels[5]

$$^{16}O + ^{24}Mg \rightarrow ^{16}O + ^{24}Mg(0^+; g.s.)$$

$$^{16}O + ^{24}Mg \rightarrow ^{16}O + ^{24}Mg(2^+; 1.37\ MeV)$$

$$^{16}O + ^{24}Mg \rightarrow ^{12}C + ^{28}Si(0^+; g.s.)$$

$$^{16}O + ^{24}Mg \rightarrow ^{12}C + ^{28}Si(2^+; 1.78\ MeV) \qquad (4.1)$$

at E_{CM} = 27.8 MeV. The results of the calculations are shown in Fig. 2.

It is important to comment on the differences between the algebraic approach and the conventional method of analysis. The major differences occur in the backward direction. At the present time, full coupled channel calculations for all four channels do not exist. The only available complete calculations are those employing an optical model code for the elastic

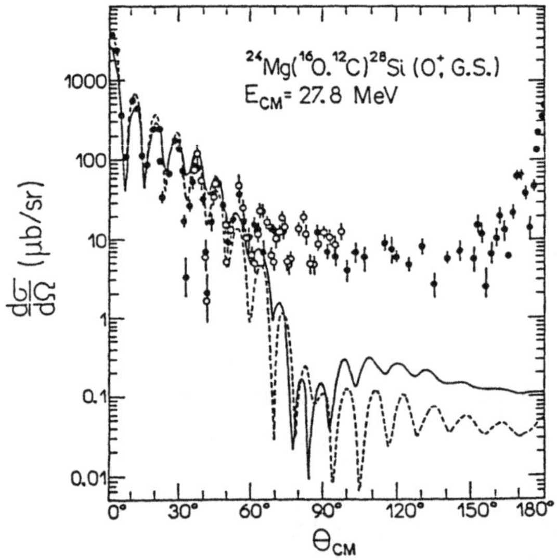

Fig. 3. A conventional calculation of the reaction $^{16}O + ^{24}Mg \rightarrow ^{12}C + ^{28}Si(0^+; g.s.)$. The dotted line represents the results od a DWUCK calculation while the continuous line represents the cross-section calculated using PTOLEMY[6].

scattering and DWBA for the inelastic and transfer channels. These calculations appear to be inadequate in the backward direction. An example is shown in Fig.3. It would be of interest to see to what extent full coupled channel calculations can reproduce the data. The algebraic approach indicates that the backward direction is dominated by coupled channel effects.

The analysis of the four reaction channels of Eq.(4.1) has been extended recently[8] to include the four additional channels

$$
\begin{aligned}
{}^{16}O + {}^{24}Mg &\rightarrow {}^{20}Ne(0^+) + {}^{20}Ne(0^+) \\
{}^{16}O + {}^{24}Mg &\rightarrow {}^{20}Ne(0^+) + {}^{20}Ne(2^+) \\
{}^{16}O + {}^{24}Mg &\rightarrow {}^{20}Ne(0^+) + {}^{20}Ne(4^+) \\
{}^{16}O + {}^{24}Mg &\rightarrow {}^{20}Ne(2^+) + {}^{20}Ne(2^+)
\end{aligned}
\tag{4.2}
$$

for which measurements have been reported. In cases in which the particles in the outgoing channel are identical as in ${}^{20}Ne(0^+)$ + ${}^{20}Ne(0^+)$, care must be taken in appropriately symmetrizing the scattering amplitudes. The results of the 8-channel calculation appear to be still of quality comparable to that reported in Fig.2.

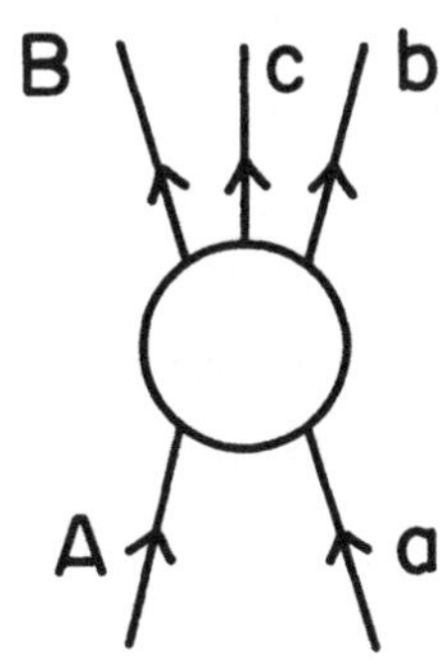

Fig.4. Schematic representation of the reaction A+a→B+b+c.

5. Multichannel problems with break-up and fragmentation.

The relatively simple form of the S-matrix (3.1) for two-body collisions allows one to construct S-matrices for more complex processes. A particularly important one is the process in which three of more particles emerge from the collision, Fig.4. Examples of these processes are

$$^{16}O + ^{24}Mg \rightarrow ^{12}C + \alpha + ^{24}Mg$$

$$^{16}O + ^{24}Mg \rightarrow ^{16}O + \alpha + ^{20}Ne \quad (5.1)$$

Work in the direction of constructing the full S-matrix for multiparticle final states is in progress. It would be of interest to obtain complete data for the reactions (5.1).

6. Combining structure and reaction theory.

The simplification introduced by the algebraic approach suggests that one may combine structure and reaction theory in order to extract some information on nuclear properties. An interesting idea here is that nuclear reactions proceed through matrix elements of an excitation operator. Consider, for example, the excitation of a rotational band. It is reasonable to assume that the excitation occurs through the matrix elements of the quadrupole operator, $\hat{T}^{(E2)}$. This operator can be represented in an appropriate nuclear structure model. For example, if the interacting boson model[7] is used to describe the rotational band, the operator $\hat{T}^{(E2)}$ has the form

$$\hat{T}^{(E2)} = \beta\,\hat{Q} = \beta\,[(d^{\dagger}\times\tilde{s}+s^{\dagger}\times\tilde{d})^{(2)} + \chi(d^{\dagger}\times\tilde{d})^{(2)}] \quad (6.1)$$

Similarly, one may consider a set of two-nucleon transfer reactions, Fig.5. Again, if the interacting boson model[7] is used to describe the pairing band, the transition operator for L=0 transfer can be written as

$$\hat{P}^{(0)\dagger} = \gamma\, s^{\dagger} \quad (6.2)$$

Assuming that indeed this is the case, one can write the coupling potentials, $V^{\alpha\alpha'}$, as

$$v^{\alpha\alpha'}(\ell,k) = \langle\alpha\|\hat{T}\|\alpha'\rangle\ f(\ell,k) \qquad (6.3)$$

where $f(\ell,k)$ is a form factor describing the ℓ and k dependence. Since the reduced matrix elements of the transition operator $\hat{T}$ are given in terms of some strength parameter, for example β, times the matrix elements of a known operator, this assumption allows one to compute cross-sections for excitation of all members of a rotational band in terms only of one parameter, β. A crucial advantage here is that, in the algebraic approach one can consider, not only simple situations, such as the excitations of a rotational band, but more complex cases, such as the excitation of bands in transitional nuclei or in nuclei with a complex geometry (for example, γ-unstable or triaxial nuclei)[8].

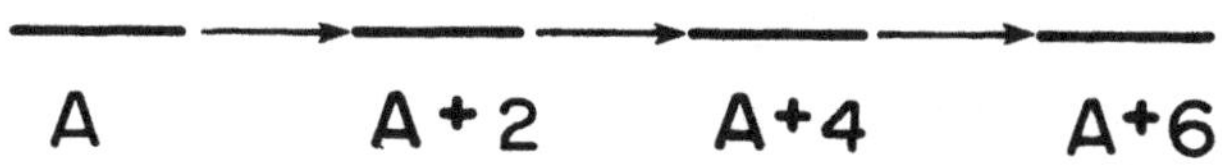

Fig.5. Excitation of a pairing band by the pair transfer operator, $\hat{P}$.

7. Conclusions

The combination of S-matrix theory and structure models offers the possibility of a detailed analysis of complex scattering processes in which several reaction channels are open and important. Heavy-ion scattering at intermediate energies are one of these situations. The algebraic method can be used here in two ways:

(i) to extract nuclear structure information if this is not known by determining the matrix elements of the transition operators $\langle\alpha\|\hat{T}\|\alpha'\rangle$ from a fit to the data;

(ii) to predict or study the outcome of a reaction if the nuclear structure is known.

8. Acknowledgements

I acknowledge useful conversations with S. Landowne and A. Vitturi. This work was performed in part under D.O.E. Contract No. DE-AC-02-76ER03074.

References

1. See for example, G.T. Chew, "S-Matrix Theory of Strong Interactions", Benjamin, New York (1961).
2. See for example, W.H. Frahn, in "Treatise on Heavy-Ion Science", Vol.I, D.A. Bromley ed., Plenum, New York (1984).
3. See for example, F. Iachello, in "Symmetries in Science II", B. Gruber and R. Lenczewski, eds., Plenum, New York(1987).
4. F. Iachello, "Algebraic Scattering Theory", Preprint YNT 87-12.
5. Y. Alhassid, F. Iachello and B. Shao, Phys. Lett. B201, 183 (1988).
6. R. Lichtenthäler, Jr., private communication.
7. F. Iachello and A. Arima, "The Interacting Boson Model", Cambridge University Press, Cambridge (1987).
8. A. Leviatan, private communication.

ABSORPTION UNDER THE COULOMB BARRIER; ITS IMPORTANCE FOR THE SCATTERING AND FUSION OF HEAVY IONS AT SUB- AND NEAR-BARRIER ENERGIES

T. Udagawa
Department of Physics, University of Texas at Austin
Austin, Texas 78712 U.S.A.

Discussions are presented concerning the absorptive part W of the optical potential for heavy ion reactions at near- and sub-barrier energies. Particularly, the information obtained from simultaneous analyses of elastic and fusion cross section data is presented, showing that the fusion part of W should have an appreciable value in the Coulomb barrier region, and thus cause a substantial absorption there. This contradicts assumptions made in all coupled-channel calculations performed in the past. Information obtained for the direct reaction part of W, as well as the real potential are also discussed.

I. INTRODUCTION

Recently, much attention has been focused on heavy ion scattering and fusion at near- and sub-barrier energies [1]. Experimentally, interesting observations have been made of the enhancement of the fusion cross section σ_F, the threshold-anomaly [2,3] (i.e., a rather striking energy dependence of the optical potentials in the above energy region), the correlation between the magnitude of σ_F and the direct reaction cross section σ_D [4,5], and so forth. Theoretically, much effort has been devoted to a study of coupled-channel (CC) effects. Thus, a number of large scale CC calculations have been done [6-10], explaining the enhancement successfully in terms of the additional attraction in the entrance channel coming from the coupling included in the calculations.

These calculations, however, often failed to reproduce the observed elastic scattering cross section σ_{EL}. Fig. 1 shows a few examples of the fit obtained. Discrepancies between the calculated and observed cross section, σ_{EL}^{cal} and σ_{EL}^{exp} respectively, are clearly seen, particularly at large angles. This seems to indicate that the absorptive potential W used in the calculation is too weak.

In fact, all the CC calculations performed so far were done by assuming a very small radius parameter $r_F = 1.0$ fm for the fusion potential W_F, i.e., the fusion part of W. Such a choice has been made by simply adopting the critical distance idea used in the usual barrier-penetration model (BPM) [11], namely that fusion takes place after the two colliding ions pass through, or go over the potential barrier, and approach within a critical distance $R_{cr} = r_{cr}(A_1^{1/3} + A_2^{1/3})$. The value of r_{cr} is usually assumed to be $r_{cr} \simeq 1.0$ fm. Now in the heavy ion reactions at near- and sub-barrier regions, there are not many direct reaction (DR) channels open. Therefore, the CC calculations performed were done by including all or most of all the important DR channels. Hence, in these calculations, W is set equal to W_F.

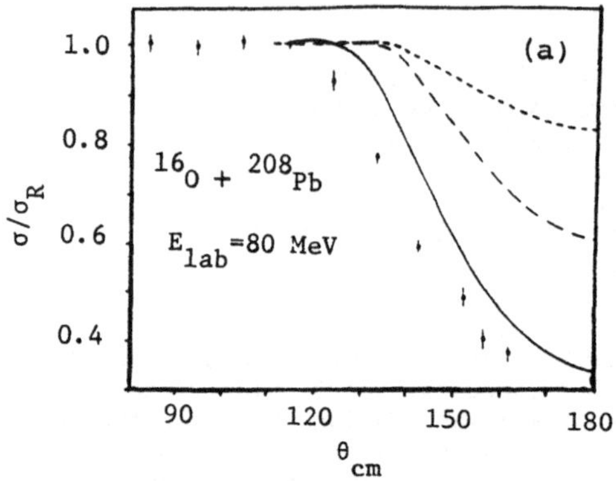

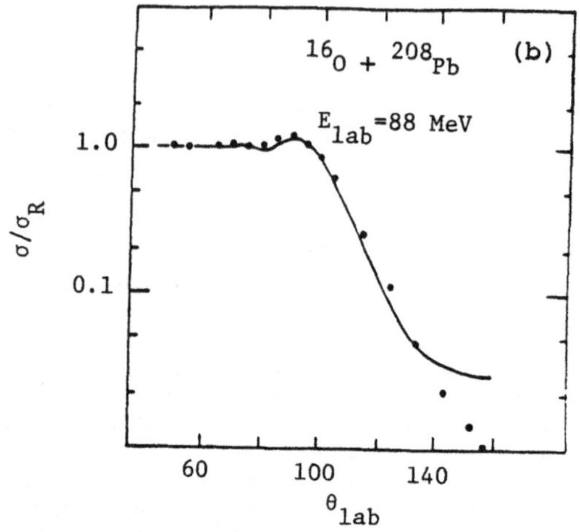

Fig. 1. Elastic scattering angular distributions for $^{16}O + ^{208}Pb$ at (a) $E_{lab} = 80$ MeV and (b) $E_{lab} = 88$ MeV. The solid lines shown in (a) and (b) are theoretical coupled-channel cross sections taken from [10] and [8], respectively.

The resultant W then becomes extremely small in the barrier region.

The failure to reproduce σ_{EL}^{exp} remarked upon above casts some doubt on the validity of the assumption $r_F = 1.0$ fm. In fact, the result of a systematic analysis [12] of fusion data has indicated that the r_F-value should be as large as $r_F \simeq 1.4$ fm. In [12], the analysis was made by using the fusion cross section formula given in terms of W_F (see Eq. 92)) [13-14], and the following simple form of W_F,

$$W_F = \begin{cases} W & \text{for} \quad r < R_F \\ 0 & \text{for} \quad r > R_F . \quad (R_F = r_F(A_1^{1/3} + A_2^{1/3})) \end{cases} \tag{1}$$

R_F (or r_F) was then treated as an adjustable parameter to be determined from experiment. The r_F-value thus determined turned out to be $r_F \simeq 1.43$ fm. This value is indeed much larger than the r_F-value used in the CC calculations [6-10] or in BPM [11]. Note that R_F with the above r_F-value can be beyond the position of the Coulomb barrier peak.

It is expected that the effect of the coupling to the DR channels obtained from the CC calculation depends strongly on the type of W_F used in the calculations. If W_F is negligible in the barrier region as assumed in [6-10], the coupling becomes effectively strong and we get a large effect there. This was indeed found to be the case in [6-10]. However, if W_F is large, the coupling will effectively be reduced and the effect will become smaller. It is thus important to have accurate information on the r_F-value, we have performed simultaneous analyses of elastic scattering and fusion data for a few systems where the data are available [15-18]. One important advantage of the method is that it makes it possible to calculate the elastic and fusion cross sections on a single footing, and thus to put together both elastic scattering and fusion data for extracting information on r_F. In what follows, after giving a brief discussion of the fusion cross section formula and the optical model potential used in the analyses, we present and discuss the results.

II. FUSION CROSS SECTION FORMULA AND OPTICAL POTENTIAL

The formula for the fusion cross section used in the analysis is given as [12-14]

$$\sigma_F = \frac{2}{\hbar v} \langle \chi^{(+)} | W_F(r) | \chi^{(+)} \rangle \quad , \tag{2}$$

where W_F is the fusion potential already introduced in the Introduction, v is the incident velocity and $\chi^{(+)}$ is the distorted wave function calculated from the usual optical model calculation. $\chi^{(+)}$ is thus generated by using the full optical potential $U = -(V + iW)$.

Note that one can write down the total reaction (or absorption) cross section σ_R similarly to Eq. (2) as

$$\sigma_R = \frac{2}{\hbar v} \langle \chi^{(+)} | W(r) | \chi^{(+)} \rangle \quad . \tag{3}$$

As is clear from the fact that W_F is a part of W, σ_F is also a part of σ_R. Actually, Eq. (2) was derived [13,14] from (3) by simply picking up the W_F part of σ_R. Similarly, we may write the total DR cross section σ_D as

$$\sigma_D = \frac{2}{\hbar v} \langle \chi^{(+)} | W_D(r) | \chi^{(+)} \rangle \quad , \tag{4}$$

where W_D is the direct reaction part in W, which may be given as

$$W_D = W - W_F \quad . \tag{5}$$

From what we have described above, it is clear that with Eq. (2) we are able to calculate the fusion cross section within the framework of the direct reaction theory (optical model). We have introduced a new ingredient W_F, which is, however, a part of W, and therefore is still confined within the framework of the direct reaction theory. In this way, we are able to put the elastic scattering and fusion data together in order to extract information about r_F as remarked in the Introduction, though the number of parameters involved in the theory has now increased.

Another important feature is that σ_F given by (2) includes DR effects through $\chi^{(+)}$. As already noted, $\chi^{(+)}$ is calculated by using the "full" imaginary potential $W = W_F + W_D$. Thus $\chi^{(+)}$ includes the effect of the absorption due to W_D, i.e., the loss of flux due to the direct reactions. It is worth remarking that because of this inclusion of the DR effects, we were able to explain successfully the observed <u>reduction</u> [19] of σ_F for very heavy systems at the above barrier energies. For these systems, the total DR cross section often exceeds 50% of σ_R. In such cases, the inclusion of flux loss due to W_D is vitally important.

In our earlier analysis [12,20,21], we took the optical potential from the literature, and then introduced W_F as the inner portion of W as given by (1). The outer portion is then identified to be W_D. In the analyses to be discussed in the following sections, however, use was made of more sophisticated forms of W_F and W_D, i.e., a volume type W_F and a surface type W_D, along with the real potential $V(r)$ of the usual Wood-Saxon form. The optical potential U used is thus given as

$$U(r) = -V(r) - i[W_F(r) + W_D(r)] \quad , \tag{6}$$

where

$$V(r) = \frac{V_R}{1 + \exp(X_R)} + V_C \quad , \tag{7}$$

$$W_F(r) = \frac{W_F}{1 + \exp(X_F)} \tag{8}$$

$$W_D(r) = 4W_D \frac{\exp(X_D)}{(1 + \exp(X_D))^2} \quad . \tag{9}$$

In (7), V_C is the Coulomb potential, while X_i (i = R, F or D) in (7)-(9) are defined as $X_i = (r - R_i)/a_i$ with $R_i = r_i(A_1^{1/3} + A_2^{1/3})$.

III. THE $^{16}O + ^{208}Pb$ SYSTEM WITH E_{lab} = 80 MeV

Let us now discuss the result of the analyses. We first summarize results obtained [22] for the $^{16}O + ^{208}Pb$ system with E_{lab} = 80 MeV [15,16]. Since the analysis made for this system is closely related to the work done by Satchler et al. [10], we first give a few remarks on that work that are relevant to the later discussion.

The authors of [10] studied the question of how well a large scale CC problem can be reduced to a one dimensional optical model problem. The authors thus searched for optical model potentials that could reproduce σ_{EL}, σ_F and σ_R obtained in such a CC calculation [9,10]. They could indeed find several such potentials, demonstrating that the reduction can indeed be done successfully. The search for such potentials was made for the case of $^{16}O + ^{208}Pb$ at E_{lab} = 80 MeV [15,16] mentioned above.

One (perhaps the best one) of the potentials has exactly the same form as that given by (2)-(9) with the following values of the parameters; V_R = 156.7 MeV, a_R = 0.584 fm, r_R = 1.179 fm, W_F = 10.0 fm, a_D = 0.25 fm, r_F = 1.0 fm, W_D = 0.0404 MeV, a_D = 0.786 fm, r_D = 1.5 fm. We call such a potential the equivalent elastic potential (EEP). Note that W_F of EEP is exactly the same as the original potential (bare potential) used in the CC calculation. (This is true for all the EEP's determined in [10].) The form of the real potential is also exactly the same, but the values of the parameters V_R and a_R are different from those of the bare potential. In particular, V_R of EEP is larger by a factor of 2.6 than that of the bare potential. Actually, the difference of the above two values can be attributed to the additional attraction due to the coupling mentioned above, which is quite large. The direct reaction part W_D of EEP is completely new in EEP; i.e., no corresponding term is present in the bare potential. It describes the absorption due to the direct reactions that are taken into account explicitly in the CC calculation.

An important claim made in [10] is that the EEP thus obtained is able to reproduce the experimental fusion cross section σ_F^{exp}. Based on this, the authors concluded that a large r_F-value is not needed in order to explain the observed large σ_F.

We point out, however, that although EEP can reproduce σ_F^{exp}, it cannot reproduce σ_{EL}^{exp}. Recall that σ_{EL}^{CC} does not agree with σ_{EL}^{exp} (see Fig. 1) and that EEP reproduces σ_{EL}^{CC}. It is thus clear that σ_{EL} obtained with EEP does not fit to σ_{EL}^{exp}. In [22], we thus investigated how this unsatisfactory feature is removed. Namely, we tried to see how the EEP of [10] is to be modified in order to remove this discrepancy, and further how this modification would affect the calculated σ_F.

We introduced first various modifications in EEP keeping, however, the r_F-value at the original value of $r_F = 1.0$ fm. Under this constraint, however, we were unable to obtain a satisfactory simultaneous fit to the data. (Since the details of the attempts are discussed in [22], we shall not repeat them here.) We thus finally carried out a χ^2-analysis including r_F as an adjustable parameter. Actually, all the six parameters in W_F and W_D were varied freely. A satisfactory fit was then immediately achieved. In Fig. 2, we present the result. As the best fit parameters, we obtained the following; $W_F = 8.9$ MeV, $a_F = 0.13$ fm, $r_F = 1.401$ fm, $w_D = 0.457$ MeV, $a_D = 0.280$ fm, and $r_D = 1.497$ fm.

A remarkable feature of the result is that the value of r_F is now 1.401 fm, which is much larger than the starting value of 1.0 fm. The value is very close to 1.43 fm, determined in [12]. The resultant W_F has now an appreciable value in the barrier, and thus causes a strong absorption there. This improved the fit of the elastic scattering data significantly as seen in Fig. 2. The modified EEP then predicted $\sigma_F = 35.7$ mb, which agrees very well with the experimental value of $\sigma_F^{exp} = 36$ mb. We were thus able to achieve a simultaneous fit to all the data.

In [22], we further evaluated with this modified potential the average spin of the compound nucleus. It turned out to be $\langle \ell^2 \rangle \simeq 142$, which is significantly larger than the value 104 obtained in [10]. Although our value 142 is not as large as the experimental value $\simeq 180$ obtained by Murakami *et al* [23], it is larger than any theoretical value quoted in [23], including the value obtained in our earlier analysis [12]. In this new analysis, the diffuseness of the fusion potential was introduced, which made the resultant spin values larger.

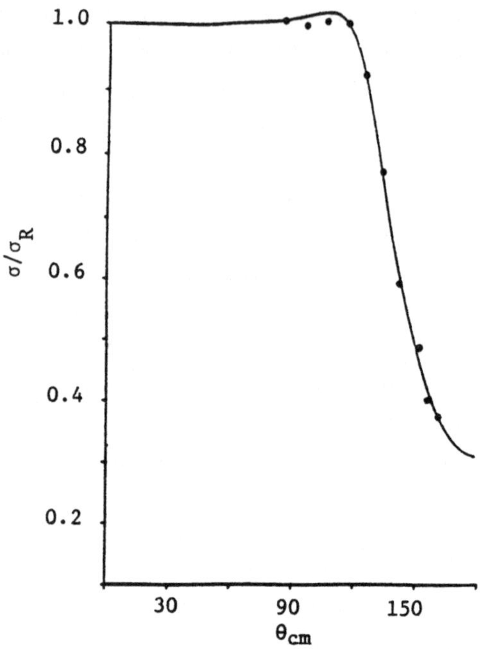

Fig. 2. Elastic scattering angular distribution for $^{16}O + ^{208}Pb$ at E_{lab} = 80 MeV. The solid line is the optical model fit with EEP.

IV. THE $^{32}S + ^{58,64}Ni$ SYSTEMS

The analysis made in [22] was then extended to data of the $^{32}S + ^{58}Ni$ and $^{32}S + ^{64}Ni$ systems taken recently by the Legnaro group [17,18]. The cross sections were measured at varied incident energies around the Coulomb barrier top ($\simeq$60 MeV). In particular, those of the elastic scattering were measured up to large angles. Therefore, the data are very suitable for our purpose of the analysis.

The analysis was performed first by treating V_R, and W_D and r_F as adjustable parameters. The primary purpose was to determine the value of r_F. V_R and W_D were, nevertheless, treated as adjustable parameters in order to achieve a reasonable fit to the data. The values of the other parameters were fixed to be r_R = 1.247 fm, a_R = 0.53 fm, W_F = 10 MeV, a_F = 0.25 fm, r_D = 1.50 fm and a_D = 0.79 fm. Those of the real and imaginary potentials are taken from [17] and [10], respectively. The values of W_F, a_F, r_D and a_D are thus the same as those of EEP discussed in the previous section. (In one case of $^{32}S + ^{58}Ni$ at E_{lab} = 88 MeV, however, use was made of W_F = 1.0 MeV, which was required to obtain a reasonable fit to the data.) The result of this analysis is presented in Subsection A.

As will be seen there, the fit to the data is reasonably good, but not complete. We thus attempted to improve the fit further. In Subsection B, we discuss the result of this attempt.

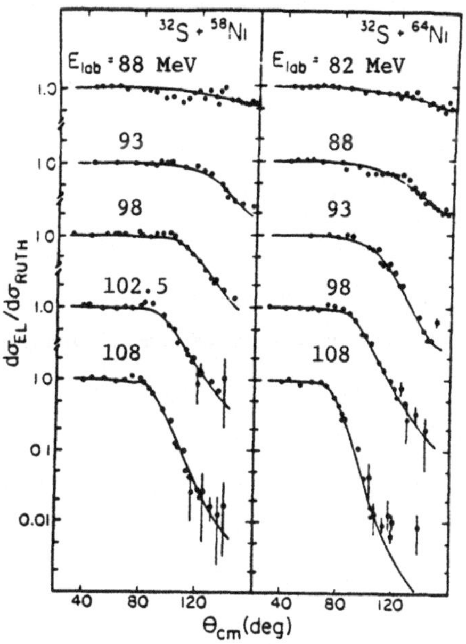

Fig. 3. Elastic scattering angular distributions. The solid lines are optical model fits.

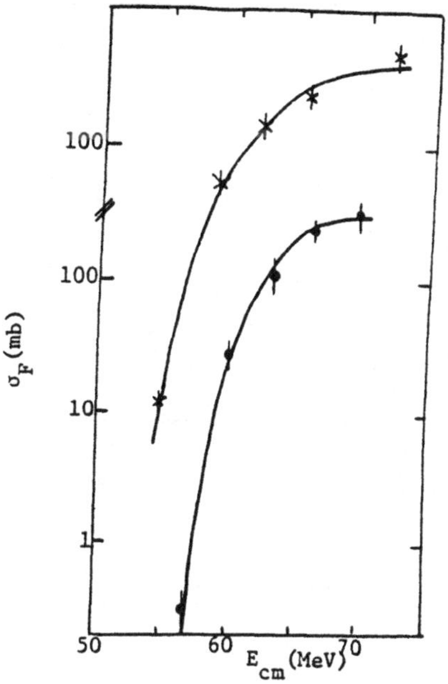

Fig. 4. Fusion cross sections. The solid lines are theoretical predictions.

A. Three Parameter Fits to Elastic Scattering and Fusion Data

In Figs. 3 and 4, we present the resultant fits to σ_{EL}^{exp} and σ_F^{exp}, respectively. As seen, the overall fit of the calculated cross sections to the data is good.

Particularly, the fit of σ_F^{cal} to σ_F^{exp} is excellent. As has been emphasized [18], the observed enhancement in σ_F^{exp} of $^{32}S + ^{64}Ni$ is much more remarkable than on that of $^{32}S + ^{58}Ni$. This feature is well reproduced in the calculation. Also, the observed angular distributions of $^{32}S + ^{64}Ni$ tend to fall off much faster than those of $^{32}S + ^{58}Ni$ do. This tendency is also well reproduced in the calculation. Since the overall fit is good, we can now discuss the significance of the parameters deduced from the analysis.

Fig. 5 shows the values of r_F, W_D and V_R determined as functions of the incident center-of-mass energy E_{cm}. It is remarkable that the r_F-values determined are all in a narrow range of 1.30 fm and 1.42 fm. The average (with the mean deviations) of the values are 1.35 ± 0.04 and 1.39 ± 02 fm for $^{32}S + ^{58}Ni$ and $^{32}S + ^{64}Ni$, respectively. These values are again very close to r_F = 1.401 fm obtained for $^{16}O + ^{208}Pb$ discussed in the previous section [12], and also the value $r_F \simeq 1.43$ fm determined in [12]. Certainly, the values are much larger than the r_F-values used in the CC calculations [6-10]. Note also the mean deviations (less than 3%) are rather small.

In order to see the role played by the elastic scattering data in the above determination of the r_F-values, we repeated the calculations without including the fusion data. The average values of the r_F-values obtained in this way turned out

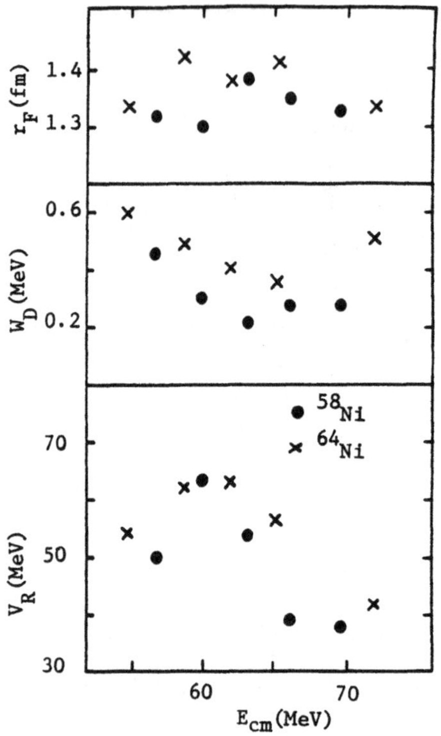

Fig. 5. The values of r_F, W_D and V_R deduced from the χ^2-analyses.

to be 1.40 and 1.39 fm for $^{32}S + ^{58}Ni$ and $^{32}S + ^{64}Ni$, respectively. These values are essentially the same as those determined before by including the data. This result shows that the large r_F-values are essentially determined by the elastic scattering data. To include the elastic scattering data in determining the r_F-value is thus vitally important.

We now turn our attention to the W_D-values. It is seen in Fig. 5 that the W_D-values of $^{32}S + ^{64}Ni$ are systematically larger than those of $^{32}S + ^{58}Ni$. The averages (with the mean deviations) are 0.31 ± 0.06 MeV and 0.47 ± 0.07 MeV for $^{32}S + ^{58}Ni$ and $^{32}S + ^{64}Ni$, respectively. W_D for the later system is thus larger by about 50% than that of the former system. This implies that direct reactions should occur more strongly in $^{32}S + ^{64}Ni$ than in $^{32}S + ^{58}Ni$. It is remarkable that this is indeed the case as observed in a recent measurement of transfer reaction cross sections [4]; there the measured transfer cross sections were found to be much larger in the ^{64}Ni target case than in the ^{58}Ni target case. Unfortunately, the data of the inelastic scattering cross sections have not been obtained yet. Therefore, it is not possible at this moment to make a quantitative comparison of our prediction of σ_D (Eq. (4)) with experiment. It is desirable that the measurement of inelastic scattering cross sections be carried out.

Finally, we turn to the value of V_R. As seen in Fig. 5, the values show a marked energy dependence, taking the maximum at about the Coulomb barrier energy

($\simeq$60 MeV). Since the radius and diffuseness parameters of V(r) are fixed as constants, the similar energy dependence is expected to appear in the values of V(r) at, say, the strong absorption radius R_A ($\simeq$10.8 fm and $\simeq$11.2 fm, respectively, for the ^{58}Ni and ^{64}Ni target cases). Note that the energy dependence seen in Fig. 5 is very similar to the "threshold anomaly" observed in the ^{16}O + ^{208}Pb [2] and ^{16}O + ^{60}Ni [3] systems. The observation of a similar energy dependence has been reported, for the systems considered here, in [17].

It may be important to give a remark here on the energy dependence of W(r); no significant energy dependence was seen in W(r) (=$W_F(r) + W_D(r)$) determined above. The situation, however, changes for the W determined in the next subsection. We shall thus return to this point later in the next subsection.

B. Improvement of Fit to Elastic Scattering Data

As already noted, there are unsatisfactory features present in the fit obtained in Fig. 3. More specifically, fits to the data at lower energies are unsatisfactory, and also for the shoulder region of the angular distributions, particularly for ^{32}S + ^{58}Ni at higher energies. In this subsection, we present the result of attempts to improve the fit. We take as an example the ^{32}S + ^{58}Ni system.

Attempts were first made to χ^2-search, varying four or five parameters freely, and finding that the fit at low energies can be improved significantly by increasing the r_D-value, and also the trouble at the shoulder can be removed by slightly reducing a_F and increasing r_F. A systematic χ^2-analysis was then performed by barying a set of parameters either (a_R,W_F,W_D and a_D) or (a_R,a_F,W_D and a_D). The value of r_F was fixed to be 1.41 fm in these anlyses. In most of the cases, the value of a_F was fixed to be 0.1 fm. Note that the values of $r_F \simeq 1.41$ fm and $a_F \simeq 0.1$ fm mentioned above are very close to $r_F = 1.401$ fm and $a_F = 0.132$ fm for ^{16}O + ^{208}Pb discussed in the previous subsection [12]. The values of the parameters deduced, and also fixed during the search, are all summarized in Table I.

Fig. 6 shows the fit thus obtained. As seen, the fit is much better than that obtained before in Fig. 3. In particular, the fit in the shoulder region of the angular distributions is greatly improved. We see, however, that the fit at E_{lab} = 88 MeV is still not good, although the overall fit is significantly improved as compared with the fit obtained before. Particularly, the oscillation observed in σ_{EL}^{exp} was not reproduced in the calculation. We could try to improve the fit further, but we have not done that, since at such low energy, CC effects may come in explicitly. Note that the fusion cross section at this energy is extremely small (σ_F^{exp} = 0.29 mb).

It is remarkable that the improvement is achieved with r_F = 1.41 fm, which is very close to the values determined in the previous subsection. The values of other parameters changed somewhat, and as consequence, the resultant V, W_F and W_D, at $r = R_A$ are slightly changed. Fig. 7 shows the values of V and W at r = 10.8 fm.

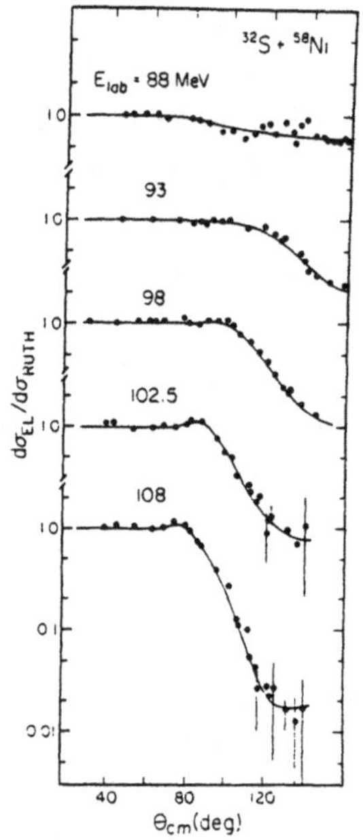

Fig. 6. Elastic scattering angular distributions. The solid lines are optical model fits.

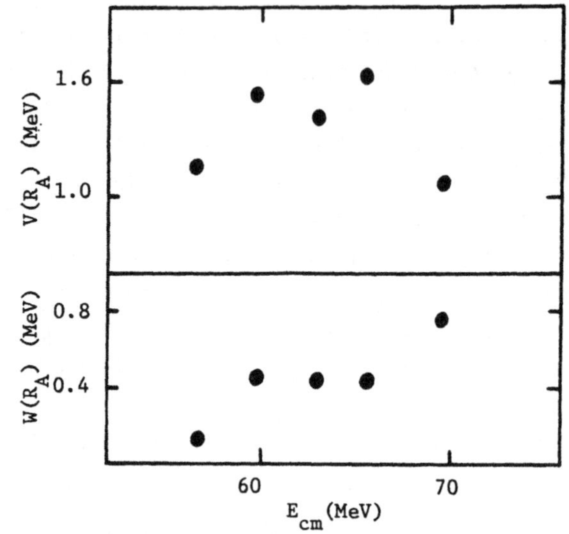

Fig. 7. The values of W(r) and V(r) at r = 10.8 fm as functions of E_{cm} for $^{32}S + ^{58,64}Ni$.

Table I

The values of the optical potential used in the calculations of cross sections shown in Fig. 6, along with the resulting σ_F, σ_R and χ^2-values. The parameters fixed during the search are underlined.

E_{lab}(MeV)	88	93	98	102.5	108
V(MeV)	<u>56.3</u>	<u>63.2</u>	<u>53.7</u>	<u>53.7</u>	<u>37.9</u>
r_0(fm)	<u>1.247</u>	<u>1.247</u>	<u>1.247</u>	<u>1.247</u>	<u>1.247</u>
a_0(fm)	0.520	0.545	0.556	0.668	0.517
W_D(MeV)	0.134	0.441	0.432	0.430	1.086
r_D(fm)	<u>1.77</u>	<u>1.518</u>	<u>1.518</u>	<u>1.518</u>	<u>1.518</u>
a_D(fm)	1.08	0.510	0.382	<u>0.300</u>	0.224
W_F(MeV)	<u>0.331</u>	10.15	17.23	<u>10.0</u>	<u>10.0</u>
r_F(fm)	1.408	<u>1.41</u>	<u>1.41</u>	<u>1.41</u>	<u>1.41</u>
a_F(fm)	<u>0.10</u>	<u>0.10</u>	<u>0.10</u>	<u>0.10</u>	0.151
σ_R	82.3	78.6	113.1	127.2	153.4
σ_F	37.6	36.4	34.6	48.2	39.5
χ^2	1.47	0.60	0.21	0.91	1.43

W now shows tendency to decrease with decreasing E_{cm}, a similar energy dependence to that seen in [2] and [3]. Note that $V(r = R_A)$ also shows an energy dependence, which is very similar to what was seen in Fig. 5.

It is important to note that the values of W presented in Fig. 7 are dominated by W_D. This means that the energy dependence of W seen in Fig. 7 is that of W_D. A question then arises to the energy dependence of W_F. In order to answer the question, we examined the value at 1.3 fm, finding that there is a tendency in the W_F-value similar to that seen in W_D at $r = R_A$. It should be emphasized, however, that the energy dependences discussed above are all based on data points which are very scarce and confined only within a narrow energy range. Therefore, the result discussed above is still too premature to draw any definite conclusion. It is thus important that further studies be made of this subject.

V. CONCLUDING REMARKS

Results are presented for simultaneous χ^2-analyses of elastic scattering and fusion cross sections within the framework of the optical model. Use is made in these analyses of an imaginary potential W consisting of a volume-type fusion and a surface-type direct reaction potential, W_F and W_D, respectively. Analyses have been made so far of data for the $^{16}O + {}^{208}Pb$ system with $E_{lab} = 80$ MeV [15,16] and the $^{32}S + {}^{58,64}Ni$ systems with $E_{lab} = 108$ MeV [17,18]. Although the cases where the analyses have been done are very limited, interesting information has been obtained for W_F and W_D, as well as the real potential V.

The most important point is that the radius parameter r_F of W_F should be as large as 1.4 fm. The value is very close to the value 1.43 fm determined before from the analysis of the fusion data [12], but is much larger than the value $r_F = 1.0$ fm, used in the coupled-channel (CC) calculations [5-10] performed so far. W_F with such a large r_F has an appreciable value in the barrier region, and thus creates a strong absorption there. This picture is quite different from those assumed in the barrier-penetration model [11], and also in the coupled-channel (CC) calculations mentioned above. It was demonstrated that the elastic scattering plays an important role in determining such a large value of r_F.

The results of the analyses mentioned above provided also interesting information on W_D and V. Thus for instance, the values of W_D deduced for the two different systems, i.e., $^{32}S + {}^{58}Ni$ and $^{32}S + {}^{64}Ni$ showed an interesting difference, namely that W_D for the latter system is larger by about a factor of 1.5 than for the former system. This indicates that direct reactions should occur more strongly in $^{32}S + {}^{64}Ni$ than in $^{32}S + {}^{58}Ni$. Such a "prediction" is in fact consistent with a recent measurement [4] of the transfer reaction cross sections in these systems. Further, an energy dependence similar to that called as threshold anomaly [2,3], was observed in $V(r)$ and $W(r)$ at $r \simeq R_A$.

The analyses we have reported here were very simple and straightforward, yet they provided much valuable information. It is thus interesting that more data will be taken and similar analyses will be done.

The author would like to thank Professor W. R. Coker for his careful reading of the manuscript. The work is supported in part by the U. S. Department of Energy.

REFERENCES

[1] See for instance, Proceeding of the Symposium on The Many Facts of Heavy-Ion Fusion Reactions, Argonne National Laboratory, March 1986, Report No. ANY-PHY 86-1.

[2] J. S. Lilley, B. R. Fulton, M. A. Nagarajan, I. J. Thompson and D. W. Banes, Phys. Lett. 151B, 181 (1985).

[3] B. R. Fulton, D. W. Banes, J. S. Lilley, M. A. Nagarajan and I. J. Thompson, Phys. Lett. 162B, 55 (1985).

[4] A. M. Stefanini et al., Nucl. Phys. A 456, 509 (1986).

[5] K.-H. Rehm, Proceeding of the Sympsoium on The Many Facts of Heavy-Ion Fusion Reactions, Argonne National Laboratory, March 1986, Report No. ANY-PHY 86-1, p. 27.

[6] C. H. Dasso, S. Landowne and A. Winther, Nucl. Phys. A 405, 381 (1983).

[7] M. J. Rhoades-Brown and M. Prakash, Phys. Rev. Lett. 53, 333 (1984).

[8] S. Landowne and S. C. Pieper, Phys. Rev. C29, 1352 (1984).

[9] I. J. Thompson, M. A. Nagarajan, J. S. Lilley and B. R. Fulton, Phys. Lett. 157B, 250 (1985).

[10] G. R. Satchler, M. A. Nagarajan, J. S. Lilley and I. J. Thompson, Ann. of Phys. 178, 110 (1987).

[11] See, for instance, a recent review article by J. R. Birkelund and J. R. Huizenga, Ann. Rev. Nucl. Part. Sci. 33, 265 (1983).

[12] T. Udagawa, B. T. Kim and T. Tamura, Phys. Rev. C 32, 124 (1985).

[13] T. Udagawa and T. Tamura, Phys. Rev. C 29, 1922 (1984).

[14] M. Hussein, Phys. Rev. C 30, 1962 (1984).

[15] F. Videbaek et al., Phys. Rev. C15, 954 (1977).

[16] E. Vulgaris, L. Grodzins, S. G. Steadman, and R. Ledoux, Phys. Rev. C33, 2017 (1986); Phys. Rev. C34, 1495 (1986), and optical parameters are in Phys. Rev. C34, 1495 (1986).

[17] A. M. Stefanini et al., Phys. Rev. Lett. 59, 2852 (1987).

[18] A. M. Stefanini et al., Phys. Lett. B 185, 15 (1987).

[19] H. Sann et al., Phys. Rev. Lett. 47, 1248 (1980); R. Bock et al., Nucl. Phys. A 388, 334 (1982).

[20] B. T. Kim, T. Udagawa and T. Tamura, Phys. Rev. C 33, 370 (1986).

[21] T. Udagawa, S.-W. Hong and T. Tamura, Phys. Rev. C 32, 1435 (1985).

[22] S. W. Hong, T. Udagawa and T. Tamura, to be published.
[23] T. Murakami, C.-C. Sahm, R. Vandenbosch, D. D. Leach, A. Ray, and M. J. Murphy, Phys. Rev. C34, 1353 (1986).

Channel Coupling Effects in Heavy-Ion Elastic Scattering and Sub-barrier Fusion

Alberto M. Stefanini

INFN - Laboratori Nazionali di Legnaro
I-35020 Legnaro (Padova, Italy)

1. Introduction

Heavy-ion interactions around the Coulomb barrier, which are the subject of this Symposium, offer the opportunity of observing spectacular effects due to the mechanisms of coupled reaction channels. The orders-of-magnitude cross section enhancements in the sub-barrier fusion were the first experimental evidences of that (see ref.1) for a review), and it was soon realized[2)] that a full understanding of those processes in terms of channel coupling can only be accomplished when "complete" sets of data are available about the reaction channels competing with fusion in the same energy range. These channnels are the quasi-elastic ones, e.g. inelastic excitations and transfer reactions of one or more nucleons or clusters. Only in a few cases[1)] it has been possible to carry out such detailed studies which imply difficult and time-consuming experiments, although quite attractive ones.

More recently, a complementary and equally interesting manifestation of the same phenomenon, e.g. channel coupling, has been revealed[3)] by analyses of elastic scattering measurements near the barrier. Here the evidence is a rather sharp energy dependence of the optical potential, leading to an enhanced attraction of the nuclear surfaces in the energy region where the non-elastic channels are becoming energetically closed. This is called "threshold anomaly" of the potential and the interpretation is based on the already known existence, on very general grounds, of a dispersion relation[4)] between the real and the imaginary parts. Neglecting the obvious radial dependence one writes the optical potential as

$$U(E) = V_0 + \Delta V(E) + iW(E).$$

The explicit energy dependence of the real potential is in the term $\Delta V(E)$, and the dispersion relation reads

$$\Delta V(E) = \frac{P}{\pi} \int_{E_t}^{\infty} \frac{W(E')dE'}{E - E'}$$

where E_t is the lowest threshold energy of the quasi-elastic reaction channels, P denoting the principal value of the integral. The term V_0 is only slowly dependent on energy due to non-locality effects. We see that near the barrier, where the absorption W changes rapidly, we get an additional attractive contribution to V, as the cooperative effect of many reaction channels.

The threshold anomaly has been observed in a few systems[5,6,7,8] and the case of $^{16}O + ^{208}Pb$ is sofar the most clear one[3,4]. This talk presents the results of elastic scattering experiments in the four near-by systems $^{32,36}S + ^{58,64}Ni$, showing evidence of strong energy dependence of the optical potential in all of them. A comparative analysis of the four cases will be done, taking into account (and relating) also the information already existent on the quasi-elastic reaction channels[9] and on the sub-barrier fusion cross sections[10].

2. Experimental

The elastic scattering measurements were carried out at the Legnaro XTU Tandem accelerator which provided the ^{32}S and ^{36}S beams at several energies in the range 82-150 MeV. A FeS sample, enriched to around 40% in mass 36, was put in the sputtering ion source to produce the ^{36}S beam. Typical beam intensities were 10-50 pnA. The Ni targets were inserted in a 60 cm diameter sliding seal scattering chamber; they were 30 $\mu g/cm^2$ evaporations on 20 $\mu g/cm^2$ carbon foils, enriched to 99.8% and 96.5% in mass 58 and 64, respectively.

The beam intensity and direction was monitored by two Si detectors placed at $\theta_{lab} = \pm 16^0$ and below the reaction plane, which insured as well proper normalization of the measured elastic yields.

The scattered Sulphur ions were detected by up to three time of flight-energy telescopes consisting of micro-channel plate detectors and 200-300 mm^2 Si surface barrier detectors, the flight path being 80-100 cm. The telescopes were external to the scattering chamber; only one micro-channel plate detector was lodged inside, at around 17 cm from the target, covering the most backward angles. The limit was anyway $\theta_{lab} \leq 135^0$.

Mass and energy resolutions were $A/\Delta A \simeq 50$ and $\Delta E \simeq$ 700-800 keV FWHM, which allowed a good separation of the elastic scattering from the quasi-elastic reaction channels.

3. Data and optical model fits

The elastic scattering angular distributions for the 4 systems at the different incident energies are shown in Fig.1. The data of ^{32}S + Ni were already published[8], and a first analysis in terms of the optical model is reported there. Here I show the results of a more comprehensive analysis of all the data, performed using the code Ptolemy[11].

The imaginary optical potential is explicitly divided[12] into two parts $W_D(r)$ and $W_F(r)$, so that the full potential is

$$U(r) = V_C(r) + V(r) + i(W_F(r) + W_D(r))$$

where $V_C(r)$ is the Coulomb interaction. $W_D(r)$ takes care of the absorption into

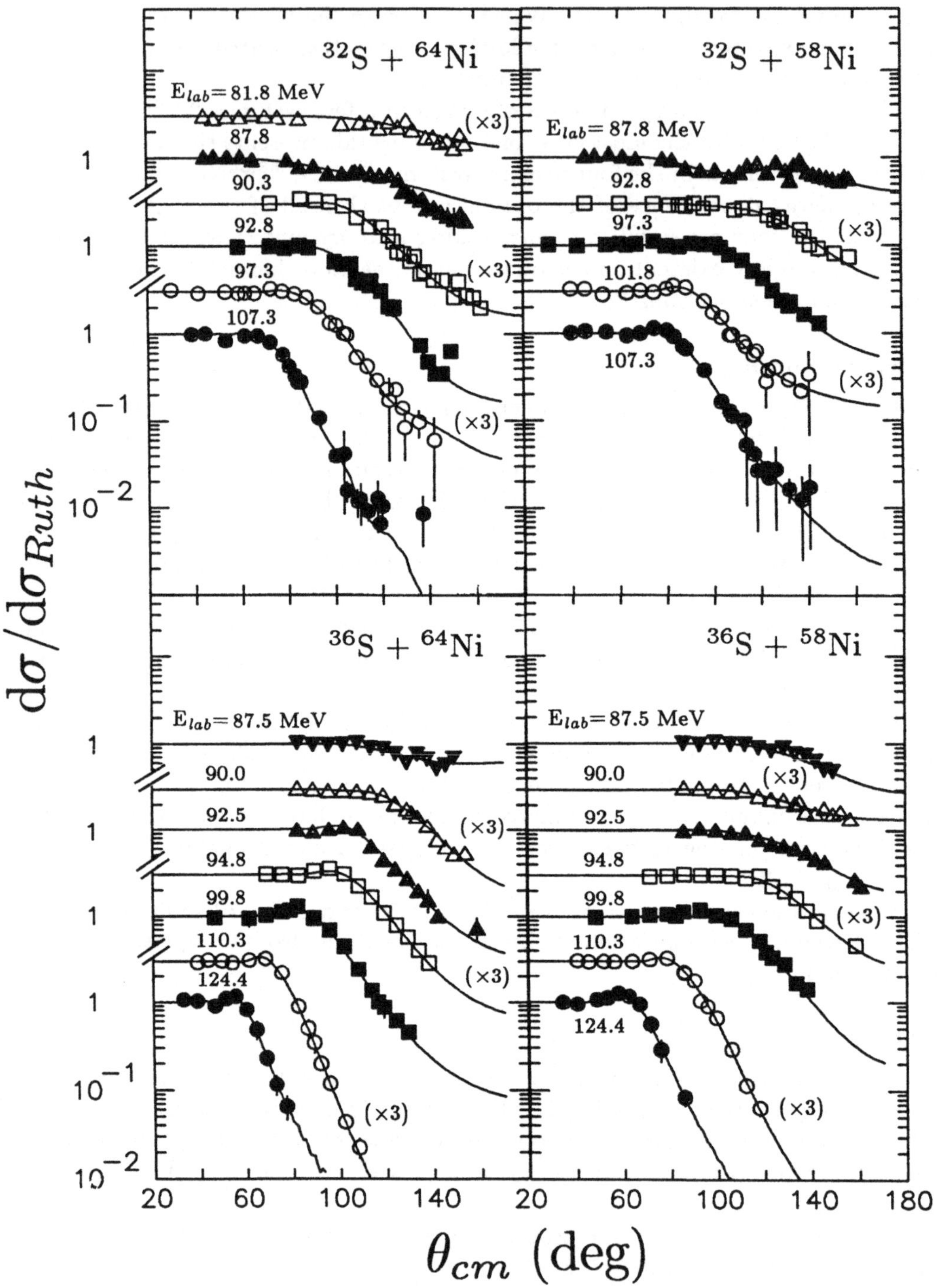

Fig.1 - *Elastic scattering angular distributions. The lines are the optical model fits described in the text.*

the quasi-elastic (direct) reaction channels and it has a surface derivative form

$$W_D(r) = 4W_D a_D \frac{d}{dr} \frac{1}{1 + exp(X_D)}$$

$$X_D = \frac{r - R_D}{a_D}, \qquad R_D = r_D(A_1^{1/3} + A_2^{1/3}).$$

W_F(r) and V(r) have usual Woods-Saxon shapes. W_F(r) is responsible for the absorption into fusion and it is chosen[12)] to be a sharp potential confined within the barrier with depth W_F=20 MeV, radius parameter r_F=1.0 fm and diffuseness a_F=0.25 fm in all cases. This assumption is equivalent to postulating that fusion takes place if, and only if, the Coulomb barrier has been tunneled or overcome. It is worth while to mention the talk of T.Udagawa in this Symposium, showing an alternative approach to the ^{32}S + Ni data, which relaxes that assumption and allows for a larger "fusion" radius.

An initial series of χ^2-searches were done, where 4 parameters of the full potential were varied, i.e. the real strength V and the three parameters of the surface interaction W_D, r_D and a_D. The real diffuseness and the reduced radius were kept fixed to a=0.5 fm and r_0=1.247 (1.277) fm for $^{32(36)}S$, respectively, following the previous analysis[8)] of ^{32}S + Ni. The 4-parameter fits showed that r_D varies little with energy for all systems above the barrier (values between 1.40 and 1.50 fm were found), whereas a general trend is observed that r_D becomes larger at the lower energies.

Then a second series of 3-parameter fits was performed, by fixing r_D to the average value resulting from the above barrier energies, separately for each system. These searches yielded the real strengths at every energy. A third series of fits was finally performed for those cases where the surface radius had the tendency to be significantly larger (or smaller) than the average values, retaining the real strengths (already determined) and varying again W_D, r_D and a_D.

The final fits are shown in Fig.1 as full lines and the potential parameters which were derived are listed in Tab.1, together with the deduced reaction cross sections and the χ^2-values. The fits are altogether quite good, except for the particular cases of the ^{32}S + Ni scattering at the lowest energies where a structure appears in the backward angular distributions, which was already noticed[8)] and whose nature is sofar unclear and deserves further investigation.

The final comment to this section is that the chosen potential parametrization gives a good account of the elastic scattering data, but this is not very significant because fits of comparable quality can be obtained by only inserting a volume absorption in the interaction. In other words, as usual elastic scattering determines 3-4 parameters of the optical potential "whatever" they are. It will be now more interesting to trace out the energy dependence of the potentials in the vicinity of the strong absorption radii, and to compare the total reaction cross sections with the experimentally known fusion and transfer cross sections.

Tab.1 - *Optical model parameters resulting from the fits to the elastic scattering data; fixed parameters are underlined. In addition, the real diffuseness and reduced radius were fixed to a=0.5 fm and r_0=1.247 (1.277) fm for $^{32(36)}S$, respectively. The "fusion" potential was W_F=20 MeV, a_F=0.25 fm and r_F=1.0 fm (see text). For 87.8 MeV ^{32} + ^{58}Ni the real diffuseness was fixed to 0.6 fm.*

System	E_{lab} (MeV)	V (MeV)	W_D (MeV)	a_D (fm)	r_D (fm)	χ^2	σ_r (mb)
$^{36}S + ^{64}Ni$	87.5	29.96	0.427	0.139	1.685	0.88	139
	90.0	69.26	0.38	0.585	1.474	0.31	221
	92.5	88.08	2.09	0.202	1.474	0.82	266
	94.8	54.30	2.11	0.200	1.474	0.34	306
	99.8	55.45	2.66	0.203	1.474	0.72	500
	110.3	48.23	2.24	0.320	1.474	0.11	920
	124.4	42.77	2.98	0.221	1.474	0.13	1183
$^{36}S + ^{58}Ni$	87.5	58.80	1.265	0.198	1.669	0.55	146
	90.0	41.61	0.56	0.152	1.694	0.47	138
	92.5	66.28	0.44	0.305	1.614	0.50	177
	94.8	68.26	0.75	0.514	1.419	0.40	181
	99.8	43.72	1.87	0.350	1.419	1.75	286
	110.3	39.54	2.73	0.340	1.419	0.38	632
	124.4	42.45	3.83	0.237	1.423	0.48	924
$^{32}S + ^{64}Ni$	81.8	33.84	1.55	0.510	1.503	1.03	160
	87.8	45.94	0.36	1.077	1.503	0.90	(489)
	90.3	88.13	3.27	0.204	1.503	1.91	286
	92.8	43.67	3.26	0.264	1.503	2.85	419
	97.3	34.62	4.30	0.243	1.503	1.08	581
	107.3	26.52	4.14	0.330	1.469	1.55	928
$^{32}S + ^{58}Ni$	87.8	81.21	0.23	0.418	1.914	1.43	(396)
	92.8	74.05	0.50	0.659	1.440	0.75	215
	97.3	77.42	4.43	0.234	1.440	0.98	279
	101.8	50.00	4.58	0.142	1.479	1.27	409
	107.3	44.93	3.43	0.271	1.440	1.26	583

4. Threshold anomalies

The analysis of the previous section has yielded us the potentials of Tab.1, that can be evaluated at any internuclear distance. One knows that rather unambiguous (i.e. fit-independent) values are only determined in the vicinity of the strong absorption radii. So the real and imaginary parts of the optical potential were calculated at $r(\theta_{1/4})$ (quarter-point) and plotted in Fig.2 vs. the bombarding energy. The different radii are listed in the caption.

The lowest energies, where the scattering data do not show any quarter point in the angular distribution, correspond to situations where the nuclei do not come close enough to each other to allow a determination of the potentials at r($\theta_{1/4}$) which is free from big uncertainties; as a consequence, it was decided not to report those data in Fig.2.

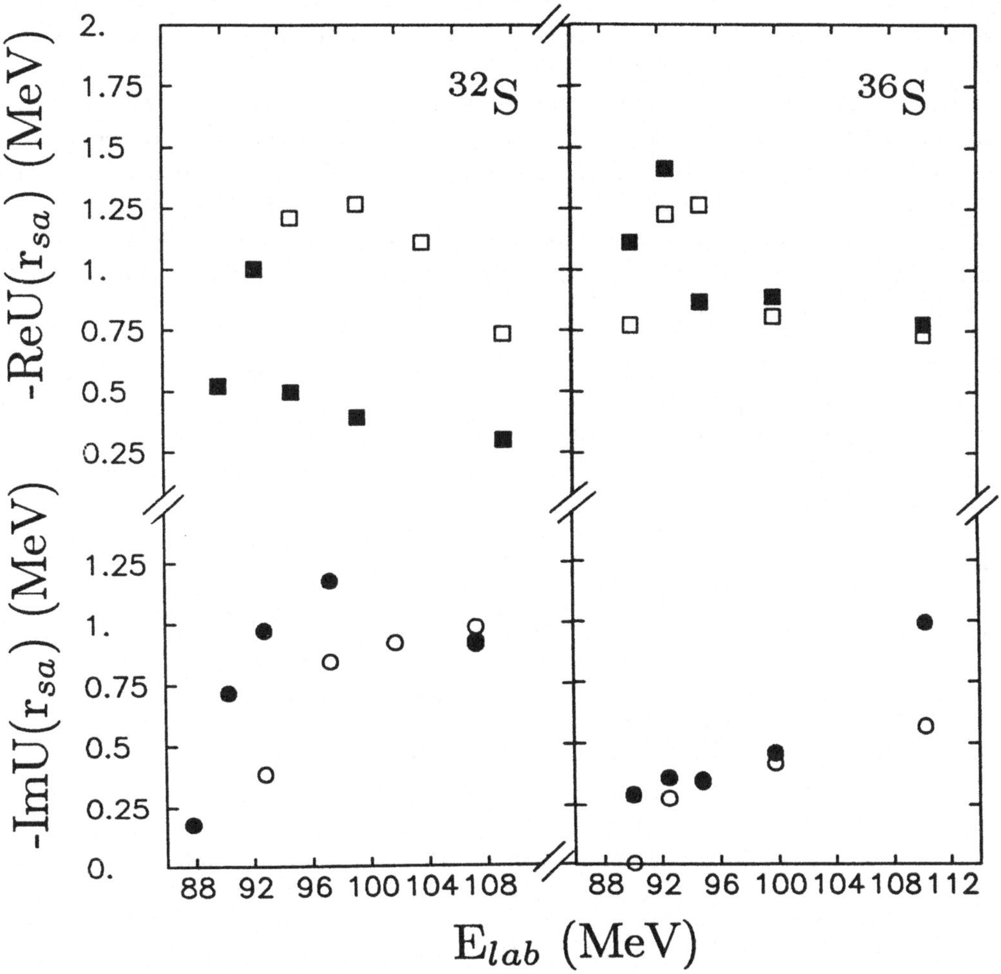

Fig.2 - *The optical potential of ^{32}S (left) and ^{36}S (right) + Ni vs. the incident energy. Full (empty) symbols refer to the $^{64(58)}Ni$ target. The adopted strong absorption radii are r_{sa}=11.18 (10.83) fm for ^{32}S + $^{64(58)}Ni$ and r_{sa}=11.38 (11.14) fm for ^{36}S + $^{64(58)}Ni$. The imaginary potentials of ^{32}S + ^{64}Ni have been scaled down by a factor 2.*

Strong variations with energy are observed in all cases, in qualitative agreement with theoretical expectations (see the introduction). As the energy is decreased and reaches the barrier region, the absorption is more and more reduced. At the same time, the real potential increases and shows a broad maximum around the barrier position. The data for $^{32}S + Ni$ are in good agreement with those obtained previously[8]. No attempt was made to quantitatively check if the two parts of the potentials actually satisfy a dispersion relation.

Rather, I would like to draw the attention on the difference existing between $^{32}S + ^{64}Ni$ and the other three cases. First, there is much more absorption in that system at all energies; second, the threshold anomaly of the real potential is more remarkable in $^{32}S + ^{64}Ni$, where in fact the potential increases by a factor $\simeq 3$ when going from above to around the barrier.

Actually, $^{32}S + ^{64}Ni$ displays the largest enhancement of sub-barrier fusion cross sections[10]. In a reduced scale, they are about one order of magnitude larger than for the other systems. In other words, the cooperative effect of the many quasi-elastic reaction channels, which produces the strongest polarization potential in $^{32}S + ^{64}Ni$, enhances considerably the sub-barrier fusion as well. It is now interesting to look for such correlations also in the transfer cross sections, which are known from previous work[9] at a number of energies. This is done in the next section.

Tab.2 - *Transfer, inelastic, fusion and total reaction cross sections. The centre of mass Coulomb barriers[13] are 60.7, 59.6 MeV for $^{32}S + ^{58,64}Ni$, and 59.6, 58.6 MeV for $^{36}S + ^{58,64}Ni$, respectively.*

System	E_{lab} (MeV)	E_{cm} (MeV)	σ_{tr} (mb)	σ_{in} (mb)	σ_{fus} (mb)	σ_{sum} (mb)	σ_r (mb)
$^{32}S + ^{58}Ni$	107.3	69.1	65±7	273	294±59	632	583
	101.8	65.6	(50)	227	236±47	(513)	409
	97.3	62.7	34±6	186	94±21	314	279
	92.8	59.8	9±5	153	26±6	188	215
$^{32}S + ^{64}Ni$	107.3	71.5	223±10	319	465±92	1007	928
	97.3	64.7	159±18	236	228±41	623	581
	92.8	61.9	121±12	194	145±29	460	419
	90.3	60.2	94±7	171	91±18	356	286
	81.8	54.5	23±6	118	0.9±0.2	142	160
$^{36}S + ^{58}Ni$	110.3	68.0	127±27	250	256±52	633	632
	99.8	61.6	106±25	144	109±24	359	286
	94.8	58.5	48±10	105	23.1±6.5	176	181
$^{36}S + ^{64}Ni$	110.3	70.6	80±18	272	453±91	805	920
	99.8	63.9	36±9	192	293±59	521	500
	94.8	60.7	32±7	150	150±30	332	306

5. The quasi-elastic reaction channels

Table 2 is a summary of the cross sections for different channels, at the incident energies where measurements of the transfer reactions have been performed, in addition to fusion and elastic scattering which are known in a wider energy range. In ref.9), the various transfer channels were identified and their Q-value integrated cross sections were individually obtained, but here the column named σ_{tr} lists the total transfer cross sections. The inelastic cross sections σ_{in} are not experimental values, as these are not available in most cases.

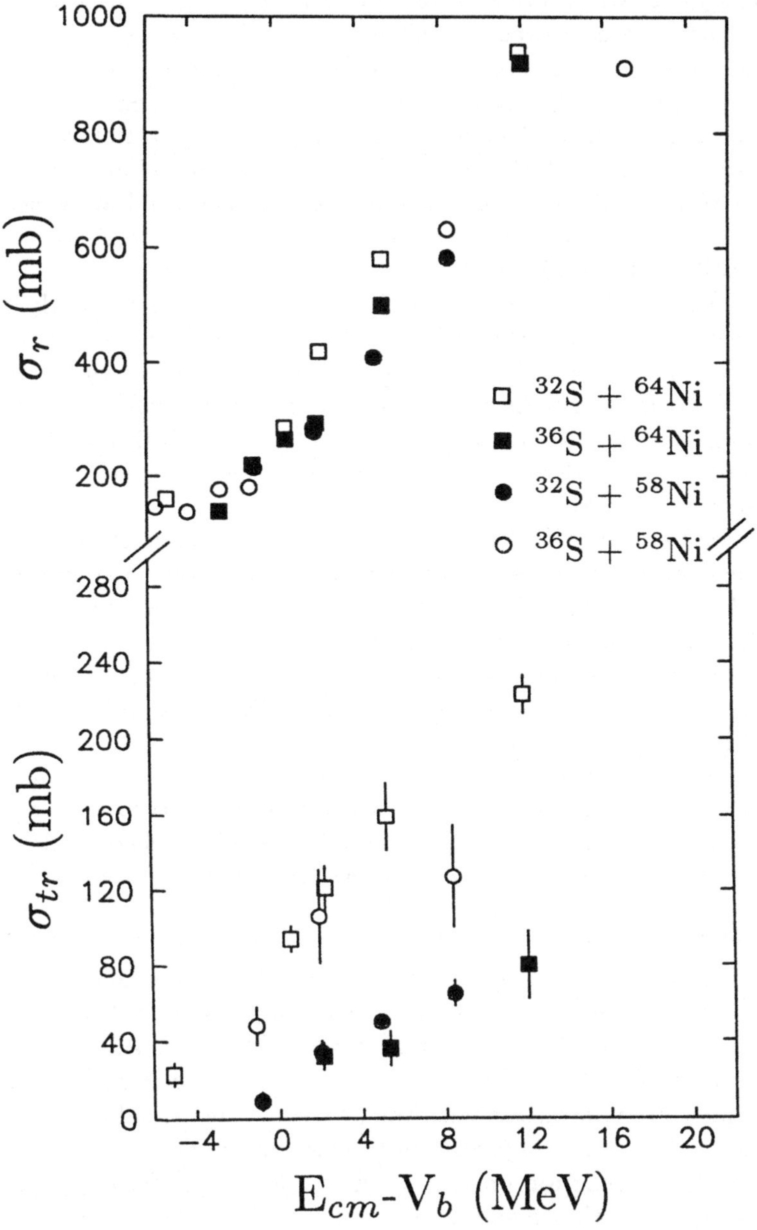

Fig.3 - *Total reaction cross sections (upper panel) deduced from the elastic scattering experiments, and total transfer cross sections vs. the energy distance from the barrier*[13].

They are the results of Ptolemy[11] coupled-channel calculations including the lowest 2^+ and 3^- excitations of both projectile and target. The coupling strengths were derived from the experimental B(E2) and B(E3) data. The "bare" potential was chosen according to the following prescriptions: 1) the real part was that determined from the elastic scattering fits at E_{lab}=107.3 (110.3) MeV for $^{32(36)}S$ (i.e. above the threshold anomaly); 2) the volume absorption was the same as described in section 3, and 3) no surface potential was introduced.

Such coupled-channel calculations for inelastic excitations are usually quite reliable, so that by summing σ_{tr}, σ_{in} and σ_{fus} we obtain the quasi-experimental cross sections named σ_{sum} which can be compared with the total reaction cross sections σ_r deduced from the elastic scattering data. The comparison, taking into account the uncertainties in the ingredients of σ_{sum}, indicates a substantial agreement between the two sets of numbers, with some tendency for σ_{sum} to overestimate σ_r in ^{32}S + Ni except for the lowest energies.

An interesting trend is observed in the σ_{tr} values, which clearly appears in Fig.3(lower panel) where the total transfer cross sections have been plotted vs. the energy distance from the barrier[13]. The four systems are divided in two pairs, ^{32}S + ^{64}Ni and ^{36}S + ^{58}Ni having cross sections larger than the other two cases by factor 3-5, depending on the energy. At variance with the systematics of the sub-barrier fusion, this isotopic effect does not correlate completely with the observations of section 4 on the various threshold anomalies (see Fig.2).

On the other hand, the (calculated) inelastic cross sections are dominated by the 2^+ excitation of the Ni targets, and those 2^+ levels have comparable excitation energies and B(E2)'s in both ^{58}Ni and ^{64}Ni. It follows that no outstanding isotopic trend in the inelastic cross sections is predicted.

The upper panel of Fig.3 shows the total reaction cross sections σ_r for the four systems. We note that the largest cross sections belong to ^{32}S + ^{64}Ni where already around 5 MeV ($\simeq$10%) below the barrier σ_r is 160 mb.

As the conclusion of this section, therefore, we are facing the open questions: which channel(s) are important in order to explain the observed trends in the cross sections and in the potential anomalies ? Should these channels explicitly taken into account in view of their possibly large coupling strengths ?

6. The sub-barrier fusion

Keeping in mind the correlations between threshold anomalies and isotopic effects in the sub-barrier fusion, which were stressed in section 4, an application of the (energy-dependent) barrier penetration model (BPM)[12] will now be presented, in order to calculate absolute fusion cross sections. At each energy where elastic scattering was measured and fitted with the procedure described above, the reaction cross section was calculated using the fitting potential, spoiled however of the surface derivative term. That reaction cross section is the (BPM) fusion cross section by definition. It is known[14] that a simple renormalization of the real potential strength at the various energies, as in this work, produces incorrect spin distributions of the compound nucleus; however, data on such spin distributions are not available for the S + Ni systems.

The results are shown in Fig.4. The BPM cross sections slightly underestimate the experimental values for $^{32}S + ^{64}Ni$ at high energies, and the gap becomes larger in the sub-barrier region; in the other three cases there is an overestimation above the barrier by factors $\simeq$1.5-2 and the BPM values get rather close to the experiment at low energies (where the interaction is increasingly adiabatic), with the exception of of $^{36}S + ^{64}Ni$ whose cross sections are somewhat underestimated. In general, the calculated excitation functions appear too steep below the barrier.

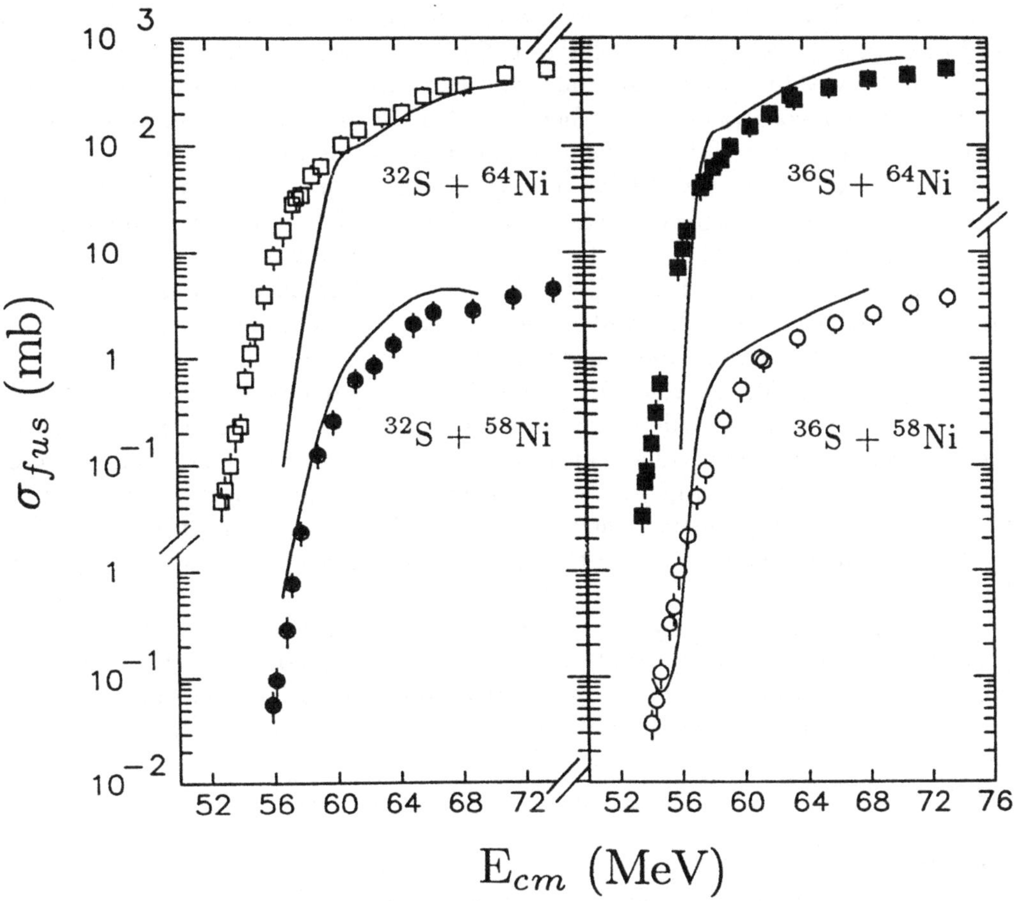

Fig.4 - *Fusion excitation functions. The lines are the (smoothed) results of the calculations within the barrier penetration model*[12]*.*

Before commenting these evidences, it is wise to put an obvious two-fold warning: the BPM cross section are model dependent, not only because of the choice of the fusion potential $W_F(r)$ which "a priori" models the fusion process, but also in view of the large ambiguities in determining the optical potential at shorter internuclear distances than the strong absorption radius (i.e. the Coulomb barrier is largely fit-dependent). Such ambiguities are largest at the lowest energies, where the elastic scattering has no quarter-point.

Then the comment is simply that the BPM picture gives a reasonable description of the fusion excitation functions with the ^{58}Ni target, while for ^{32,36}S + ^{64}Ni the disagreement below the barrier is noteworthy.

7. Conclusions

The measurement of elastic scattering in the four systems ^{32,36}S + ^{58,64}Ni at various energies close to the barrier has provided systematic information on those low energy interactions, which supplement what was already known on the competing reaction channels such as quasi-elastic transfer and fusion. The scattering data were analysed within the optical model and good fits were obtained in most cases, separating the absorption in a surface part (simulating the quasi-elastic reaction channels) and in a more internal volume part responsible for the fusion.

The threshold anomaly of the potential shows up in all cases, being more conspicuous in ^{32}S + ^{64}Ni where the sub-barrier fusion has the largest enhancement as well. The comparison between total reaction cross sections and fusion + quasi-elastic cross sections shows a good overall agreement, with fluctuations which need to be understood better. ^{32}S + ^{64}Ni and ^{36}S + ^{58}Ni have comparable total transfer cross sections, and larger than in the other two systems.

Finally, it was shown that the barrier penetration picture of the sub-barrier fusion, employing an energy dependent nuclear interaction consistent with the elastic scattering data, is able to yield an "average" fair description of the experimental cross sections. This is possibly a good starting point for more refined analyses within the explicit coupled-channel formalism, trying to single out the important channels in the different cases, which produce the observed and yet sofar unexplained isotopic effects.

It is a pleasure to acknowledge the friendly collaboration of all members of our group, i.e. A.Tivelli, G.Montagnoli, D.Bonamini, D.R.Napoli, G.Fortuna, R.Pengo, S.Beghini, C.Signorini, F.Soramel and F.Scarlassara. Precious help during the experiments came from A.De Rosa, G.Inglima, M.Sandoli, G.Cardella, F.Rizzo and G.Pappalardo.

References

1) - M.Beckerman, Rep.Prog.Phys. **51** (1988) 1047
2) - C.H.Dasso, S.Landowne and A.Winther, Nucl.Phys. **A405** (1983) 381; R.A.Broglia et al., Phys.Lett. **B133** (1983) 34
3) - J.S.Lilley et al, Phys.Lett. **B151** (1985) 181
4) - M.A.Nagarajan, C.Mahaux and G.R.Satchler, Phys.Rev.Lett. **54** (1985) 1136
5) - A.Baeza et al., Nucl.Phys. **A419** (1984) 412
6) - B.R.Fulton et al., Phys.Lett. **B162** (1985) 55
7) - B.Bilwes et al., Nucl.Phys. **A449** (1986) 519
8) - A.M.Stefanini et al., Phys.Rev.Lett. **59** (1987) 2852
9) - A.M.Stefanini et al., Phys.Lett. **B185** (1987) 15
10) - A.M.Stefanini et al., Nucl.Phys. **A456** (1986) 509

11) - M.H.Macfarlane and S.C.Pieper, Argonne Nat. Lab. Report ANL-76-11 (Rev.1), 1978; M.J.Rhoades-Brown et al., Phys.Rev. **C21** (1980) 2417
12) - G.R.Satchler et al., Ann.Phys. **178** (1987) 110
13) - L.C.Vaz, J.M.Alexander and G.R.Satchler, Phys.Rep. **69** (1981) 373
14) - M.A.Nagarajan and G.R.Satchler, Phys.Lett. **B173** (1986) 29

REACTIONS IN THE COLLISION OF OXYGEN IONS WITH LEAD AT ENERGIES CLOSE TO THE COULOMB BARRIER

M.A. Nagarajan
SERC Daresbury Laboratory
Warrington WA4 4AD, England

1. Introduction

The revival of renewed interest in the collisions of heavy ions at energies close to the Coulomb barrier is due to the evidence of strong energy-dependence of the optical potential that was observed during recent years[1-3]). This energy dependence termed the 'threshold anomaly' is believed to reflect the very strong coupling between the elastic and non-elastic (inelastic and nucleon transfer) channels at these energies.

To achieve a more complete understanding of the phenomenon of the threshold anomaly, experiments were performed at Daresbury Laboratory on the collision of ^{16}O and ^{17}O isotopes with ^{208}Pb target. I shall discuss the progress made in the analyses of ^{16}O-^{208}Pb collision in Section 2 and that of the ^{17}O-^{208}Pb system in Section 3. A simple three channel model for the transfer reaction is described in Section 4. The summary and conclusions are presented in Section 5.

2. The Interactions of ^{16}O with ^{208}Pb

The evidence of the threshold anomaly[2]) and its interpretation in terms of the dispersion relation[4,5]) have been published. The elastic scattering of ^{16}O by ^{208}Pb at 80 MeV revealed the fact that the real part of the optical potential, at the strong absorption radius (12.4 fm), $V(R_{SA})$ was much more attractive than the double folding potential[6]). The latter had to be renormalized by a factor of approximately 1.6. Upon reanalysing the available data on this system at higher energies, it was observed that $V(R_{SA})$ progressively increased in magnitude as it approached the Coulomb barrier. At the same time, the imaginary part of the optical potential at the strong absorption radius, $W(R_{SA})$, was found to become smaller as the energy approached the Coulomb barrier. The reduction of the imaginary part can be understood as due to the closing of the available open channels. The increase in the real potential arises from causality which imposes a relation between the real and imaginary parts[5]).

To test the source of the energy dependence, coupled-channels[7] including the inelastic channels and coupled-reaction-channels[8,9] which considered both inelastic and single nucleon transfer channels, were performed. These calculations suggest that a large part of the discrepancy between the predictions of the double folding model and the experimental data on the elastic scattering cross section can be accounted for by the coupled-channels effect. In table 1, the effect of the coupled channels on the inelastic, transfer, fusion, and total reaction cross sections are shown. The experimental cross sections[10] are also shown. It can be seen that all the cross sections except the total reaction cross section are well reproduced. Recent calculations indicate that this missing cross section can be accounted for by the inclusion of the $(^{16}O,^{12}C)$ α transfer channel. This will be discussed by Dr. Lilley[10].

Table 1
Effect of α-transfer on integrated cross sections for $^{16}O + ^{208}Pb$ at 80 MeV

Channel	Couplings included				
	No coupling	Inelastic	Inelastic + nucleon transfer	Inlastic + nucleon + alpha transfer	Expt.
Fusion	1.1	14	38	37	1 ± 6
Inelastic	-	8	14	10	45 ± 5
Nucleon transfer	-	-	31	39	
Alpha transfer	-	-	-	18	19
Total	1.1	22	83	104	105 ± 10

The elastic scattering data at several energies in $^{16}O + ^{208}Pb$ collision could thus be understood in terms of coupled-reaction-channels (CRC) effects[9]. It has been suggested that the observed energy dependence of the real part of the optical potential can also explain the enhancement of sub barrier fusion cross section in the framework of the barrier penetration model[5,11,12].

It is of interest to study whether similar 'threshold' anomaly is also observed in non-elastic channels. Evidence of the threshold anomaly in the inelastic channel in the reaction $^{208}Pb(^{16}O,^{16}O)^{208}Pb$ (3^-, 2.64 MeV) is presented by Dr. Lilley[10] in this conference. One should expect similar effects also in the transfer channels.

3. Reactions in the $^{17}O + ^{208}Pb$ system

The last neutron binding energy in ^{17}O is very weak. The study of ^{17}O collision with ^{208}Pb allows us to investigate the effect of the weakly bound neutron. The investigation of the inelastic excitation of ^{17}O to its first excited $1/2^+$ state by ^{208}Pb [13]) clearly presented the effects of multistep processes. The nuclear inelastic excitation was dominated, at 78 MeV projectile energy, by neutron transfer

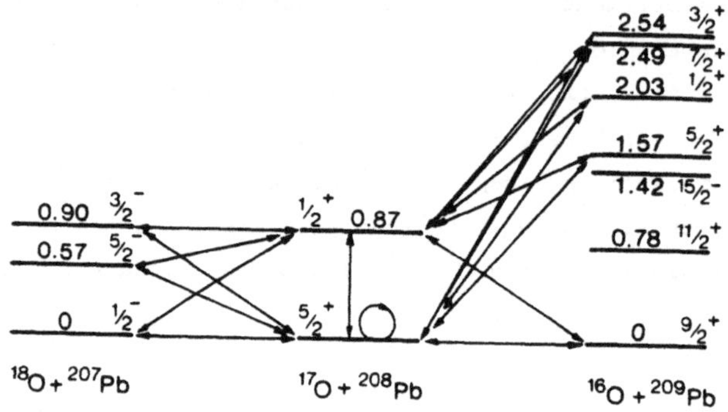

Fig. 1. Channel couplings used in the multistep CRC calculations of the ^{17}O-^{208}Pb interaction.

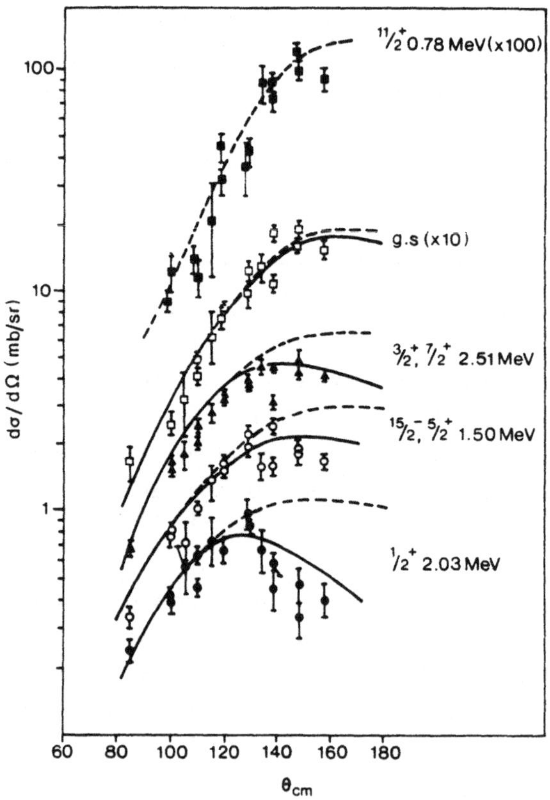

Fig. 2. Differential cross sections for the $^{208}Pb(^{17}O,^{16}O)^{208}Pb$ reaction at 78 MeV projectile energy. The short dashed curves are DWBA predictions and the solid curves are the full CRC predictions.

followed by the neutron pickup process. Subsequently, the entire set of nucleon transfer cross sections and the elastic cross section were measured for the 78 MeV $^{17}O + ^{208}Pb$ system. The transfer cross sections leading to the weakly bound, low angular momentum states of ^{209}Pb revealed the inadequacy of DWBA. DWBA predicted these cross sections to peak at 180°, typical of sub Coulomb transfer reactions, while the experimental data showed them to peak at smaller angles. In order to account for the 'shift of the grazing peak', coupled-reaction-channels calculations, with the couping scheme of fig. 1, were performed. The results of the CRC calculations for the transfer reactions[14]) are shown in fig. 2.

To investigate the dominant channel which causes the shift in the DWBA peak, further test calculations were performed. In fig. 3, we first show the effect of omitting the $1/2^+$ state of ^{17}O from the coupling scheme on the transfer cross section leading to the $1/2^+$ state of ^{209}Pb. These are denoted by crosses. Next, we included the $1/2^+$ state of ^{17}O in the coupling scheme but omitted the direct coupling between the ground state and $1/2^+$ state of ^{17}O. The results are shown by open circles and are seen to account for most of the multichannel effects. This is consistent with the observation that at 78 MeV, the dominant contribution to the inelastic excitation of the $1/2^+$ state of ^{17}O was from the stripping-pickup mechanism. In figs. 4 and 5, the predictions of CRC on the transfer cross sections at 86 and 102 MeV are shown. It can be seen that the discrepancy between DWBA and CRC persists but progressively becomes smaller. In table 2, we show the

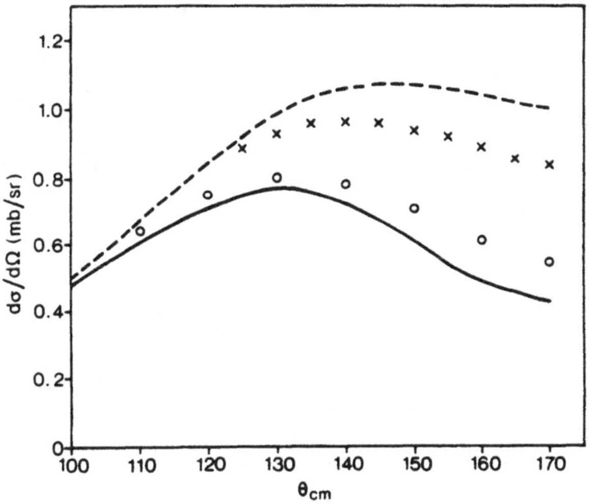

Fig. 3. Differential cross section for the $^{208}Pb(^{17}O,^{16})^{209}Pb$ reaction to the $1/2^+$ state of ^{209}Pb at 78 MeV. The short dashed curve is the DWBA prediction and the full line the CRC prediction. The crosses (x) are the predictions of the CRC calculations when the excited $1/2^+$ state is omitted from the coupling scheme, and the circles (o) are the predictions when the direct coupling between the ground and the first excited state of ^{17}O is neglected.

Table 2

Shift in the DWBA grazing angle and ratio of peak magnitudes for the reaction $^{208}Pb(^{17}O, ^{16}O)^{209}Pb\ 1/2^+$

Projectile Energy (MeV)	$\Delta\theta^{grazing}$	$\sigma_{DWBA}/\sigma_{CRC}$
78	40°	1.43
86	15°	1.25
102	7-8°	1.15

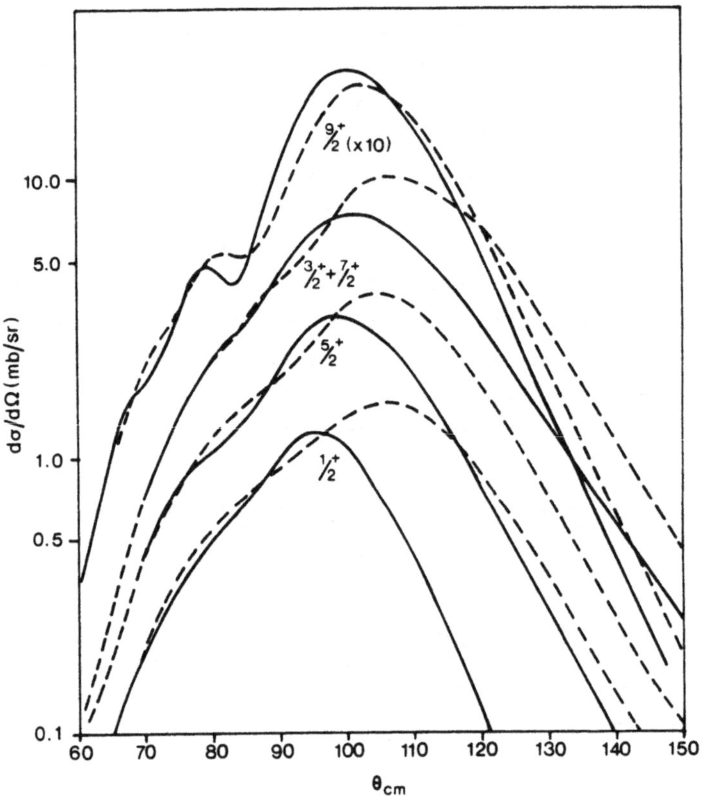

Fig. 4. The comparison of DWBA (short dashed curve) and CRC (solid curve) predictions for the transfer cross sections in $^{208}Pb(^{17}O, ^{16}O)^{209}Pb$ reaction at 86 MeV. The spins and parities of the states of ^{209}Pb are shown on the figures.

shift in the DWBA grazing angle due to CRC effects as well as the ratio of the peak CRC and DWBA cross sections for the three energies. In all the cases, it was verified that most of this CRC effect on transfer is accounted for by transfer

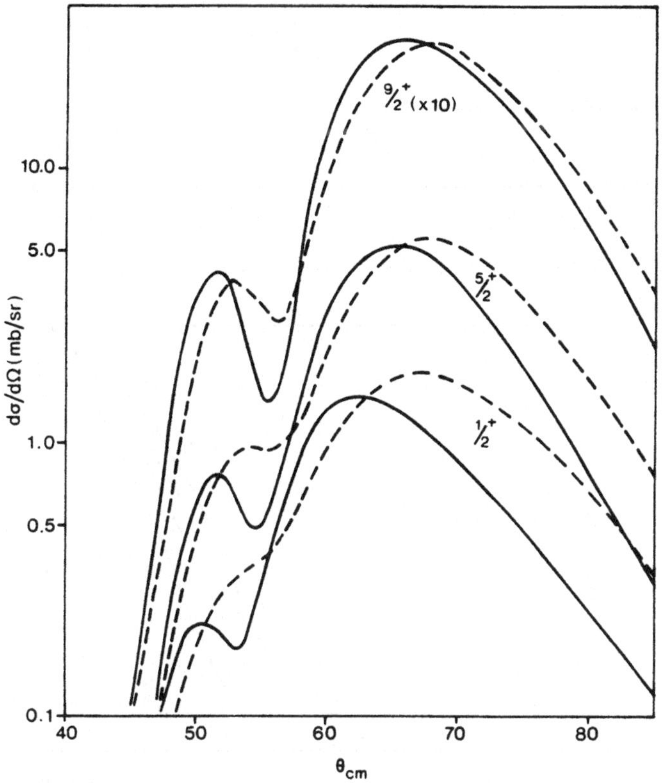

Fig. 5. Same as fig. 4 for 102 MeV projectile energy

through the excited $1/2^+$ state of ^{17}O. It should be noted that even when the shift in grazing angle is large, the change in the peak magnitude is not too large. Thus, inference of the importance of CRC effects by deviations in spectroscopic factors can be misleading.

4. A three channel model for (^{17}O,^{16}O) transfer

The coupling scheme shown in fig. 1 does not include any direct couplings between the transfer channels. We had also shown from CRC calculations that the main CRC effect on the transfer channels arises from the coupling to the $1/2^+$ state of ^{17}O. Thus, as far as the trasnfer reactions are concerned, we considered a simple three channel model comprising of the elastic, the inelastic (to the $1/2^+$ state of ^{17}O) and the specific transfer channels interested in. Instead of doing of a full CRC calculations, we consider only one way coupling (see fig. 6). The aim was to determine the polarisation effects on the inelastic transition potential. The

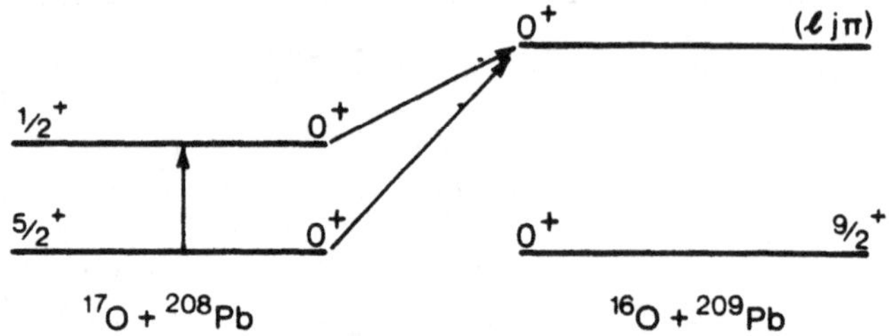

Fig. 6. Coupling scheme used in the three channel model.

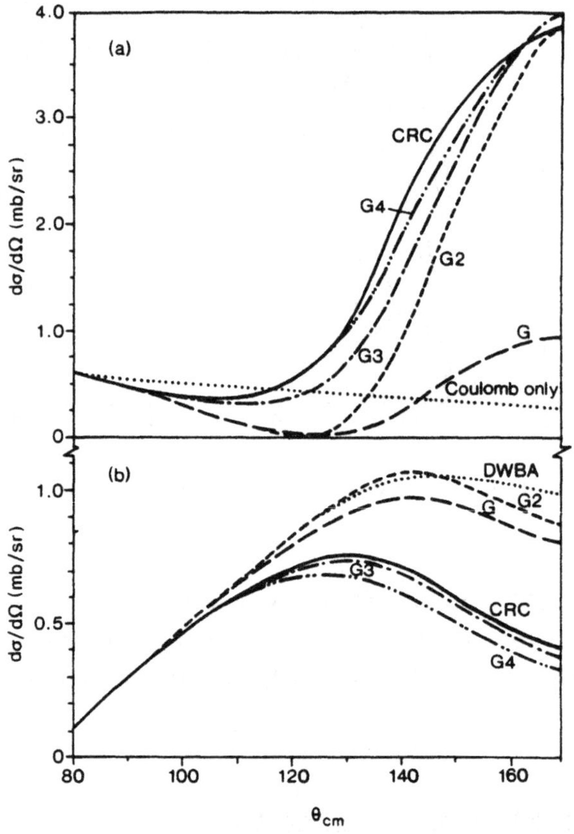

Fig. 7. a) The dependence on the inelastic form factor of the inelastic cross section to the $1/2^+$ state of ^{17}O by ^{208}Pb at 78 MeV. The dotted curve represents pure Coulomb excitation, and the CRC is shown by the full line curve. The others curves are designated by the symbols G (δ_2 = 0.476 fm, R_I = 1.238 fm), G2 (δ_2 = 2.25 fm, R_I = 1.238 fm), G3 (δ_2 = 0.95 fm, R_I = 1.44 fm) and G4 (δ_2 = 1.25 fm, R_I = 1.40 fm).
b) The dependence on the inelastic form factor of the cross section for the reaction $^{208}Pb(^{17}O,^{16}O)^{209}Pb$ ($1/2^+$). The dotted curve represents the DWBA prediction and the full curve the CRC prediction. The other curves are labelled by the symbols which are given in fig. 7a.

elastic optical potential was chosen to fit the elastic scattering and the inelastic form factor was given the form

$$V_{inel}(r) = - \delta_2 \frac{d}{dR} [V(r_R,a_R) + iW(r_I,a_I)] + M(E2) \frac{4\pi Z_{Pb}e}{5 R^3}$$

where M(E2) is the Coulomb E2 matrix element. The potentials $V(r_R,a_R)$ and $W(r_I,a_I)$ are of Woods-Saxon forms with (r_R,a_R) and (r_I,a_I) their respective radius and diffuseness parameters. δ_2 is a deformation length which was treated as a free parameter. If the excited $1/2^+$ state of ^{17}O were a true collective state, one would expect V and W to represent the real and imaginary parts of the optical potential and δ_2 the deformation length consistent with M(E2). In fig. 7a, the attempts at fitting the inelastic scattering cross section are shown. The choice of the nuclear form factor identical to the one used in the CRC calculations is seen to grossly underpredict the large angle cross section. Simply scaling δ_2 to fit the back angle cross section does not improve the predictions at smaller angles. Two other choices of form factors are shown where both δ_2 and the imaginary radius parameter, r_I were varied. These are denoted G_3 and G_4 and yield acceptable fits. The corresponding transfer cross sections to the $1/2^+$ state of ^{209}Pb are shown in fig. 7b. It is seen that the form factor G_3 yields the best fit to the CRC cross section. Similar effects were seen at 88 MeV (figs. 8a and 8b) and 102 MeV (fig. 9). The inelastic form factor which fits the inelastic CRC cross section also gave reasonable fit to the transfer cross sections. These results indicate that it is possible to construct a three-channel model which reproduces all the features of the full CRC, provided the inelastic form factor and the elastic potential are independently normalized to fit their respective cross sections.

The main value of this anlysis is to gain a deeper understanding of these renormalisation effects. In earlier calculations[14]), we had found that the effect of the inelastic and transfer channel couplings was to introduce a dominantly imaginary polarization potential to the elastic potential. In fig. 10, we show the renormalized form factors compared with collective form factors (which are defined as the bare deformation length times the deritative of the optical potential) at 78 and 102 MeV. The optimum inelastic form factors are seen to be dominantly imaginary in the region of the nuclear surface. Furthermore, calculations indicated that transfer cross sections were more strongly affected by the imaginary part of the inelastic form factor than by its real part. The polarisation of the bare inelastic form factor arises from coupling to all the transfer states except the one specifically included in the three channel model. One thus finds that the effect of the neutron transfer channels on both the elastic potential and inelastic transition potential is to yield a polarisation potential which is dominantly imaginary in the region of the nuclear surface.

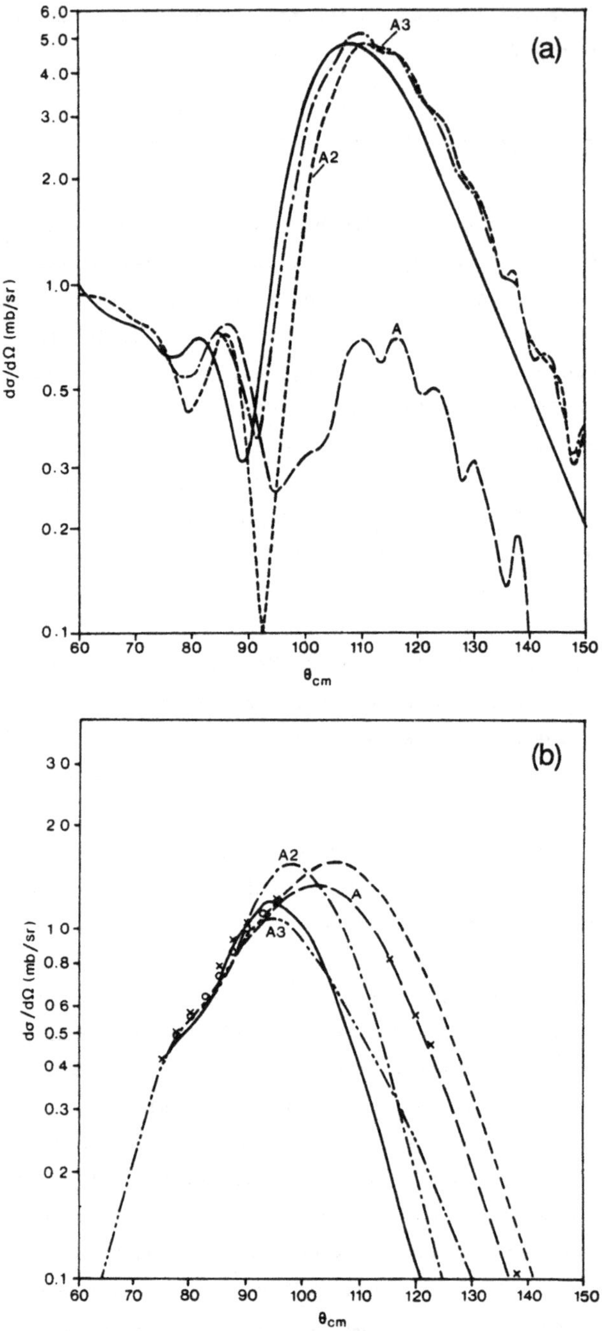

Fig. 8. a) Same as fig. 7a for 86 MeV projectile energy. CRC predictions are shown by the full curve. The form factors used were A (δ_2 = 0.476 fm, R_I = 1.12 fm), A2 (δ_2 = 1.65 fm, R_I = 1.12 fm) and A3 (δ_2 = 0.476 fm, R_I = 1.29 fm).
b) Cross sections for the reaction $^{208}Pb(^{17}O,^{16}O)^{209}Pb$ ($1/2^+$) at 86 MeV. The dashed line represents the DWBA and the full curve the CRC predictions. The other curves are the predictions of the three channel model using the form factors labelled on them.

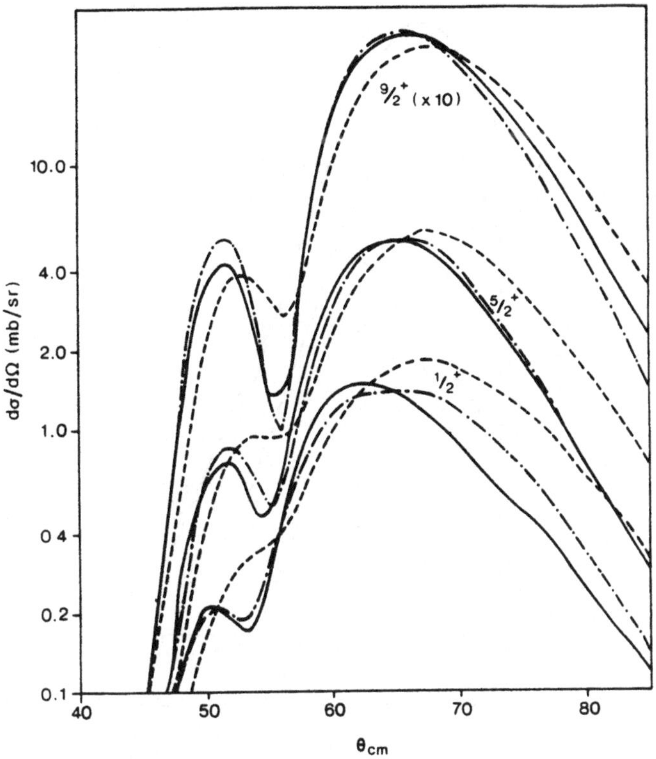

Fig. 9. Cross sections for the reaction $^{208}Pb(^{17}O,^{16}O)^{209}Pb$ leading to the different states of ^{209}Pb at 102 MeV. The predictions of the three-channel model are shown by the dash-dotted curve and used the inelastic form factor B3 (δ_2 = 0.70 fm, R_I = 1.35 fm).

5. Summary and Conclusions

Coupled-reactions-channels (CRC) calculations using the program FRESCO[15]) were performed to analyse the elastic scattering of ^{16}O by ^{208}Pb as well as to analyse elastic, inelastic and transfer reactions in the collision of ^{17}O by ^{208}Pb. In the case of ^{16}O + ^{208}Pb, it was shown that the CRC calculations were able to reproduce the elastic cross sections at near barrier energies and at 80 MeV, they were able to predict the total inelastic, transfer, fusion and reaction cross sections in addition. We were thus able to conclude that the observed threshold anomaly was caused by the strong coupling of the elastic to the non elastic channels. The channel coupling was also seen to become weaker as the energy increased beyond the Coulomb barrier.

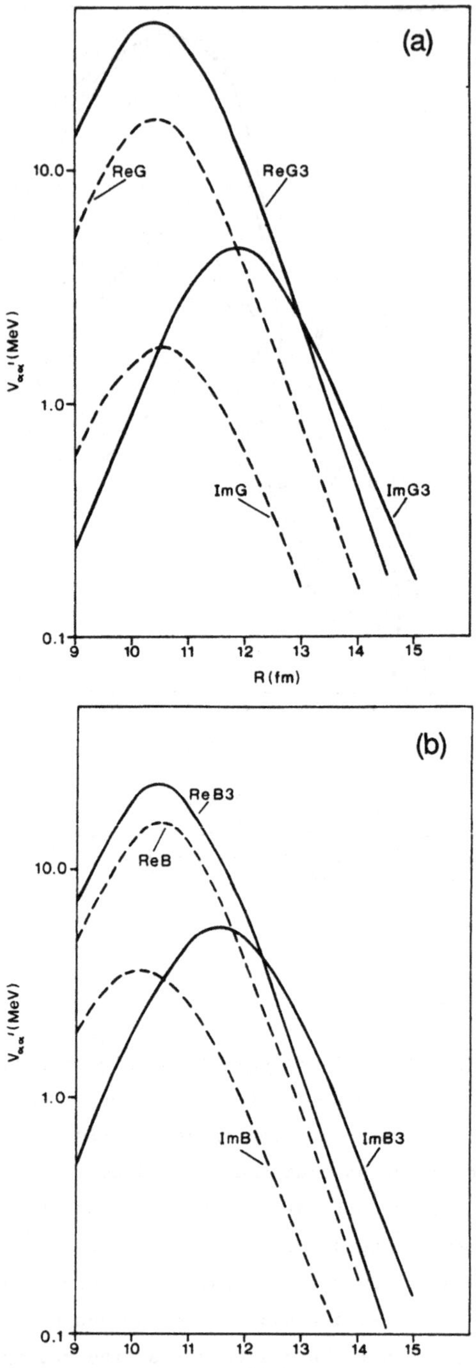

Fig. 10. a) The radial dependence of the inelastic form factors for 78 MeV projectile energy. The labels G and G3 correspond to those in fig. 7a. b) Same as fig. 10a for 102 MeV energy. The form factor B is the collective model approximation (δ_2 = 0.476 fm, R_I = 1.19 fm) while B3 is the same as in fig. 9.

In the case of the collision of ^{17}O by ^{208}Pb, the weakly bound neutron in ^{17}O and the low lying $1/2^+$ state of ^{17}O had dramatic effect on the reactions. The effect of the nonelastic channels on the elastic scattering was to primary influence the imaginary part of the optical potential. The weakly bound neutron in ^{17}O provided for a large cross section for transfer reactions which took flux away from the elastic channel. At energies close to the Coulomb barrier, the two-step mechanism of stripping and pickup was the dominant one for the nuclear inelastic excitation of ^{17}O. With increasing energy, the CRC calculations predicted an increase in the direct inelastic excitation cross section.

The shift of the grazing angle in the transfer reactions due to the CRC effects was found to persist at higher energies even though it gradually became smaller as one progressed beyond the Coulomb barrier. The main source of the CRC effects in this case was verified to emerge from transfer following the inelastic excitation of ^{17}O.

The effect of the coupling of the transfer channels on the inelastic form factor was found to provide for a polarisation from factor which was dominantly imaginary at large internuclear separations. Thus, flux was being taken out of both the elastic and inelastic channels into the transfer channels.

These analyses were focused on the specific system of oxygen isotopes on lead. However, the effect of strong channel coupling on all the reactions at energies close to the Coulomb barrier would be expected to manifest itself in all the systems which exhibit the threshold anomaly. Detailed experiments and theoretical analysis of other systems would throw light on the effect of nuclear structure on the reaction mechanism.

Acknowledgement

This review is a summary of the work done in collaboration with J.S. Lilley, B.R. Fulton, D.W. Banes, I.J. Thompson and G.R. Satchler. I would like to thank all of my collaborators for involving me in these interesting studies.

References

1. A. Baeza, B. Bilwes, R. Bilwes, J. Diaz and J.L. Ferrero, Nucl. Phys. A419 (1984) 412.
2. J.S. Lilley, B.R. Fulton, M.A. Nagarajan, I.J. Thompson and D.W. Banes, Phys. Lett. B151 (1985) 181.
3. B.R. Fulton, D.W. Banes, J.S. Lilley, M.A. Nagarajan and I.J. Thompson, Phys. Lett. B162 (1985) 55.
4. M.A. Nagarajan, C. Mahaux and G.R. Satchler, Phys. Rev. Lett. 54 (1985) 1136.
5. C. Mahaux, H. Ngo and G.R. Satchler, Nucl. Phys. A449 (1986) 354; A456 (1986) 134.
6. G.R. Satchler and W.G. Love, Phys. Rep. 55 (1979) 183.
7. M.J. Rhoades-Brown and M. Prakash, Phys. Rev. Lett. 53 (1984) 333.
8. I.J. Thompson, M.A. Nagarajan, J.S. Lilley and B.R. Fulton, Phys. Lett. B157 (1985) 250.
9. S.C. Pieper, M.J. Rhoades-Brown and S. Landowne, Phys. Lett. B162 (1985) 43.
10. J.S. Lilley, see the proceedings of this conference.
11. M.A. Nagarajan and G.R. Satchler, Phys. Lett. B173 (1986) 29.
12. G.R. Satchler, M.A. Nagarajan, J.S. Lilley and I.J. Thompson, Ann. Phys. 178 (1987) 110.
13. J.S. Lilley, B.R. Fulton, D.W. Banes, T.M. Cormier, I.J. Thompson, S. Landowne and H. Wolter, Phys. Lett. 128B (1983) 153.
14. J.S. Lilley, M.A. Nagarajan, D.W. Banes, B.R. Fulton and I.J. Thompson, Nucl. Phys. A463 (1987) 710.
15. I.J. Thompson, to appear in Computer Physics Reports, Code FRESCO.

DYNAMICAL ASPECTS OF HEAVY ION REACTIONS A MICROSCOPIC APPROACH

G. POLLAROLO
Dipartimento di Fisica Teorica dell'Universitá di Torino and INFN Sezione di Torino
via P. Giuria 1, 10125 TORINO, Italy

1. INTRODUCTION

In the past few years we had a renewal of interest for the understanding of the dynamics of heavy ion reactions at energies close to the Coulomb barrier. This interest has been prompted by the discovery that the cross section for fusion considerably exceeds the theoretical predictions based on single channel elastic potentials[1]. These large enhancements have been ascribed [2] to the coupling of the elastic channel to reaction channels like inelastic excitation and particle transfers. These coupling strongly influence the position of the barrier in the effective potential and thus the fusion cross section.

In the Optical Model analysis of elastic scattering at energies close to the Coulomb barrier has been noticed[3] an *anomaly* in the real part of the optical potential in that it was necessary to increase sharply its strength for these low energies. Also this behaviour has been ascribed[4] to the coupling of reaction channels to the elastic one and it has been correlated to the behaviour of the imaginary potential through a dispersion relation[5].

At the same time has been possible to provide calculations of the imaginary potential[6] in term of the inelastic and one-particle transfer channels. These calculations, performed in the semiclassical approximation, have demonstrated that, at least in the surface region, the imaginary potential is determined by coupling of these channels with the elastic one. This potential is weakened at energies close to and below the Coulomb barrier since the non-elastic channels are effectively closed at these low energies.

In this talk I will briefly review the formalism and the results of microscopic calculations both for the imaginary potential and for the correction to the real part (polarization potential) and I will also discuss how the same *anomaly* has to be expected in the coupling matrix elements to well defined states. In particular I will discuss the microscopic origin of the imaginary part of the formfactor for the inelastic excitation.

2. IMAGINARY POTENTIAL

In the Optical Model description of elastic scattering one has to introduce an imaginary potential to account for the loss of flux in reaction channels that are not explicitly included in the calculation. If the depopulation of the elastic channel is due to a large number of independent transfer reactions and to nuclear inelastic excitation to collective vibrations, one can write the imaginary potential as:

$$W(r) = W_{trans}(r) + W_{inel}(r) \tag{1}$$

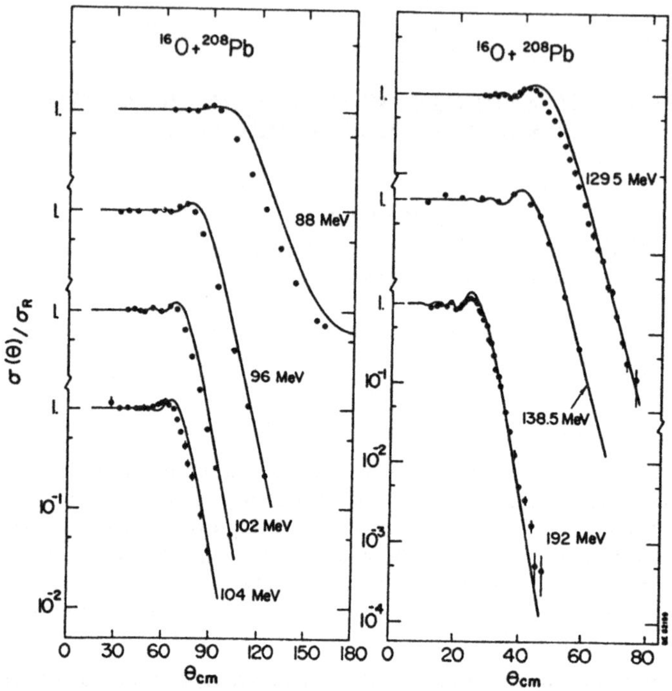

Fig. 1: *The ratio of elastic to Rutherford angular distributions for the $^{16}O + ^{208}Pb$ at different bombarding energies, see ref* [6].

The first component can be written as:

$$W_{trans} = \sum_{a_1,a_2,\lambda} \sqrt{\frac{a_{tr}(a_1,a_2)}{16\pi\hbar^2\ddot{r}_o}} \Big\{(2j_1+1)U(a_2)^2 V(a_1)^2 \mid f_\lambda^{(a_2,a_1)}(r) \mid^2 \Big\} g_\lambda(Q) \tag{2}$$

where the summation has to be extended over all one-particle transitions between projectile and target. The single-particle levels connected by the transition are labelled by $a \equiv (n,l,j)$, with n indicating the number of nodes, while l and j are the orbital and total angular momenta. The quantity a_{tr} is the diffuseness of the formfactor ($a_{tr} \simeq 1.2 fm$) associated with the reactions connecting the single-particle states a_1 and a_2. The parameters U^2 and V^2 are the spectroscopic factors. Thus, V^2 gives the probability that a given orbital is occupied and $U^2 = (1 - V^2)$ is the corresponding probability that the orbital is empty. The adiabatic cut-off function $g_\lambda(Q)$ weights the probability with which the different transfer channels contribute, at a given bombarding energy, to W_{trans}.

The second component, arising from the nuclear excitation of collective states can be written as:

$$W_{inel}(r) \simeq -\sqrt{\frac{\pi a}{\hbar^2\ddot{r}_o}} \left(\frac{\partial U}{\partial r}\right)^2 \sigma^2 \tag{3}$$

with

$$\sigma^2 = \frac{<\alpha>^2 R_0^2}{4\pi} exp\Big\{-\frac{a\omega^2}{\ddot{r}_o}\Big\} \tag{4}$$

representing the zero-point fluctuation amplitude in the nuclear radius weighted by the adiabatic cut-off function that takes into account the finite frequency of the mode. The parameter a is the diffuseness of the nuclear potential ($a \simeq 0.6 fm$).

The first term in (1) leads to a long-range component of the absorption with a diffuseness of the order of $0.6 fm$. The second to a short range part with a diffuseness of the order of $0.3 fm$. The $W(r)$ so calculated depends on energy mostly through the adiabatic cut-off function $g_\lambda(Q)$. As the bombarding energy increases, the collision time become smaller leading, for a given channel, to a larger value of the function $g_\lambda(Q)$ and thus to $W(r)$.

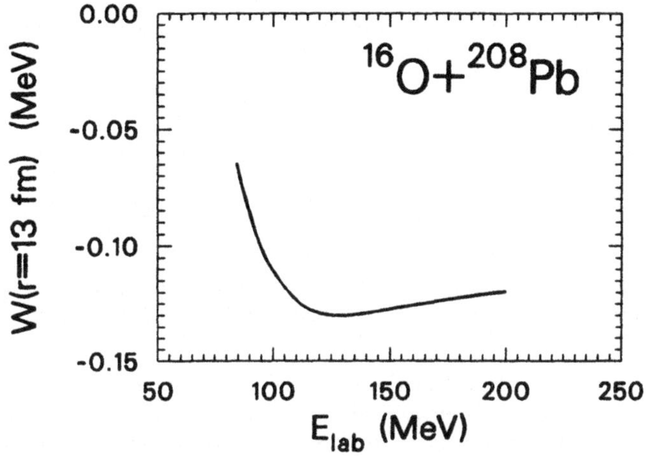

Fig. 2. *Energy dependence of the absorptive potential due to particle transfer as calculated in ref.* [6]. *The magnitude is taken at* $r = 13 fm$.

The above formalism has been applied to the study of the energy dependence of elastic scattering of several heavy-ion reactions. Here I just recall the results for the $^{16}O + ^{208}Pb$ elastic scattering that are relevant for the following discussions. The calculated angular distributions (ratio to Rutherford) are shown in fig. 1., while in fig. 2. is shown the energy dependence of the imaginary potential calculated at a distance $r = 13 fm$. The transfer component has been calculated including all the one particle transfer channels (stripping and pick-up) compatible with the bound states of target and projectile. It is interesting to notice that the imaginary potential displays a rapid rise in the low energy regime and it stabilizes at around $\simeq 150 MeV$.

3. POLARIZATION POTENTIAL

The elastic scattering data are most commonly analysed in the framework of the Optical Model. The potential used in these analysis has the general form:

$$U = V + \Delta U \tag{5}$$

with

$$\Delta U = \Delta V + iW \tag{6}$$

where W is the absorptive potential and ΔV is the polarization potential that corrects the bare potential V.

The experimental identification of ΔV is difficult since its presence is masked by the uncertainties in the determination of V and W. However detail analysis of elastic scattering data at energies close to the Coulomb barrier have shown an *anomalous* behaviour in the energy dependence of the real part of the potential[3]. As the energy approaches the Coulomb barrier, this part has to be increased considerably respect to the one needed at higher energies. In Fig.3. are shown the results obtained for the strength of the optical potential for the reaction $^{16}O + ^{208}Pb$.

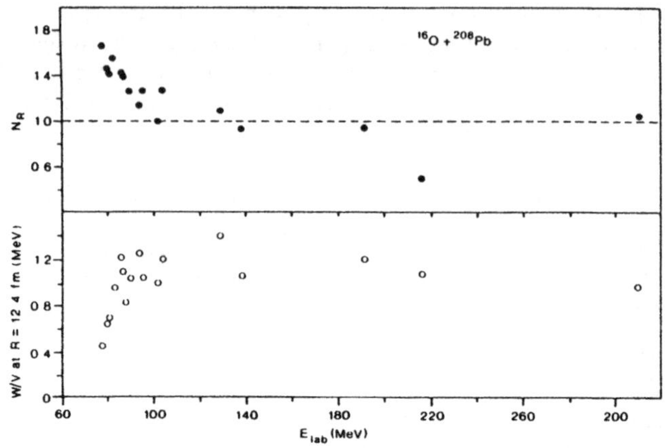

Fig. 3. *Variation of (top) normalization of the real folded potential and (botton) imaginary potential at $r = 12.14 fm$ from the optical model fits of elastic scattering data at various energies. From ref.* [3].

This behaviour of the real part of the potential has been correlated to the behaviour of the imaginary potential through a dispersion relation[5]. Here I will aim in calculating ΔU [7] within a semiclassical description of the collision process, focusing on the role played by the intermediate transfer channels. It will be shown that the present derivation will contains, for the imaginary potential, the result briefly recalled above.

3.1 A semiclassical expression for the polarization potential

Our approach is based on using semiclassical perturbation theory to obtain the amplitude for second order transitions. In order to avoid the complication of the angular momentum algebra I will start discussing the derivation for spinless transitions, and later on I will show results in actual cases.

The elastic scattering amplitude for an even-even system can be written in general as

$$f_{\alpha\alpha} = \frac{i}{2\kappa} \sum_{\ell} (2\ell + 1) P_\ell(\cos\theta)(1 - e^{-2i\beta_\ell} a^\ell_{\alpha\alpha}) \tag{7}$$

where β_ℓ is the real wave phase shift due to the bare elastic potential and $a^\ell_{\alpha\alpha}$ is the reaction amplitude for the elastic channel. Up to second-order semiclassical perturbation theory one obtains

$$a^\ell_{\alpha\alpha} = 1 - \frac{1}{\hbar^2} \sum_{\gamma} \int_{-\infty}^{+\infty} dt V_{\alpha\gamma}(r(t)) e^{-i\omega_\gamma t} \int_{-\infty}^{t} dt' V_{\gamma\alpha}(r(t')) e^{+i\omega_\gamma t'} \tag{8}$$

where $V_{\alpha\gamma}(r)$ specifies the coupling, i.e. the form factor, to an intermediate state with energy $E_\gamma = \hbar\omega_\gamma$ $(E_\alpha \equiv 0)$ and the integration is carried out along the trajectory for the impact parameter $\rho = \ell/k$.

The effect of the coupling can be taken into account by modifying the potential by an amount $\Delta U = \Delta V + iW$ in such a way that the phase shifts $\tilde{\beta}_\ell$ pertinent to the new potential satisfy the relation:

$$e^{2i\tilde{\beta}_\ell} = e^{2i\beta_\ell} a^\ell_{\alpha\alpha} \tag{9}$$

In the spirit of the semiclassical approximation one can write

$$e^{2i\tilde{\beta}_\ell} = e^{2i\beta_\ell} \left\{ 1 - \frac{i}{\hbar} \int_{-\infty}^{+\infty} dt \Delta U \right\} \tag{10}$$

thus comparing (10) with (7) and (8) one can identify:

$$\int_{-\infty}^{+\infty} dt \Delta U = -\frac{i}{\hbar} \sum_{\gamma} \int_{-\infty}^{+\infty} dt V_{\alpha\gamma}(r(t)) e^{-i\omega_\gamma t} \int_{-\infty}^{t} dt' V_{\gamma\alpha}(r(t')) e^{+i\omega_\gamma t'} \tag{11}$$

As it stands this formula is inadequate to specify the function $\Delta U(r)$. For each partial wave ℓ we only know the integral of ΔU. So to be useful this information as to be supplemented by additional constraint. In this choice we are guided by the previous experience on the imaginary potential so we require that the function ΔU is independent of the partial wave ℓ (this constraint is born out from the empirical observation that the elastic scattering can be described by ℓ-independent potentials).

This additional constraint is enforced by noticing that due to the exponential decay of the matrix element $V_{\alpha\gamma} \sim e^{-r/a}$ the main contribution to the integrals comes from a region very close to the turning point. Using a parabolic approximation to the classical trajectory

$$r = r_o + \frac{1}{2} \ddot{r}_o t^2 \tag{12}$$

with $\ddot{r}_o$ the acceleration at the classical turning point r_o, it is possible to transform (10) into an integral equation. This can be solved approximately to give

$$\Delta U(r) = -\frac{i}{\hbar} \sum_{\gamma} | V_{\alpha\gamma} |^2 \frac{1}{\sqrt{\pi \sigma_\gamma^2}} I(\omega_\gamma) \tag{13}$$

with

$$I(\omega_\gamma) = \int_{-\infty}^{+\infty} dt \exp \left\{ -\frac{1}{2\sigma_\gamma^2} t^2 - i\omega_\gamma t \right\} \int_{-\infty}^{t} dt \exp \left\{ -\frac{1}{2\sigma_\gamma^2} t'^2 + i\omega_\gamma t' \right\} \tag{14}$$

Here $\sigma_\gamma = a_\gamma / \ddot{r}_o$. Notice that $\tau = 2.3\sigma_\gamma$ gives a measure of the collision time. From the analytic property of the $I(\omega)$ in the complex ω-plane it is easy to see that the function (14) satisfies the dispersion relation:

$$ImI(\omega) = \frac{1}{\pi} \wp \int_{-\infty}^{+\infty} \frac{ReI(\omega')}{\omega' - \omega} d\omega' \tag{15}$$

$$ReI(\omega) = \frac{1}{\pi} \wp \int_{-\infty}^{+\infty} \frac{ImI(\omega')}{\omega - \omega'} d\omega' \tag{16}$$

An important observation on the radial dependence of the ΔU can be made by considering that in the case of particle transfer the matrix elements $V_{\alpha\gamma}$ have a diffuseness of $a_\gamma \simeq 1.2 fm$ thus the correction to the elastic potential has a diffuseness of $\simeq 0.6 fm$ that is the typical range of the empirical potential. For transfer reactions the $\hbar\omega_\gamma$ frequency has to be related to the difference $Q - Q_{opt}$, between the Q-value and the optimum Q-value of the transition. Notice that with this definition negative frequencies correspond to positive effective Q-value transfer reaction channels.

The real and imaginary part of $I(\omega_\gamma)$ are plotted in fig. 4. While the real part peaks at $\omega = 0$ and drops-off rapidly for large values of $| \omega |$, the imaginary part is antisymmetric around $\omega = 0$ and deceases slowly for large ω. From this behaviour we can make some interesting points considering some possible distribution in energy for the intermediate states (in Fig 4. these are represented with shaded areas). For a uniform distribution over positive and negative frequencies (Fig. 4a) there will be absorption but no polarization, while for a distribution over positive (Fig. 4b) or negative frequencies $\omega > 0$ ΔV and W receive comparable contributions. However for intermediate states localized at high frequency (Fig. 4c) we have only polarization but no absorption. Note that inelastic excitation and negative effective Q-values for reaction channels lead to a negative contribution to ΔV and thus to an increase of the nuclear attraction.

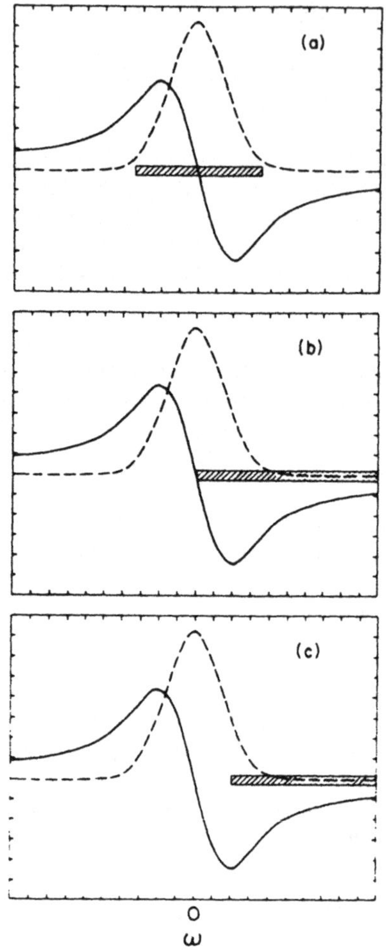

Fig 4. *Behaviour of the real (dotted) and imaginary (full) parts of the integral $I(\omega)$ and three type of energy distribution.*

The energy dependence of ΔV and W is essentially contained in the collision time. Besides the energy dependence brought by the term $1/\sqrt{(\pi\sigma_\gamma^2)}$ the collision time determines the width of the curves shown in Fig 4. At low energy these curves become narrow. This cuts off the higher frequency contribution to W leaving ΔV favored. For transfer channels the effective Q-value depends on bombarding energy through the Q_{opt}. Thus the distribution of intermediate states shown in Fig. 4. can vary as the energy changes. In the next section examples of this energy dependence will be shown in actual cases.

It should also be noticed that while the summation over intermediate states has a natural cut-off for W there is an apparent difficulty for ΔV. This is because at large ω

$$ImI(\omega) \rightarrow -\frac{\sqrt{\pi\sigma_\gamma^2}}{\omega} , \tag{17}$$

as can be seen from the integral expression (15). This slow rate of convergence does not pose a fundamental problem since the coupling strengths at high frequencies effectively limit the number of active channels. In the application that follows we have restricted the sum over intermediate steps to the bound states of the $A \pm 1$ systems. While this practical truncation leaves uncertain the magnitude of the polarization potentials, their energy dependence can be calculated more reliably, in fact the contribution of the intermediate channels to the derivative $d\Delta V/dE$ would have a better convergence.

3.2 Applications

Considering the case of a spinless projectile and target and focusing on the stripping reaction where a particle from the state a_1 in the projectile is transferred to a state a_2 in the target, one obtains the following expression for the effective potential,

$$\Delta U(r) = \sum_{a_1,a_2} \sqrt{\frac{a_{tr}(a_1,a_2)}{16\pi\hbar^2 \mid \ddot{r}_o \mid}} (2j_1+1) U^2(a_2, I_A) V^2(a_1, I_a) \mid f_\lambda^{a_2,a_1}(r) \mid^2 \times (-i)[g_\lambda(Q) + i\, p_\lambda(Q)] \quad . \tag{18}$$

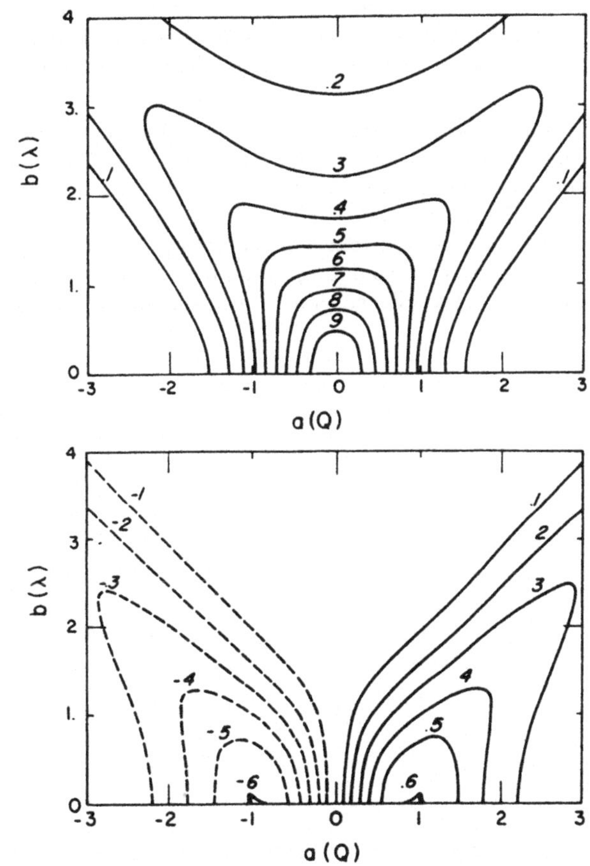

Fig. 5. *Contour plots of the adiabatic cutoff function (top) and the polarization function (bottom).*

A similar expression holds also for pick-up reactions. Here f_λ denote the transfer formfactor of multipolarity λ while the U and V are the occupation numbers of the orbitals. The quantity in brackets is given by the expression

$$|g_\lambda(Q) + ip_\lambda(Q)| = \frac{1}{\pi}\int_0^\pi d\theta \int_{-\infty}^{+\infty} dt e^{-t^2/2+i(a-b\cos\theta)t} \int_{-\infty}^{t} e^{-t'^2/2-i(a-b\cos\theta)t'}, \tag{19}$$

where

$$\begin{aligned} a &= \sqrt{\frac{a_{tr}(a_1,a_2)}{\hbar^2 \ddot{r}_o}}(Q - Q_{opt}), \\ b &= \sqrt{\frac{a_{tr}(a_1,a_2)}{\hbar^2 \ddot{r}_o}}\frac{(\ell_g + \frac{1}{2})\lambda}{m_{aA} r_o^2} \quad . \end{aligned} \tag{20}$$

This expression generalizes the function $I(\omega)$ introduced in the previous section.

The quantity $g_\lambda(Q)$ is the adiabatic cut-off function introduced in chapter 2.. The new feature introduced here is the polarization function $p_\lambda(Q)$ that determines the real part of ΔU. The plots in Fig. 5 show the contours of g and p in the (a, b) plane. These plots can be used to construct the effective potential once the formfactor for one-particle transfer are known. Notice that each channel corresponds to a well defined point in the (a, b) plane, its position is a function of the bombarding energy.

We have applied the above formalism to the elastic scattering of ^{16}O on ^{208}Pb and ^{60}Ni concentrating on the energy dependence of the effective potential. The calculations have been done including only the bound states of projectile and target and using the formalism of ref.[6] to calculate the one-particle transfer formfactors. The results for the $^{16}O + ^{208}Pb$ are shown in fig.6 where with the broken line we indicated the values of the absorptive potential for $r = 13$ fm, while with the full line we have plotted the corresponding values for the polarization potential ΔV. This last calculation displays an energy dependence that is very similar to the one extracted from the analysis of ref. [3].

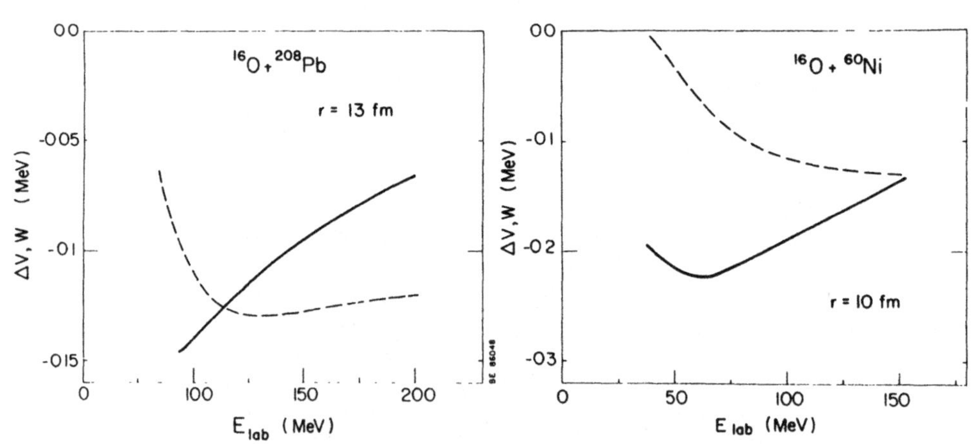

Fig. 6. *Energy dependence of the polarization potential (full line) and absorption (dashed line) due to one-particle transfer channels for the $^{16}O + ^{208}Pb$ reation (left) and for the $^{16}O + ^{60}Ni$ reaction (right).*

As discussed before the absolute magnitude of the polarization potential is somewhat uncertain due to the effective truncation of intermediate channels with high Q-values. The sum were here included all the bound state with $| Q | \leq 25$ MeV, a range for which the enerrgy dependence shown in fig. 6 was found to be quite stable .

For the system $^{16}O + ^{60}Ni$ has been found[8]) the same *anomaly* in the real part of the Optical Potential at energies close to the Coulomb barrier. Our results for this system are shown in fig. 6.

4. RENORMALIZATION OF THE OFF-DIAGONAL MATRIX ELEMENTS

The transitions to well defined final states, inelastic scattering or few particles transfer reactions, are usually analysed in the Distorted Wave Born Approximation (DWBA) formalism. The matrix elements used to calculate these transitions should also be renormalized due to the presence of channels not included explicitly in the calculation, as in the case of the elastic scattering. A nice example of this is provided by description of inelastic transitions where one has to include an imaginary part of the formfactor in order to have a consistent description of the data. In the macroscopic model this is taken to be the derivative on the imaginary part of the Optical Potential. Here I will show how the imaginary part of the inelastic formfactor can be derived microscopically from the reaction channels (in particular one-particle transfer) not included in the calculation[9]). This will be achieved by generalizing the formalism used to calculate the polarization potential.

4.1 Inelastic single-particle transitions.

In this section I will derive, in the semiclassical approximation, an explicit expression for the effective interaction for single-particle transitions in terms of one-step inelastic and two-step inelastic and and transfer processes. The extension of the model to surface collective mode can be done by utilizing the formalism of ref. [10]).

Consider a collision between two heavy ions where an inelastic transition takes place between the two shell model states $| j_\alpha, m_\alpha >$ and $| j_\beta, m_\beta >$ of the target due to the residual interaction V. In first-order semiclassical perturbation theory the amplitude for this transition is given by:

$$c_{\beta\alpha}^{(1)} = -\frac{i}{\hbar} \int_{-\infty}^{+\infty} dt < j_\beta m_\beta \mid V(\mathbf{r}(t)) \mid j_\alpha m_\alpha > e^{i\omega_{\beta\alpha} t} \quad , \tag{21}$$

where $\mathbf{r}(t)$ is the trajectory of relative motion and $\hbar\omega_{\beta\alpha} = E_\beta - E_\alpha$ is the Q-value of the transition. The corresponding second-order contribution is

$$\begin{aligned} c_{\beta\alpha}^{(2)} = -\frac{1}{\hbar^2} \sum_\gamma \int_{-\infty}^{\infty} & dt < j_\beta m_\beta \mid V(\mathbf{r}(t)) \mid j_\gamma m_\gamma > e^{i\omega_{\beta\gamma} t} \\ \times \int_{-\infty}^{t} & dt' < j_\gamma m_\gamma \mid V(\mathbf{r}(t')) \mid j_\alpha m_\alpha > e^{i\omega_{\gamma\alpha} t} \quad . \end{aligned} \tag{22}$$

The strategy to construct the imaginary part of the formfactor is to recast the second-order amplitude $c_{\beta\alpha}^{(2)}$ into the same form of $c_{\beta\alpha}^{(1)}$ and thereby obtain an effective first-order interaction which account for the second-order processes.

The intermediate states $| \gamma >$ may be of two kinds, either two-step inelastic transitions within the target itself or two step-transfer reactions which go back and forth between the collision partners. For the transfer formfactors the no-recoil approximation will be assumed so that the transfer formfactors will have the same form as for the inelastic transitions. Of course the transfer formfactor is characterized by a larger diffuseness. The calculation will be carried out introducing the radial formfactors through the multipole expansion:

$$< j_\beta m_\beta \mid V(\mathbf{r}) \mid j_\alpha m_\alpha > = \sum_{\lambda\mu} \frac{(-1)^{\lambda+\mu}}{\sqrt{4\pi}} (j_\alpha m_\alpha \lambda - \mu \mid j_\beta m_\beta) f_\lambda^{\beta\alpha}(r) Y_{\lambda\mu}(\hat{r}) \tag{23}$$

and using the focal coordinate system[11] where the z-axes is perpendicular to the plane of the orbit and the x-axes bisects the trajectory. Furthermore I will restrict myself to backward scattering trajectories where $\phi(t) \approx 0$ through the motion. After some angular momentum algebra one gets the result:

$$c^{(1)}_{\beta\alpha} + c^{(2)}_{\beta\alpha} = \sum_{\lambda\mu} \frac{(-1)^{\lambda+\mu}}{\widehat{\lambda}} (j_\alpha m_\alpha \lambda - \mu \mid j_\beta m_\beta) Y_{\lambda -\mu}(\tfrac{1}{2}\pi, 0) \Big[- i I_{\beta\alpha}(\omega_{\beta\alpha}) \\ - \sum_{\gamma} \sum_{\lambda_\alpha \lambda_\beta} \frac{\widehat{\lambda j_\gamma}}{\sqrt{4\pi}} (\lambda_\alpha 0 \lambda_\beta 0 \mid \lambda 0) W(j_\alpha \lambda_\alpha j_\beta \lambda_\beta; j_\gamma \lambda) I_{\beta\gamma\alpha}(\omega_{\beta\gamma}, \omega_{\gamma\alpha}) \Big] \tag{24}$$

where the first and second orbital integrals are given by:

$$I_{\beta\alpha}(\omega_{\beta\alpha}) = \frac{1}{\hbar} \int_{-\infty}^{+\infty} dt\, f^{\beta\alpha}_{\lambda}(r(t)) e^{i\omega_{\beta\alpha} t} \\ I_{\beta\gamma\alpha}(\omega_{\beta\gamma}, \omega_{\gamma\alpha}) = \frac{1}{\hbar^2} \int_{-\infty}^{+\infty} dt\, f^{\beta\gamma}_{\lambda_\beta}(r(t)) e^{i\omega_{\beta\gamma} t} \int_{-\infty}^{t} dt' f^{\gamma\alpha}_{\lambda_\alpha}(r(t')) e^{i\omega_{\gamma\alpha} t'} \tag{25}$$

The quantity $I_{\beta\alpha}$ is a pure real number while $I_{\beta\gamma\alpha}$ is complex. Its real and imaginary parts may be separated introducing the function

$$\epsilon(t - t') = \frac{i}{\pi} \wp \int_{-\infty}^{+\infty} \frac{dq}{q} e^{-iq(t-t')} \tag{26}$$

as was done in ref.[11] for the Coulomb excitation processes. In this way one finds:

$$Re\ I_{\beta\gamma\alpha} = \frac{1}{2} I_{\beta\gamma}(\omega_{\beta\gamma}) I_{\gamma\alpha}(\omega_{\gamma\alpha}) \\ Im\ I_{\beta\gamma\alpha} = \frac{1}{2\pi} \wp \int_{-\infty}^{+\infty} \frac{dq}{q} I_{\beta\gamma}(\omega_{\beta\gamma} - q) I_{\gamma\alpha}(\omega_{\gamma\alpha} + q) \tag{27}$$

By bringing these terms into the same form as $I_{\beta\gamma}$ one sees that the real part of $I_{\beta\gamma\alpha}$ gives rise to the imaginary formfactor while the imaginary part gives rise to a correction to real formfactor. The reduction of (27) to the same form as (25) is done by exploiting the exponential form of the formfactors $f^{\beta\alpha}(r) = f^{\beta\alpha}_0 \exp[-(r - r_o)/a_{\beta\alpha}]$ and by using the procedure outlined in the previous chapter to solve (11). After some manipulation one can write:

$$I_{\beta\gamma\alpha} = \sqrt{\frac{\pi a_{\beta\alpha}}{\hbar^4 \ddot{r}_o}} \int_{+\infty}^{+\infty} dt\, f^{\beta\gamma}_{\lambda_\beta} f^{\gamma\alpha}_{\lambda_\alpha} e^{i\omega_{\beta\alpha} t} \Big\{ \exp\Big(- \frac{\omega^2_{\beta\gamma\alpha} a_{\beta\gamma\alpha}}{2\ddot{r}_o} \Big) \\ + \frac{i}{\pi} \wp \int_{-\infty}^{+\infty} \frac{dq}{q} \exp\Big[- \Big(q + \omega_{\beta\gamma\alpha} \Big)^2 \frac{a_{\beta\gamma\alpha}}{\ddot{r}_o} \Big] \Big\} \tag{28}$$

where I defined:

$$\omega_{\beta\gamma\alpha} = - \frac{\omega_{\beta\gamma} a_{\beta\gamma} - \omega_{\gamma\alpha} a_{\gamma\alpha}}{a_{\beta\gamma} + a_{\gamma\alpha}} \\ a_{\beta\gamma\alpha} = \tfrac{1}{2}(a_{\beta\gamma} + a_{\gamma\alpha}) \tag{29}$$

It is interesting to notice that when the final state $\mid \beta >$ coincides with the initial $\mid \alpha >$ the terms in curl brackets reduce the the adiabatic cut-off function $g_\lambda(Q)$ and to the "polarization" function $p_\lambda(Q)$ introduced in the previous chapter.

Using (28) one obtains the following result for the imaginary part of the effective formfactor between two shell-model states.

$$Im\ \tilde{f}^{\beta\alpha}_{\lambda}(r) = - \sum_{\gamma} \sum_{\lambda_\alpha \lambda_\beta} \widehat{\lambda j_\gamma} (\lambda_\alpha 0 \lambda_\beta 0 \mid \lambda 0) W(j_\alpha \lambda_\alpha j_\beta \lambda_\beta; j_\gamma \lambda) \\ \times \sqrt{\frac{a_{\beta\gamma\alpha}}{\hbar^2 \ddot{r}_o}} f^{\beta\gamma}_{\lambda_\beta} f^{\gamma\alpha}_{\lambda_\alpha} \exp\Big[- \frac{\omega^2_{\beta\gamma\alpha} a_{\beta\gamma\alpha}}{2\ddot{r}_o} \Big] \tag{30}$$

The term proportional to the principal part in (28) will renormalize the real part of the formfactor.

The results obtained above apply directly to the inelastic excitation of a system with a nucleon outside the closed shell, like ^{209}Pb. To generalize the results for collective states one may use the formalism of ref.[10] noticing that the matrix element for the particle-hole transition may be easily obtained from the one describing the particle-particle transition(for mode detail cf. sect. 2 of ref.[9]).

4.2 Applications

To calculate the imaginary formfactor for a collective excitation using the formalism of ref. [10] one needs the RPA wave function of the state. To have a good RPA wave functions one has to use a very large shell-model space and thus one needs to calculate a very large number of single-particle transfer formfactors and these are easy to calculate only for bound states. To avoid all these difficulties we have only calculated the imaginary formfactor for selected individual particle-hole components of the low-lying 3^- state in ^{208}Pb in the collision with ^{16}O.

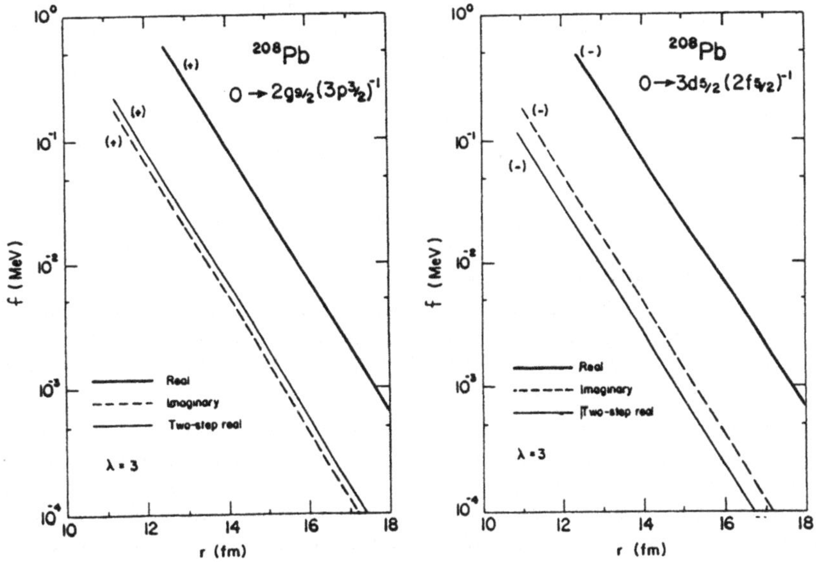

Fig. 7. *Microscopic formfactors for the indicated particle-hole excitation in* $^{208}Pb(\lambda = 3^-)$. *Indicated are the direct real formfactor and the effective real and imaginary formfactors coming from two-step transfer reactions.*

Let first consider the neutron particle-hole configuration $(2g_{9/2}, (3p_{3/2})^{-1})$ coupled to angular momentum $\lambda = 3$ (cf. fig. 7). This is the strongest component in the RPA wave function for the low-lying 3^- state. The direct excitation formfactor and the imaginary and real formfactors resulting from the two-step transfers via ^{17}O, are shown in fig. 7a. The single-particle transfer formfactors have been calculated in the no-recoil approximation using the formalism of ref.[6] and using the set of bound states shown in fig. 8.

The ratio of the real to imaginary formfactor is approximately equal to 0.1. The analysis of the inelastic cross section in the DWBA has been found that this ratio is about $0.2 - 0.4$, which is not too far from our estimation. To check if this ratio may be influenced by the coherent superposition of individual particle-hole excitations the calculation has been repeated for a weaker particle-hole configuration of the 3^- that has a much larger Q-value. The result for the configuration $(3d_{5/2}, (2f_{5/2})^{-1})$ is shown in fig. 7b. Also in this case the ratio is about 0.1.

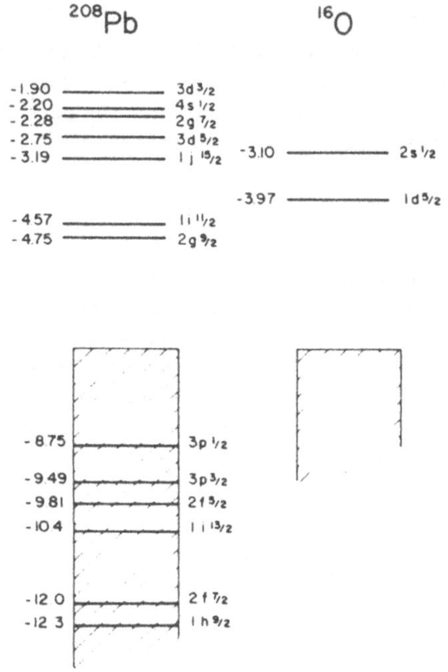

Fig. 8. *Neutron single particle levels used to construct the formfactors. See text.*

These microscopic calculations show how the main features of the usual macroscopic prescription can be understood. The imaginary coupling interaction should be proportional to and of the same order of magnitude as the imaginary part of the Optical Potential in the tail region. The empirical diffuseness of the imaginary formfactor is explained by the fact that the two-step transfer contribution dominates in the tail region.

It is interesting to notice that the second-order contribution to the real formfactor has the same strength as the imaginary formfactor itself. One thus expects to find in the inelastic scattering the same need to increase its strength for bombarding energies close to the Coulomb barrier as has been found for the Optical Potential. To check this idea a systematic analysis of the inelastic excitation of the low lying 3^- state in ^{208}Pb in the collision with ^{16}O at very low bombarding energies[12)] has been performed.

Since is very difficult to do a full microscopic calculation both for the Optical Potential and the inelastic formfactor the analysis of the data has been carried out with a conventional DWBA calculation utilizing the collective model to calculate the formfactor. We start with a set of Wood-Saxon parameters and Coulomb and nuclear deformation parameters which give a reasonably good description of the data at $104MeV$ $(V_o = -68.4MeV, W_o = -39MeV, r_o = r_w = 1.178fm, a_o = 0.758fm, a_w = 0.568 and \beta_N = \beta_C = 0.1227$ see ref [6)]). For energies below $104MeV$ one allows the strength of the real potential to increase and the imaginary potential to decrease according to the linear relations:

$$V(E) = V_o + \alpha(E_{lab} - 104)$$
$$W(E) = W_o + 0.85(E_{lab} - 104)$$

The latter slope parameter has been chosen to account the energy dependence of the theoretical calculation (cf. fig. 2. and ref. [6)]). The parameter α will be varied to fit the data.

In fig. 8a, are shown the results of the calculation for the excitation function at $\theta_{cm} = 172^o$ in comparison with the experimental data and with the pure Coulomb excitation (dotted line). This calculation was performed using $\alpha = 2.54$. The full angular distribution is shown in fig. 9b, for

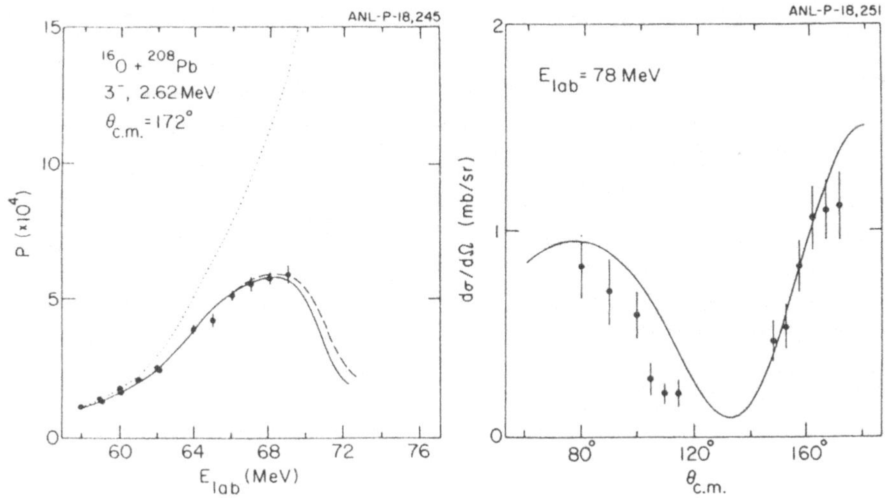

Fig. 9. *Ratio to Rutherford of inelastic cross section at a fixed angle (left). The Coulomb excitaion is shown with the dotted line. On the right is shown the angular distribution at $E_{lab} = 78MeV$.*

the bombarding energy of $78MeV$[13]. This energy dependence of the off-diagonal matrix elements may also be reconciled with a dispersion relation[14].

5. CONCLUSIONS.

In this talk I have shown that working in the semiclassical approximation it is possible to account for the strong coupling effects seen in the vicinity of the Coulomb barrier in terms of one-particle transfer channels and inelastic excitations. This formalism not only provides the right energy dependence of the Optical Potential but also allows us to calculate the r-dependence of the potential itself and of the off-diagonal matrix elements.

REFERENCES.

1) M. Beckerman, M. Salomaa,A. Sperduto, J.D. Molitoris and and A. DiRienzo, Phys. Rev. C25(1982)837.
W. Reisdorf et al, Phys. Lett. 49(1983)1811.

2) H. Esbensen, Nucl. Phys. A352(1981)281
R.G. Stokstad and E.E. Gross, Phys. Rev. C23(1981)281
R.A. Broglia, C.H. Dasso, S. Landowne and Aage Winther, Phys. Rev. C27(1983)2433.

3) A. Baeza, B.Bilwes, R. Bilwes, J. Diaz, and J.L. Ferrero, Nucl. Phys. A419(1984)412
J.S. Lilley, B.R. Fulton, M.A. Nagarajan, I.J. Thomson and D.W. Banes, Phys. Lett. 151B (1985) 181

4) I.J. Thompson, M.A. Nagarajan, J.S. Lilley and B.R. Fulton Phys. Lett. 157B(1985)250.

5) M.A. Nagarajan, C. Mahaux and G.R. Satchler, Phys. Rev. Lett. 54 (1985) 250,
C. Mahaux,H. Ngo and G.R. Satchler, Nucl. Phys. A449 (1986) 354.

6) R.A. Broglia, G. Pollarolo and Aa. Winther, Nucl.Phys. A361(1981)301,
G. Pollarolo, R.A. Broglia and Aa. Winther, Nucl. Phys. A406(1983)369.

7) C.H. Dasso, S. Landowne, G. Pollarolo and Aa. Winther, Nucl. Phys. A459 (1986) 134.

8) B.R. Fulton, D.W. Banes, J.S. Lilley, M.A. Nagarajan and I.J. Thompson, Phys. Lett. 162B (1985) 55.

9) C.H. Dasso, G. Pollarolo and S. Landowne, Nucl. Phys. A443 (1985) 365.

10) R.A. Broglia. C.H. Dasso, G.Pollarolo and Aa. Winther, Phys. Reports 48 (1978) 351.

11) K. Alder and Aa. Winther, Electromagnetic excitation (North-Holland, Amsterdam, 1975).

12) W.J. Vermeer, M.T.Esat, J.A. Kuehner, R.H. Spear A.M. Baxter ans S.Hinds, Aust. J. Phys. 37 (1984) 123.
C.S. Lim, M.D. Rebgetz. R.H Spear. W.J. Vermeer, M.P. Fewell, R.A. Robson and R. Smith, Phys. Lett. 200B (1988) 263.

13) J.S. Lilley and B.R. Fulton, private communication.

14) G.R. Satchler, Nucl. Phys. A472 (1987) 591.

A MICROSCOPIC NUCLEUS-NUCLEUS POTENTIAL* °

M. V. Andrés[a] and E. G. Lanza[b]

Departamento de Fisica Atómica y Nuclear, Universidad de Sevilla, 41080 Sevilla, Spain
b) Istituto Nazionale di Fisica Nucleare, Sez. di Catania, 95129 Catania, Italy

1. Introduction

The energy independent double-folding potential [1], together with a phenomenological imaginary part, has been successfully used to reproduce elastic differential cross sections. Recent reanalisys of experimental data at energies around the coulomb barrier [2], show the need of introducing an energy dependence, due to coupled channel effects [3]. The coupling does not only introduce absorption but changes also the real part of the optical potential. The complex polarization potential due to the transfer mode has been studied in ref. [4]. Here we construct the one arising from the coupling of the elastic channel to the ones corresponding to the excitation of collective vibrational degrees of freedom. These collective modes are described in the framework of the Random Phase Approximation (RPA). The probability amplitudes for exciting these modes are evaluated by integration along classical trajectories. In the next section we introduce the model and study the contribution of low and high lying modes to the polarization potential. In the last section comparisons between calculated and experimental elastic differential cross section are made.

2. The Model

The Hamiltonian of the system is given by the sum of the intrinsic Hamiltonians of the two colliding nuclei. For each of the nuclei we have

$$H_{intr} = \sum_k E_k B_k^\dagger B_k + \sum_k V_{k0}(t) B_k^\dagger + h.c. \tag{1}$$

We checked that the inclusion in the Hamiltonian of the term $V_{kk'}$ did not change our results. They affect mainly the inelastic cross section at low energy [5], by enhancing the population of the collective high lying states which, however, remains much smaller than the population of the low lying modes.

In eq. (1) $B_k^\dagger$ is the boson creation operator corresponding to the RPA k-mode $|\varphi_k\rangle$, whose energy is given by E_k. The $V_{k0}(t)$ are the matrix elements

$$V_{k0}(t) = < \varphi_k | U(t) | 0 > \tag{2}$$

* Talk given by M.V. Andrés.

° This work has been done in collaboration with F. Catara (Catania, Italy) and Ph. Chomaz (Orsay, France).

of the mean field of the other nucleus. This mean field is the Hartree- Fock potential.

The state of the system $|\Psi, t>$ is the product of two coherent states [6]. The probability amplitude for the system to remain in its ground state is then

$$< 0|\Psi, t >= e^{-i\phi(t)} e^{-\frac{1}{2}\mathcal{N}(t)} \quad , \tag{3}$$

where

$$\phi(t) = Im \sum_k \int_{-\infty}^{t} dt' e^{-iE_k t'} V^*_{k0}(t') \int_{-\infty}^{t'} dt'' e^{iE_k t''} V_{k0}(t'') \tag{4}$$

and

$$\mathcal{N} = \sum_k < \Psi, t|B_k^\dagger B_k|\Psi, t >= \sum_k N_k(t) \quad , \tag{5}$$

with N_k the average number of k- phonons in $|\Psi, t>$.

Semiclassicaly one can define the complex polarization potential [7] as

$$< 0|\Psi, t >\equiv e^{-i\int_{-\infty}^{t} dt' [\Delta V(t') + iW(t')]} \quad , \tag{6}$$

where the integration is done along a classical trajectory. From eq.s (3) and (6) we can get microscopic expressions for ΔV and W,

$$\Delta V = \dot{\phi}(t) \tag{7}$$

$$W = -\frac{1}{2}\dot{\mathcal{N}}(t) \tag{8}$$

which depend on energy and relative angular momentum.

The dependence on time appears through the relative distance $R(t)$ which evolves in time according to classical equations of motion. The latters are governed by a nucleus-nucleus potential which is the sum of the monopole- monopole coulomb part and the nuclear part obtained by folding the ground state densities of the nuclei with the M3Y effective nucleon- nucleon interaction. The energy and angular momentum loss as well as the modification of the nucleus- nucleus potential due to the real part of the polarization potential are taking into account in the description of the trajectory.

The functions ΔV and W can not be interpreted directly as the real and imaginary parts of the polarization potential. Their values, at any time t, depend on the history of the system till that instant. At a relative distance R one can find two different values for W (or ΔV) because R can be reached twice, in the approaching and in the outgoing phases (see figs. 1 and 2).

In order to define a local potential we assume that each trajectory (i.e. each L) defines the value of the potential at the distance of closest approach, and then fit the collection of values thus obtained with Saxon- Woods forms, whose radii is fixed according to ref. [8].

We have done calculations for the system $^{40}Ca+^{40}Ca$. All the RPA modes contributing to the energy weighted sum rule more than 5% have been considered [9]. At energies close to the coulomb barrier, ΔV and W arise mainly from the excitation of the low lying 3^- collective state. The situation changes drastically at higher energies. In fig. 1 W is given, at $E/A = 44\,MeV$, for different significant angular momentum values. In the right insert of the figure it is shown W obtained when only the contribution from the 3^- state was considered. The absoption strongly increases as soon as one takes into account the Giant Quadrupole Resonance (GQR) states. The other modes slightly modify this result. As can be seen in fig. 2, the main contribution to ΔV comes also from the GQR, while the low lying 3^- state does not give any sizeable contribution.

2. Elastic Differential cross section

In order to test our microscopic optical model, we have calculated elastic differential cross sections at several energies for the system $^{40}Ca+^{40}Ca$, and compared them with the experimental data [10]. Figures 3 and 4 show the elastic angular distribution for energies in the center of mass system between once and twice the height of the coulomb barrier. The dashed line corresponds to a calculation done by using a nuclear potential whose real part is given by the folding potential and its imaginary part by the absorption calculated as described in the previous section. The agreement with the experimental data is improved when the real part of the polarization potential is included in the calculation, solid line. At the lowest energy, $E_{cm} = 55.45\,MeV$, the angular distribution essentially agrees with Mott differential cross section. As the energy increases the effect of the nuclear potential shows up by decreasing the amplitudes of the oscillations, while preserving their positions.

The impressive agreement with the data, in spite of the absence of free parameters, gives us confidence in the way we are constructing the polarization potential.

We also have done calculations at higher energy, $E/A = 44\,MeV$, in order to see whether the important role played by the GQR in heavy ion collisions [5] shows also in the elastic angular distribution. In figure 5 the dotted line corresponds to Mott result. The dot- dashed line includes only the contribution of the low lying 3^- state. All the states were considered in the calculations whose results are given by the dashed and solid line, but only in the last one the real part of the polarization potential was taken into account.

The three calculations in which the nuclear potential has been considered show a similar behaviour: There is first a strong decrease of the cross section with the scattering angle, till some strong oscillations of *small period* appear, after that the cross section changes slope and decreases smoother than before. That can be understood when a near-far side decomposition of both nuclear and coulomb scattering amplitude is done [11]. The result at the small angles is near side dominated, the strong oscillations arise from the interference between the near and the far side contributions, the last one being the dominant at bigger angles.

We would like to point out the large modifications of the cross section due to the inclusion of the GQR states, as well as the ones due to the real polarization potential. Furthermore, we remark that both these effects are sizeable and hence can be experimentaly detectable.

References

[1] G. R. Satchler and W. G. Love, Phys. Rep. **55** (1979) 183;

[2] J. S. Lilley, B. R. Fulton, M. A. Nagarajan, I. J. Thompson and D. W. Banes, Phys. Lett. **151** (1985) 181;

[3] I. J. Thompson, M. A. Nagarajan, J. S. Lilley and B. R. Fulton, Phys. Lett. **157** (1985) 250;

[4] C. H. Dasso, S. Landowne, G. Pollarolo and Aa. Winther, Nucl. Phys. **A459** (1986) 134;

[5] F. Catara, Ph. Chomaz and A. Vitturi, Nucl. Phys. **A471** (1987) 661;

[6] Ph. Chomaz and D. Vautherin, Phys. Lett. **B139** (1984) 244; F. Catara and U. Lombardo, Nucl. Phys. **A455** (1986) 156; F. Catara and Ph. Chomaz, Nucl. Phys. **A482** (1988) 271c;

[7] R. A. Broglia, S. Landowne, R. A. Malfliet, V. Rostokin and Aa. Winther, Phys. Rep. **11** (1974) 1;

[8] R. A. Broglia and Aa. Winther, *Heavy Ion Reactions*, Lecture notes in Physics, (Benjamin, New York, 1981), vol 1, pag. 114;

[9] M. V. Andrés, F. Catara, Ph. Chomaz and E. G. Lanza, to be pubblished in J. of Phys. G;

[10] H. Doubre, J. C. Jacmart, E. Plagnol, N. Poffé, M. Riou and J. C. Roynette, Phys. Rev. **C15** (1977) 693;

[11] R. C. Fuller, Phys. Rev. **C12** (1975) 1561; M. V. Andres and E. G. Lanza, in preparation.

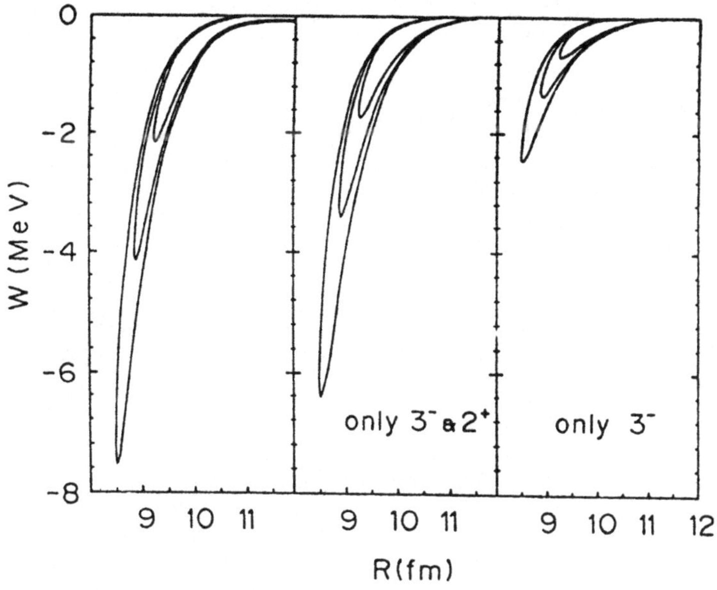

Fig. 1- Imaginary part W of the polarization potential for the $^{40}Ca+^{40}Ca$ system at $E_{cm} = 880\,Mev$, as function of the relative distance R. In the three sectors are reported the results obtained by including (from the right to the left): Only the low lying 3^- state, the 3^- state and the three 2^+ states (i.e. the GQR), and all the states of ref. [9]. In each sector, the three curves refer to three different values of the angular momentum: $L = 240, 250$ and $260\hbar$.

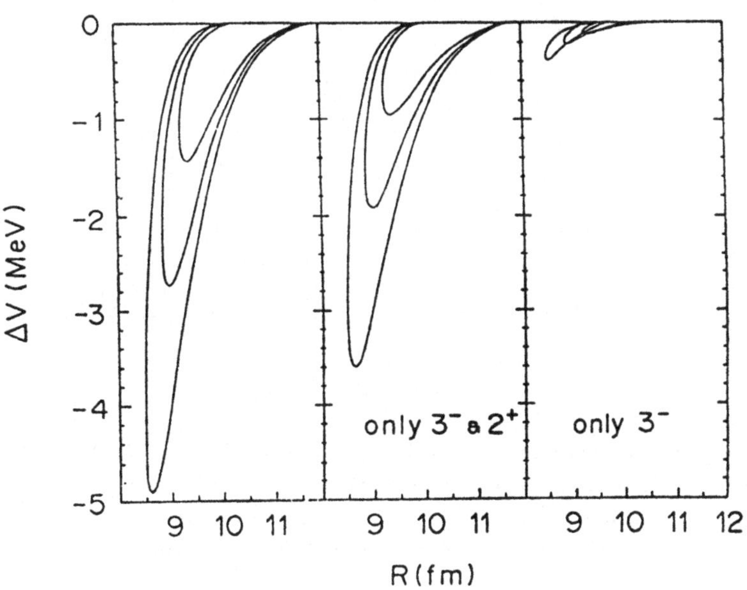

Fig. 2- Same as fig. 1 but for the real part ΔV.

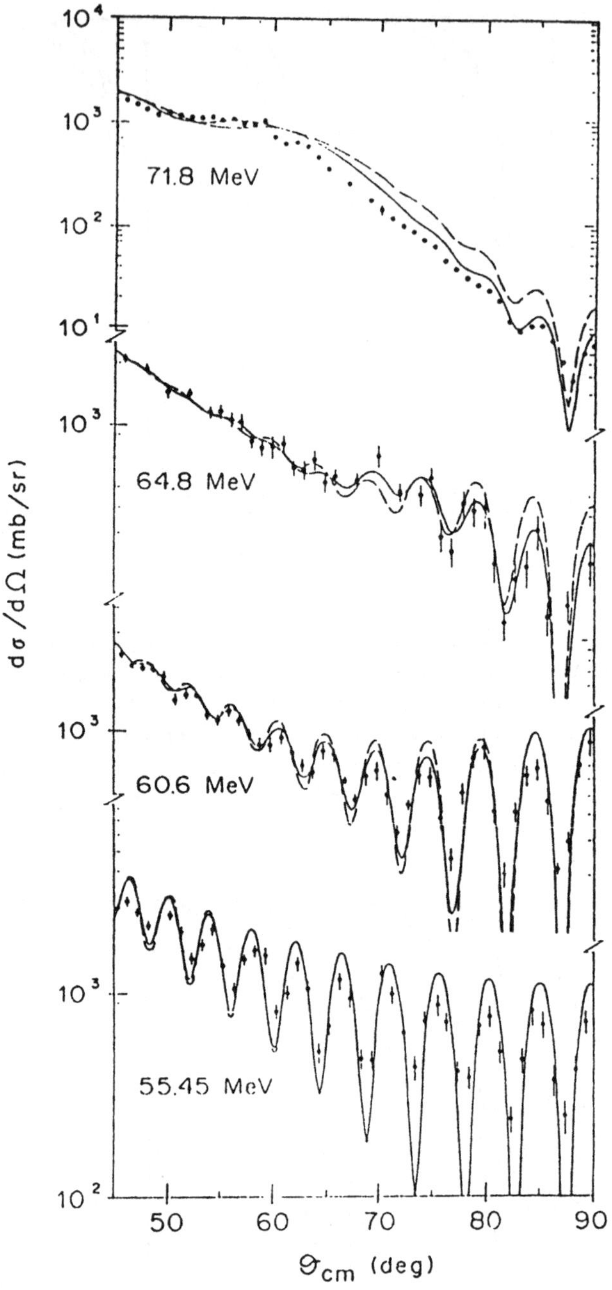

Fig. 3- Comparison between experimental data [10] and theoretical results for the elastic differential cross section for the $^{40}Ca + ^{40}Ca$ system at several center of mass energies as indicated. The dashed line is the result obtained using $V_{fold} + iW'$, while the solid line using $V_{fold} + \Delta V + iW$.

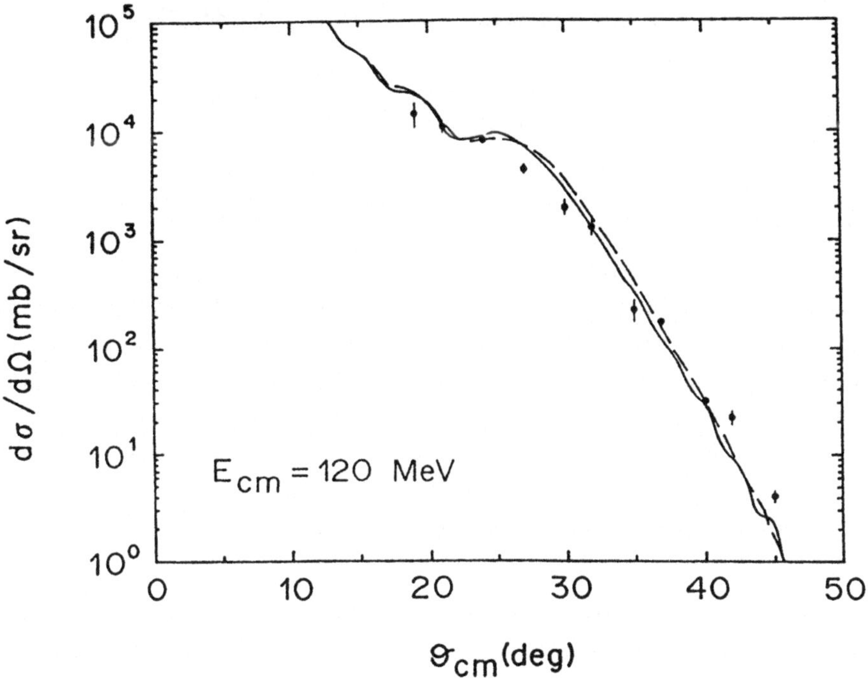

Fig. 4- Same as fig. 3 but at $E_{cm} = 120\,Mev$.

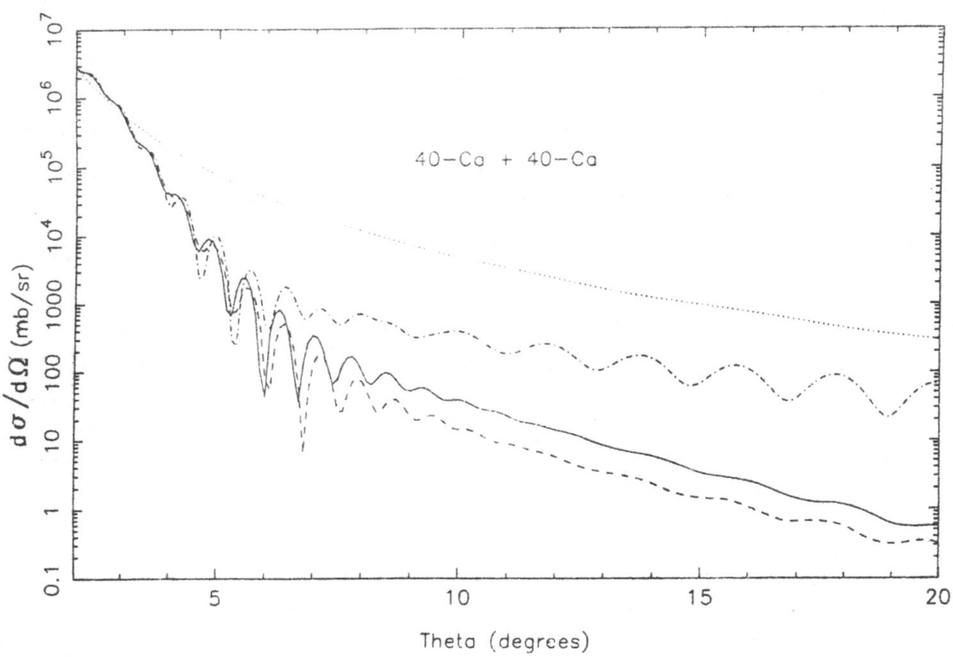

Fig. 5- Elastic differential cross section for the ^{40}Ca+^{40}Ca system at $E_{cm} = 880\,Mev$ (see text).

FUSION AND SCATTERING OF POLARIZED ^{23}NA*)

D. Fick and H. Leucker
Philipps-Universität, Fachbereich Physik
3550 Marburg, FR Germany

1. Introduction

Despite an intensive investigation in the past fusion at the barrier in particular and heavy ion interaction in general are still topics of wide interest for heavy-ion energies around the Coulomb barrier. Presently heavy ion (quasi)elastic scattering around barrier energies draws the attention of an increasing number of experimentalists since it became quite obvious that the enhancement of the subbarrier fusion cross section is not an anomaly peculiar to the fusion channel. The underlying mechanism - coupling of the entrance channel to other degrees of freedom, mainly to inelastic and transfer channels - manifests itself in any channel including the (quasi)elastic one.

Aligned heavy ion beams are a particular useful tool to investigate such phenomena, since in addition to the cross section three further quantities, the socalled 2nd rank analyzing powers T_{2q} are accessible experimentally /1/. In an aligned beam the quadrupole moments of the projectiles are aligned as well, and therefore the analyzing powers tell about the sensitivity of the reaction on the spatial alignment of such a beam. Depending on the internal structure of the projectile (mainly the spin sequence) certain alignment axes can be identified for which it is easier to transfer orbital angular momenta into internal angular momenta and opposite (this is a well known effect in Coulomb excitation) /2/. Thus, 2nd rank tensor analyzing powers are a very sensitive tool for coupling to other degrees of freedom. This was the very reason why we started to investigate heavy ion interaction around the Coulomb barrier using aligned beams.

For the experiments to be described the polarized ^{23}Na beam at the Heidelberg MP-Tandem accelerator plus post-accelerator was used /3,4/. With this combinations of accelerators also low energy ^{23}Na beams stripped to the 9^+ charge state can be produced. Among the accessable charge states this is the one which displays after stripping a minimum of depolarization. By intensity reasons it has to be generated with an as high as possible terminal voltage (U_T=12MV). In order to obtain beam energies smaller than 120MeV, the post-accelerator is used just as a decelerator.

---

*) supported partly by the Bundesministerium für Forschung und Technologie, Bonn, under contract 06 MR 853 I

Using such a beam the fusion cross section and T_{20}^{fus} for fusion have been measured at energies around the fusion barrier for $^{23}Na + ^{48}Ti$ and $^{23}Na + ^{206}Pb$ /5/. Fig.4 displays the energy dependence of the cross section and the analyzing power T_{20}^{fus} for $^{23}Na+^{206}Pb$. The analyzing power T_{20}^{fus} is large and positive below the barrier and decreases to zero far above the barrier. Qualitatively this bahaviour is well understood in terms of the prolate deformation of the ^{23}Na projectiles. For a quantitative understanding the authors of these data defined the fusion cross section as the difference between the total reaction cross section and the sum of all direct cross sections, now restricted to inelastic scattering, which occurs at the outside of the barrier. They calculated S-matrix elements with the coupled channel code ECIS 79. In order to assess the coupling effects later they started with a "bare" ion potential, chosen as the Woods-Saxon parametrization of the Akyüz-Winther potential, which however was adjusted slightly for the $^{23}Na+^{206}Pb$ case in order to reproduce the experimentally determined barrier.

The one dimensional calculations reproduced the fusion cross section and the corresponding analyzing power T_{20} well above the barrier, but, as expected, did not describe both below the barrier. Coupling to inleastic states of projectile and target increased the subbarrier fusion cross section somewhat, but not enough to explain the observed enhancement. But it definitely decreased the tensor analyzing powers in a way, to describe them in remarkable detail. The failure to describe the cross section in this way was interpreted as the need to include transfer channels which couple to the fusion channel. The fact that the inclusion of the inelastic channels alone yields the observed 2nd rank tensor analyzing power T_{20}^{fus} while the cross section itself needs an additional enhancement process showed that these additional processes must not change the analyzing powers. In view of the delicate balance of couplings which conspired to give the correct T_{20}^{fus} this appeared to be a severe constraint to the choice of additonal degrees of freedom necessary to describe the cross sections for fusion for both systems under investigation.

2. Coupled channel calculations

Coupled channel calculations are rather computer time consuming and become increasingly boring and inpracticable with increasing number of coupled channels. Moreover, the transparence of the underlying physics fades with increasing complexity of such calculations. Therefore, since long in a series of theoretical studies the socalled semiclassical or adiabatic approximation /6,7/ was developed which allows to diagonalize these equations if the following presuppositions are fulfilled:

i) the coupling form factors have identical radial form,

ii) the coupled states are degenerated,

iii) differences in centrifugal potentials can be ignored.

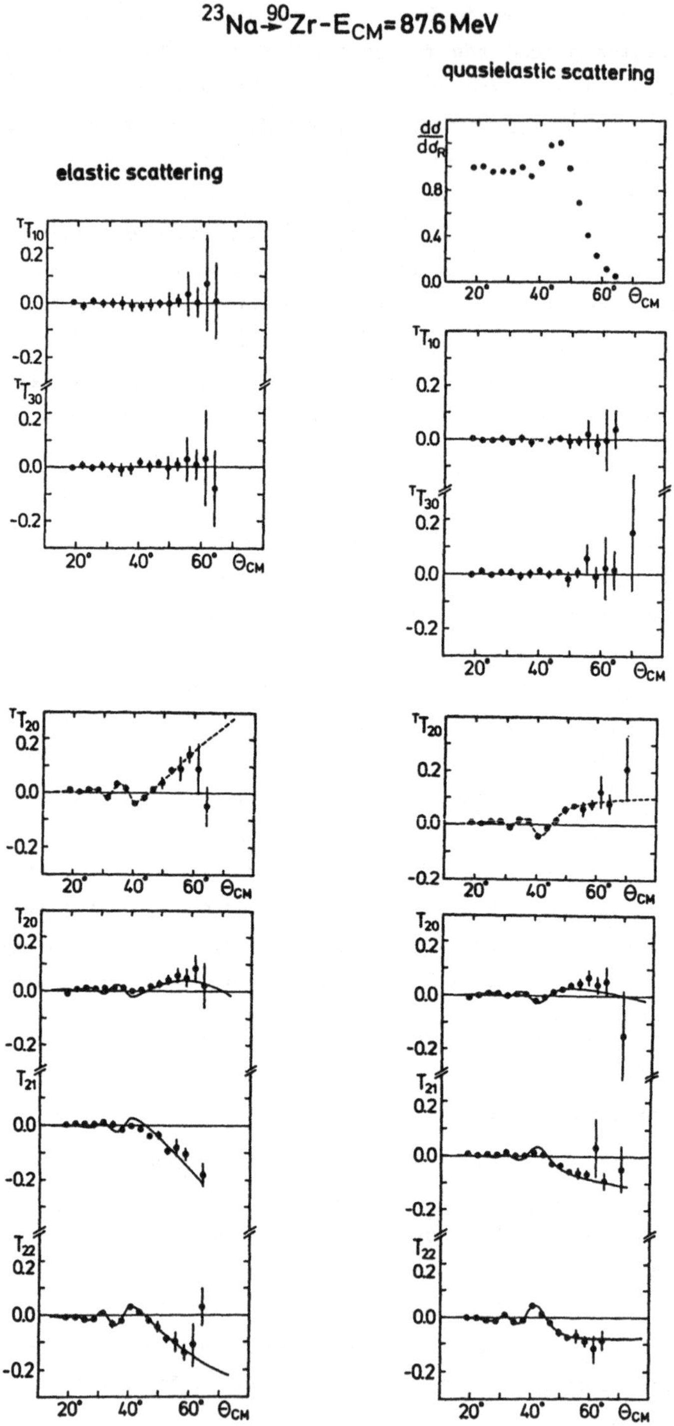

$^{23}Na \to ^{90}Zr - E_{CM} = 87.6$ MeV
quasielastic scattering
elastic scattering
$^{T}T_{10}$
$^{T}T_{30}$
$^{T}T_{20}$
T_{20}
T_{21}
T_{22}
$\frac{d\sigma}{d\sigma_R}$
Θ_{CM}

Fig.1

It has been shown /8/ that the latter presupposition is fulfilled, if the socalled "isocentrifugal approximation" can be used, which leads to a conservation of the socalled "tidal spin". Polarized beams are an excellent tool to proof this approximation since it can be shown that conservation of tidal spin requires that all odd rank analyzing powers must vanish identically and the 2nd rank analyzing powers must obey the socalled shape effect relations /9/

$$T_{2q} = -\sqrt{16\pi/5}\ Y_{2q}\{(\pi-\theta)/2,0)\}\,{}^{T}T_{20}(\theta).$$

The conservation of tidal spin was investigated for the interaction of polarized ^{23}Na with ^{90}Zr at a bombarding energy of 110MeV (Fig.1). (On very general grounds /10/ it can be shown that conservation of tidal spin in this system means conservation of tidal spin in the systems $^{23}Na+^{206}Pb$ under investigation in sect.3.) Angular distributions of analyzing powers for elastic and inelastic scattering (sum of elastic scattering and inleastic excitation to the first excited state of ^{23}Na) were determined. Obviously within the error bars the odd rank analyzing powers vanish. To test the second requirement the data for ${}^{T}T_{20}$ were connected in Fig.1 by a dashed line. The solid lines in the plots of the other analyzing powers T_{2q} are calculated by the above equation "inserting" for ${}^{T}T_{20}$ the dashed line of Fig.1. Altogether it becomes appearent that tidal spin must be conserved in such a system and the socalled "isocentrifugal approximation" must be an excellent approximation. Therefore, coupled channel calculations can be facilitated quite a lot by just using the socalled adiabatic or semiclassic approximation in solving them.

3. Threshold anomaly and enhancement of the fusion cross section

To learn further on the interaction potentials for $^{23}Na+^{206}Pb$ the energy dependence for quasielastic scattering was investigated. Fig.2 shows such data for the differential cross section and the tensor analyzing power T_{20} to which the investigation could be restricted because of the validity of the shape effect relations. The solid line through the data points displays optical model fits to the quasielastic scattering which was achieved by using the coupled channel code ECIS 79. As it is well known for such systems optical potentials are only determined in a very narrow region, here around 12.6fm. However, the additional use of aligned beams determines not only the value of the potential but also its slope at this sensitive distance. Fig.3 displays the strength of the real (V) and imaginary (W) part of the optical potential at this distance as they emerged from the fit. Even with this limited amount of data the socalled threshold anomaly is clearly visible /11/. Following previous publications /12/ we parametrized the energy dependence of the imaginary

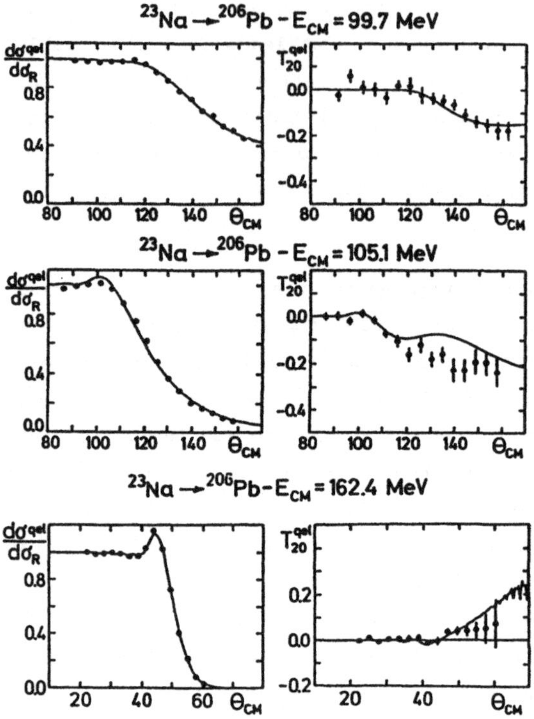

Fig.2

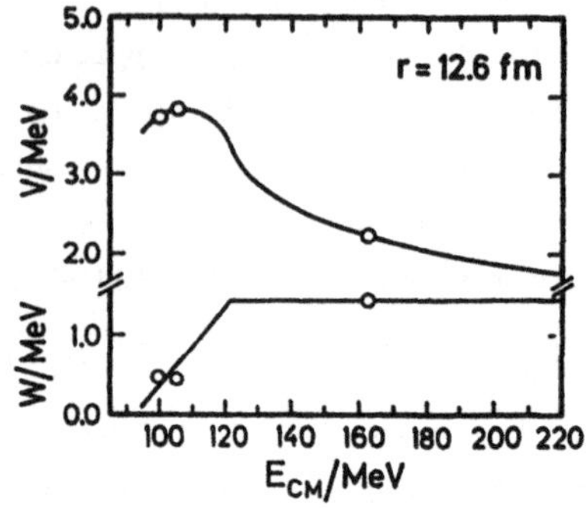

Fig.3

part trapizium like. Using now the well known dispersion relation between the real and imaginary part in its local form

$$V(r,E) = V_o(r) + \pi^{-1} P\int_{-\infty}^{+\infty} dE'W(r,E')/(E'-E),$$

from the energy dependence of W the one for V can be infered. It describes beautifully the determined potential strengths V at this distance (Fig.3). This agreement indicates that the increasing strength of the real potential around barrier energies is caused by an increasing strength of the imaginary part of the optical potential.

Even though the real part of the optical potential determined in quasielastic scattering may necessarily not be usable to describe the fusion process, we used this potential for that purpose. Certainly the fusion barrier for Na with Pb is at an about 1fm smaller distance than the region of sensitivity in quasielastic scattering. But since we determined not only the value of the real potential but also its slope around 12.6fm, there exists the hope that we may be able to describe excitation functions for fusion of ^{23}Na with ^{206}Pb by the energy dependent real part of the optical potential of Fig.3. According to the findings in sect.2 the

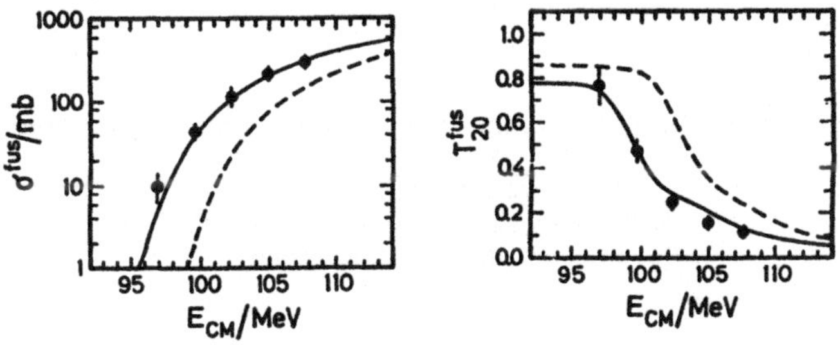

Fig.4

calculations were performed with a semiclassical code written by Grawert and coworkers /7/. Fig.4 displays as a solid line the results of such a calculation. It turns out that the energy dependent optical potential of Fig.3 describes in remarkable detail the observed excitation function for fusion. The enhancement of the fusion cross section becomes very clear. The dashed line in Fig.3 displays the results of a calculation with an energy independent optical potential as determined by the high energy value of V in Fig.3. Obviously quasielastic scattering of aligned ^{23}Na determines value and slope of the real part of the optical potential well enough to use it at an about 1fm smaller radius for the description of the fusion cross section.

References

1 M. Simonius in 'Polarization Nuclear Physics', ed. D. Fick, Lecture Notes in Physics, Vol.30, Springer 1974

2 K. Alder and A. Winther, Electromagnetic Excitation; North-Holland 1975

3 D. Krämer et al., Nucl.Instr.Meth. **220** (1984)123

4 H. Jänsch et al., Nucl.Instr.Meth. **A254** (1987) 7

5 R. Butsch et al., Phys.Rev. **C36** (1987) 1351

6 P. Jacobs and U. Smilansky, Phys.Lett. **127B** (1983) 313
C.H. Dasso et al., Nucl.Phys. **A405** (1983) 381
H. Esbensen et al., Nucl.Phys. **A441** (1983) 275
R. Lindsay and N. Rowley, J.Phys. **G10** (1984) 805

7 D. Mukhopadhyay et al., Phys.Rev. **C35** (1987) 1324

8 J. Gomez-Camacho and R.C. Johnson, J.Phys. **G12** (1986)235

9 Z. Moroz et al., Nucl.Phys. **A381** (1982) 284

10 W. Ott et al., J. Phys. **G14** (1988) L7

11 P.R. Christensen and A. Winther, Phys.Lett. **65B** (1976) 19

12 M.A. Nagarajan et al., Phys.Rev.Lett. **54** (1985) 1136

THRESHOLD ANOMALIES IN THE SCATTERING OF OXYGEN BY ^{208}Pb

J.S. Lilley
SERC Daresbury Laboratory
Warrington WA4 4AD, England

In recent work at Daresbury Laboratory, a study of the elastic scattering of ^{16}O by ^{208}Pb revealed a threshold anomaly in the energy dependence of the effective interaction[1]). An analysis of the elastic data near the barrier indicates a large (70-80%) increase in the nuclear attraction compared with the bare potential given by the double-folding model. This increase is consistent with the measured fusion cross section, which is more than a factor of thirty larger than the bare potential prediction. The results also indicated that the effect is energy dependent and related to the behaviour of the quasi-elastic reactions, which are changing rapidly at near barrier energies.

A dispersion relation interpretation of these data, using several simplifying assumptions, relates the surface absorption potential and the real potential[2]). This analysis reproduced the observed energy dependence of the real potential above the barrier, and indicated that it should peak near the lowest energy for which data exist and then decrease at lower energies (see fig. 1).

Coupled reaction channel calculations confirm that a significant polarization potential arises from coupling to both inelastic and transfer channels[3]). The latest CRC calculations for E_{lab} = 80 MeV are presented in fig. 2 which shows the major improvements obtained by successively including couplings to inelastic scattering, nucleon transfer and, finally, α particle transfer. The open squares in fig. 1 are the surface potentials obtained from the CRC calculations using an inversion procedure[5]). They are consistent with the phenomenological potentials. Total cross sections are listed in Table 1. Evidently the major features of the interaction are well accounted for, although many details are not reproduced. For example, inter-channel couplings which hitherto have been neglected, may be important. The magnitude of the polarization potential appears to be highly sensitive to a number of factors, such as the reaction Q-value and the projectile energy. The radial dependence of the polarization potential also is expected to depend on these factors and would be changing most rapidly near the barrier.

It now is important to obtain more precise data to apply more critical tests to the theory. For example, the dispersion relation interpretation, which reproduces the observed increase in V as the bombarding energy approaches the barrier from

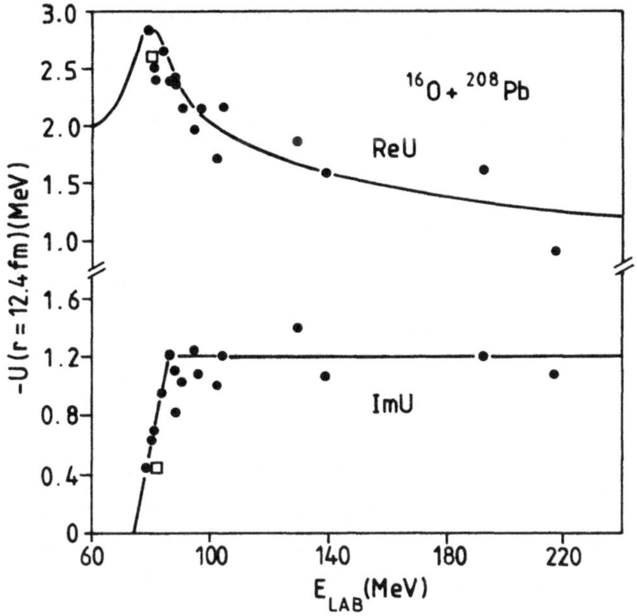

Fig. 1. Dispersion relation fits to empirical ^{16}O-^{208}Pb optical potentials evaluated at 12.4 fm (ref. 2).

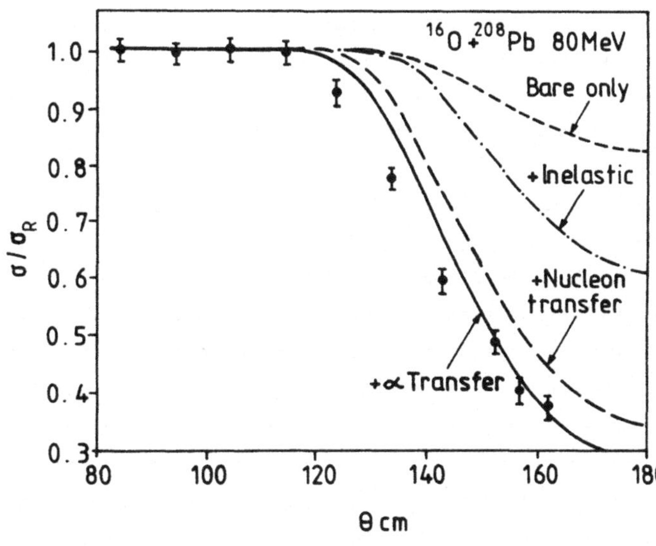

Fig. 2. Elastic scattering of ^{16}O-^{208}Pb at 80 MeV (ref. 4). The curves are CRC calculations showing the improvement obtained by successively including the coupling effects of inelastic scattering, nucleon transfer and α-transfer.

Table 1. Integrated cross sections at 80 MeV.

	CRC Couplings				
	None	Inelastic only	Inelastic + n,p transfer	Inelastic + n,p,α transfer	EXP (mb)
Fusion	1	14	38	37	41 ± 6
Inelastic	-	8	14	10	45 ± 5
n,p transfer	-	-	31	39	
α transfer	-	-	-	18	19 ± 2
TOTAL	1	22	83	104	105 ± 8

above, predicts that the potential should decrease as the energy is further reduced. There is evidence for this in ^{16}O-Ni, and ^{32}S-Ni measurements[6,7]). In ^{16}O-^{208}Pb collisions, it is difficult to obtain information below 80 MeV from elastic scattering data alone, because it rapidly tends towards Rutherford scattering, and optical model analyses are ambiguous.

Several approaches have been considered to try to overcome this limitation. For example, the fusion cross section σ_f is very sensitive to the barrier at low energies, and values for the real potential V(r) could be obtained from the data using the barrier penetration model (BPM). Unfortunately, σ_f is also sensitive to the shape of V(r), which is not well determined by elastic scattering. Thus, while values for V(r) can be obtained which agree well with those of fig. 1, they are by no means unique. Furthermore, the fact that the validity of the BPM is open to question casts further doubt on the utility of this procedure.

In a different approach, excitation functions of the total reaction cross section and elastic scattering taken at 155° (lab) have been used in a global optical model analysis. Including both sets of data enables the V/W ambiguity to be lifted to some extent. Both excitation functions are shown in fig. 3. The solid lines are predictions of an energy independent optical potential derived by fitting the complete elastic differential cross section at 78 MeV. Deviations from the data, particularly above this energy, are further evidence of the energy dependence of the ^{16}O-^{208}Pb nuclear potential in this region. Both $\sigma_{el}(155°, E)$ and $\sigma_R(E)$ data were fitted simultaneously at each energy by adjusting the surface strengths of V and W. The results are summarized in fig. 4. They reproduce the results of the elastic scattering analysis above 80 MeV and indicate that V begins to decrease at lower energies according to the dispersion relation, which is shown by the solid curve. Below 77 MeV, uncertainties due to errors in the data are very large and these results have not been included.

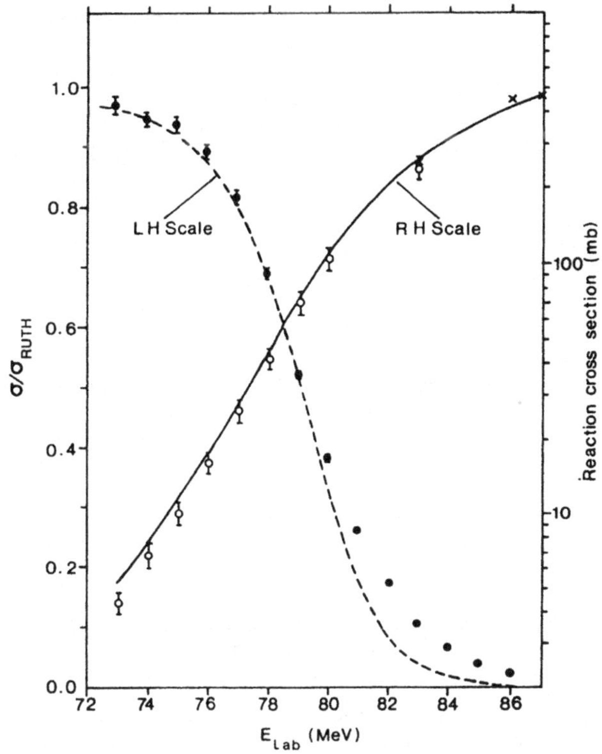

Fig. 3. Excitation functions for $^{16}O-^{208}Pb$ elastic scattering at 155° lab (solid circles) and σ_{Total} (open circles). The curves are predictions of an optical potential which fits 78 MeV data.

A third approach makes use of the generality of the dispersion relation which suggests that threshold anomalies also should manifest in non-diagonal as well as diagonal channels[8]). For example, inelastic excitation of the collective (3^-) state at 2.614 MeV in ^{208}Pb may show anomalous behaviour, and, because of Coulomb-nuclear interference, it should be possible to study it at energies significantly below the present limit. Some information on this already exists but it is based mainly on single-angle, excitation function data which are insufficient to enable both V and W to be determined.

Using the NSF at Daresbury Laboratory, we have measured angular distributions for the $^{208}Pb(^{16}O,^{16}O')^{208}Pb(3^-)$ reaction at a number of bombarding energies between 70 and 94 MeV. These data were taken using a magnetic spectrometer; great care being necessary to minimise the effects of slit-edge scattering and target deterioration. A preliminary analysis has been carried out within the DWBA framework to determine the effective potentials. Results obtained so far for the real transition potential (V_{tr}) at r = 13 fm are given by the squares in fig. 5. They indicate a strong peak

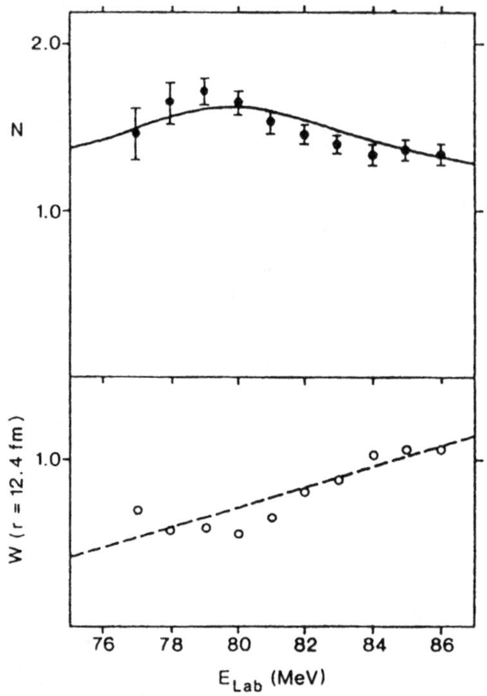

Fig. 4. Folded potential scaling factors (N) and surface imaginary strengths required to fit the data of fig. 3. The curves show a linear fit to W and a dispersion relation dependence for N (ref. 2)

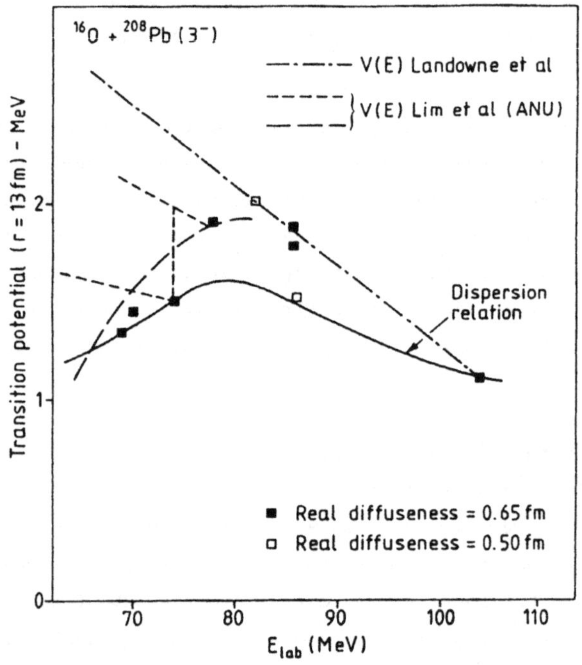

Fig. 5. Surface transition potentials for inelastic excitation of the $^{208}Pb(3^-)$ state. Energy dependences quoted in refs. 9 and 10 are shown. The solid curve shows the energy variation of the elastic dispersion relation.

in the energy dependence near threshold, which is similar in position, but possibly even more pronounced than that given by the elastic dispersion relation. The dot-dashed line is the energy dependence given in ref. 9, obtained from fitting 78 MeV data and a 172° excitation function from 60-70 MeV. The solid, long-dashed, and short-dashed lines are, respectively, the dependences of the dispersion relation (fig. 1), and those used in ref. 10 to reproduce the 172° excitation function extended to 79 MeV.

At low energies (< 78 MeV), the data are insensitive to the optical potentials and the imaginary transition potential. Thus these values for V_{tr} are reasonably well determined. At higher energies, as expected, both the real and imaginary parts play a role, especially in the vicinity of the interference minimum. Within the framework of the collective model, equivalent elastic potentials predict similar inelastic distributions, but at 82, 86 and 94 MeV the experimental data at large angles are not well reproduced by the calculations. The behaviour is reminiscent of that observed in transfer reactions near the Coulomb barrier[11]), which was evidence of the importance of multistep transfer processes.

I wish to thank my colleagues, M.A. Nagarajan, I.J. Thompson and B.R. Fulton for contributions to all aspects of this work, D.W. Banes for his help on the elastic scattering and M.J. Smithson, P.V. Drumm and R.A. Cunningham for their work on the inelastic scattering experiment.

References

1. J.S. Lilley et al., Phys. Lett. 151B (1985) 181.
2. M.A. Nagarajan, C. Mahaux and G.R. Satchler, Phys. Rev. Lett. 54 (1985) 1136; C. Mahaux, H. Ngo and G.R. Satchler, Nucl. Phys. A449 (1986) 354.
3. I.J. Thompson et al, Phys. Lett. 157B (1985) 250.
4. F. Videbaeck et al., Phys. Rev. C15 (1977) 954.
5. A. Ioannides and R. Mackintosh, Phys. Lett. 161B (1985) 43.
6. B.R. Fulton et al., Phys. Lett. 162B (1986) 55.
7. A.M. Stefanini et al., Phys. Rev. Lett. 59 (1987) 2852.
8. G.R. Satchler, Nucl. Phys. A472 (1987) 591.
9. S. Landowne, C.H. Dasso and G. Pollarolo, Phys. Lett. 178B (1986) 336.
10. C.S. Lim, et al., Phys. Lett. 200B (1988) 263.
11. J.S. Lilley et al., Nucl. Phys. A463 (1987) 710.

CONCLUDING REMARKS

Concluding Remarks

R.A. Broglia
Dipartimento di Fisica, Università di Milano, and INFN Sez. Milano, Milano, Italy, and
The Niels Bohr Institute, University of Copenhagen, Copenhagen, Denmark

The study of sub-barrier fusion reactions shares with electron assisted tunneling and SQUID, among others, the realm of tunneling of quantal system with many degrees of freedom. While the conditions for using concepts like friction and thermodynamic fluctuations are marginally satisfied, the different phenomena can be understood in terms of selected quantal states. This unique feature of sub-barrier fusion reactions was obvious already from the first talk presented by Landowne, where full coupled channel studies were presented , and the long awaited possibility of carrying out a simultaneous analysis of elastic, transfer, and inelastic data was exploited to the present state of the art.

Such a treatment aims at a full description of tunneling processes, taking into account all the degrees of freedom of the interacting ions, thus dispensing of an absorptive potential. The reason why the present results may not be able to completly accomplish this goal can probably be traced back to channels not included in the present calculations, e.g. channels where heavy particles are exchanged, which in turn may lead to short range, sharp diffusivity absorptions. Similar situations have already been encountered in the analysis of the backward rise in the elastic scattering of ^{16}O on ^{28}Si. In this sense, although not only, is the work presented by Udagawa well taken.

Non-linear coupling terms leading to inelastic processes are found to play an important role in some coupled channel calculations. To better asses the consequences of these results, it is likely that one has to await for a consistent treatment of the associated anharmonicities of the nuclear spectrum. They may lead to a breaking of the two-phonon strength with the possible consequence that multiple excitation of collective modes becomes less prolific.

The power of spin distributions in testing the models was demonstrated, among others, the case of the reaction where ^{16}O impinges on ^{232}Th. The discrepancy between theory and experiment found at 90 MeV remains a challenge, let alone the low energy behaviour. One may speculate to which extent the underestimation of the spin distribution and the overestimation of the fusion cross section are not partially connected with a somewhat poor choice of the ion-ion potential.

The role of macroscopic variables were emphasized by Wohlter and Fröbrich. Naturally, coherent modifications of the surface are also produced by the excitation of collective states, which may lead to neck-like configurations. However, there are limits to how many channels can be included in the calculations. Furthermore, the friction arising from non-collective degrees of freedom, like uncorrelated particle-hole excitations will be very difficult to treat in detail. Most important, statistical fluctuations, as required by the fluctuation-dissipation relation are likely to be important, as indicated by the surface friction model. On the other hand, quantal fluctuations are also needed in a consistent description. To make a parallel: in the study of the damping of giant resonances in excited nuclei, both quantal and thermodynamic fluctuations are essential to account for the observed changes in the damping widths. In particular, because the two contributions behave very differently as a function of the excitation energy.

In any case, judgement has to be excersized in comparing the results of the friction model, and those of the coupled-channel calculation. They carry some amount of complementary information but they are essentially very different: one is dominated by quantum tunneling, while in the other, statistical fluctuations allows the system to go over the barrier in a classical fashion. Of course, prescriptions can be worked out to extend the classical trajectory into the forbidden region. The problem becomes formidable, however, if the actual manifold of trajectories has to be considered.

A fully microscopic frictional model was presented by Rapisarda. It shares with that of Fröbrich, the classical trajectory problem below the barrier. On the other hand, both the transfer and inelastic degrees of freedom are treated in this model, microscopically. This is done making use of a shell- model basis, the starting point of essentially all nuclear structure calculations. In this model, as well as in the coupled-channel model, the calculation of the cross section associated with any particular channel is well defined, a possibility that goes beyond the macroscopic models. The model discussed by Rapisarda also contains the diabatic limit, a limit invoqued by Reissdorf to discuss the reduction in the sub-barrier fusion cross section in the shoulder region found in reactions like $^{90}Zr + ^{90}Zr$, as compared with potential predictions.

Concurrent with these observations, a very sharp increase of the transfer channel was found, as reported by Bellwied. This effect, is likely to be the clearest proof that transfer reactions are a very important doorway channel in the sub-barrier fusion process. Increasing the product of the charges of the interacting ions, forces the transfer processes to give rise to real transitions instead of virtual. This is similar to what happens in the direct neutron decay of giant resonances. Decreasing the nuclear mass forces the virtual particle-hole excitation into a real transition which populates levels of the nucleus with one less particle, interrupting the evolution of the vibration into itself or into the compound nucleus.

The far distant contact between very heavy systems through particle tunneling shown by Werth was a beautiful example of quantal processes, and a contribution to a very exciting field of research.

The power of the recoil mass spectrometer (RME) as a tool to study transfer as well as fusion processes, was pointed out by Betts and Lee, and we expect to learn much physics using such devices, in particular from the one due to operate in Legnaro.

In his report, Betts showed that the distance of closest approach behaviour of one-particle transfer process is the one expected from optimum Q-value arguments, but only in the case of reactions on speherical target nuclei. In fact, it goes wrong for deformed systems. The reason for this behaviour is not clear, and once found will shed light on the sub-barrier transfer mechanism. Nonetheless, one may speculate that the deformed nucleus is excited into a variety of states in the approaching phase, and transfer takes place from them, thus changing the tail behaviour of the form factors. An elegant attempt at treating some of these questions was discussed by Lotti.

Simultaneous studies of fission, elastic, inelastic and transfer processes were presented by Rehm, who demonstrated that one- particle transfer processes among heavy ions are well understood, the situation being quite different in the case of two-particle transfer reactions. It is likely that this last conclusion is associated to the use, in the analysis of the data, of less than state of the art reaction theory and spectroscopic information. At low bombarding energies, the transfer process is essentially dominated by the successive transfer of two nucleons which, to get correlated, need the phase space of large single-particle configurations.

Studies similar to those discussed by Rehm were presented by Corradi, results of a Garchin-Legnaro collaboration. From data of such quality will come the decisive tests to the current ideas on sub-barrier fusion reactions.

Very important progress has taken place in the study of the polarization corrections to the ion-ion potential. This was made clear by the talks presented by Stefanini and Lilley, which conveyed the experimental situation, and by Nagarajan and Pollarolo, as well as by Andrés who discussed the theoretical developments.

The consistency between the microscopically calculated absorptive and polarization potentials reflects the role played by transfer processes, shedding light on the energy dependence of the two-particle transfer process. In spite of the many successes, much work still remains to be done in this field. In particular, no microscopic calculation has ever produced the sharp change in the absorptive potential around the Coulomb barrier needed to give rise to a peak in the polarization potential.

A similar situation is known from the case of the single-particle motion in nuclei, where the effective mass and the mean free path arising from the coupling of the nucleons to surface vibrations, are connected with the real and imaginary part of Dyson's operator, related by standard dispersion relations. Both positive and negative (Pauli principle) contributions to the effective mass are obtained, in much the same way as the coupling to positive Q-value channels tends to reduce the real part of polarization potential. Fittingly, the calculated values of the effective mass fall short of the value required by the observed level density around the Fermi energy. It is an open question how to force the imaginary single-particle potential to change more rapidly and to produce larger values of the energy-dependent effective mass.

A delicate probe of the reaction mechanisms which controls the fusion process, is provided by polarization measurements, as illustrated by Fick. While sub-barrier fusion cross section can be well reproduced invoquing a variety of degrees of freedom, the tensor analysing power responds very selectively to the detailed nuclear structure input. It is tempting to connect this result with the predictions presented by Fernandez Nielo, on the dependence of sub-barrier fusion cross-section on hexadecapole deformations. It is likely that an even clearer fingerprint of the effect discussed by him will emerge from eventual polarization studies.

An elegant attempt at tackling the full complexity of coupled situations was presented by Iachello. He discussed the general properties to be fulfilled by the S-matrix describing the heavy ion collision process, with the help of dynamical symmetries. While the ability of the model to fit the data with few parameters is clear, it is less transparent at this early stage how to relate the parameters to the motion of the nucleons and their interaction. Another question which deserves attention is to which extent the S-matrix associated with reactions between complex nuclei of interwearing degrees of freedom can be treated in terms of analytic functions.

Although many successes can be claimed during the last years in the study of sub-barrier fusion reactions, a variety of open problems still remain. In particular, as emphasized by Vandebosch, no final understanding of the spin distribution exists as yet. He also called the attention to the possible analogy which may exist between the mechanism involved in sub-barrier fusion and exotic decay. In particular concerning the inertial mass. This point is very well taken and will most likely affect future studies.

While it is true that the meeting has not offered any big surprises, it was an outstanding forum where data of the highest quality and versatility were presented, and theoretical analysis was found to match, although not quite, the experimental findings. The steady growth of activity and insight which has taken place during the last years has made sub-barrier fusion reactions a well established field of research, and this meeting proved it quite clearly. The organizers deserve much credit for the success of the Symposium, and are to be warmly congratulated.

On the other hand, the only way to proceed from here is to identify new exciting problems. I have no doubt that one frontier to be opened is that of the extreme sub-barrier fusion processes, where the adiabatic picture should essentially be correct. In the process of running exotic decay backwards, one may learn about the energy intervals in which the slope of the cross sections change, thus reflecting the presence of specific channels. Paying attention to details, we may reconstruct the history of the process, which eventually may be of interest in connection with astrophysical studies. In this context it is timely to remember that important tests on unified gauge theories provided by double β-decay seem to crucially depend on a delicate nuclear structure interplay, namely the interweaving between the particle-particle and particle-hole interaction between protons and neutrons. To bring sub-barrier fusion reactions to a similar level of accuracy may be penalizing, but in all likelehood it is a very exciting task.

Another frontier to lie within reach of present day techniques is the one recently opened by studies of deep inelastic reactions, where rapid oscillations in the excitation function have been observed. It is unlikely, for example, that the spin distribution shows a smooth decreasing behaviour around the limiting angular momentum,much in the same way as the sharp triangular behaviour was too simplistic an approximation.

At high bombarding energies, the fusion cross section saturates. Considerable freedom still exists in the selection of the ion-ion potential needed to fit the data. This uncertainty reflects itself with vengeance at the low bombarding energies. Polarization measurements of elastic scattering, which can switch the absorption on and off may prove valuable in trying to solve this question.

In summary, there were intense and enjoyable long working days at Legnaro, which provided much information and shed light on a variety of topics at the forefront of heavy ion research. It is in place to thank the speakers, chairmans and organizers, and in particular the attendees, who actively participated in the discussions, for their contribution to the Symposium.

RECOIL MASS SPECTROMETERS

The Design of Recoil Separators

H. Wollnik

2. Physikalisches Institut der
Justus–Liebig–Universität Giessen
6300 Giessen, W–Germany

ABSTRACT

For the separation of energetic recoil ions several arrangements of sector field mass analyzers are discussed. Included are laterally mass–dispersive systems consisting of magnetic and electrostatic sector fields alone or systems consisting of purely magnetic sector fields and quadrupoles employing, however, an energy loss foil at an intermediate image position and finally longitudinally dispersive systems that again use magnetic fields only.

1 Introduction

To separate energetic recoil ions of 1 MeV/amu or more, large and strong electromagnetic fields are required. Thus beam lengths of 10 m or 20 m are common for recoil separators and there are even larger systems. All these separators must achieve at least mass resolving powers $m/\Delta m$, i.e. about 500 or 1000, despite the fact that the recoil energies vary drastically. There are four ways how this has been achieved for multiply charged recoil ions:

1. Laterally dispersive mass separators for medium energy ions where the separators use magnetic as well as electrostatic sector fields. Such systems may be arranged in three versions:
   (a) an angle– and energy–focusing system with both fields having the same plane of deflection (double–focusing mass analyser[1–6])
   (b) an angle– and charge–focusing system with both fields having the same plane of deflection (angle– and charge–focusing velocity filter[7–10])
   (c) an angle–focusing system with perpendicular directions of energy and momentum dispersions (angle–focusing "parabola" analyzer[11–14])
2. Time–of–flight mass separators that use magnetic sector fields, where ions of all energies and masses are focused laterally and the ion flight time is independent of the ions' energies, their start points or their initial angles of inclination[15–19].
3. Gas–filled magnetic separators[14–16] which analyze the ions proportional to mv/q where due to collisions with residual gas atoms the ion charge q on average is proportional to the ion's velocity v so that the system achieves a pure mass separation[20–22].
4. Laterally dispersive mass separators for fully stripped energetic ions. Such separators use a double magnetic analysis of the ions' momenta, where the ions loose a part of their kinetic energy in an intermediate foil but where ions of slightly different energies are still focused to one point as long as they have the same masses and nuclear charges[23,24]. Also such systems may be arranged in three versions:

(a) a system with two magnetic analyzing stages where the energy loss foil has a wedge shape and where the ion formation limits the initial energy spread considerably[23,24]

(b) a system with two magnetic analyzer stages where the energy loss foil has a wedge shape and where the initial energy spread of the ions under investigation is limited by an electrostatic sector field[25]. As a small improvement I would propose to arrange this electrostatic sector field perpendicularly to the plane of deflection of the two magnetic stages so that ions of initially different electrostatic rigidities are finally focused to neighbouring mass spectra. Preferably this electrostatic sector field should follow the magnetic stages.

(c) As a third system, I propose here a system with two magnetic analyzer stages that are arranged perpendicularly to each other, where the intermediate energy loss foil has a constant thickness and where additionally to the measurement of the final ion's position its final velocity must be determined.

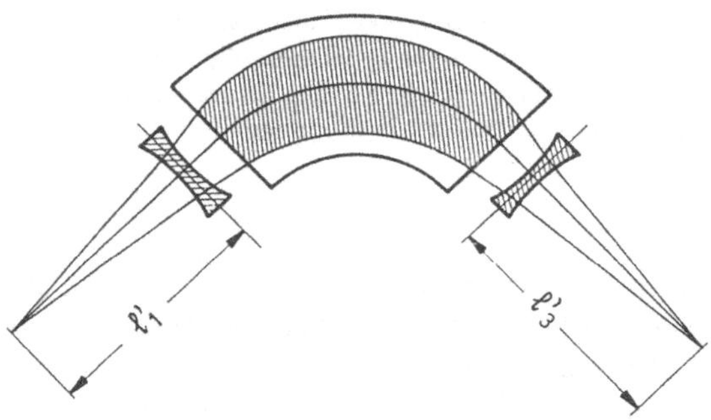

Figure 1: A sector field analyzer preceded and followed by a quadrupole lens. Note that the second quadrupole varies both the magnification and the dispersion such that the resolving power stys constant. Note also that the first quadrupole varies the resolving power but leaves the dispersion unchanged.

2 Medium Energy Recoil Separators

All mass separators consist of magnetic (and electrostatic) sector fields as well as of quadrupole lenses. If such quadrupole lenses follow a sector field they change its mass dispersion D_m but leave its mass resolving power R_m unchanged. If such quadrupole lenses precede the sector field on the other hand they change its mass resolving power R_m but leave its dispersion D_m unchanged[8,26]. More generally one can define a Q–value[8,26]

$$Q_m \quad = \quad R_m(2x_{00}2a_{00}) = R_m[2x_{00}2(p_{x0}/p_{00})]$$

which is the product of the achievable mass–to–charge resolving power R_m and the lateral phase–space area (see Figure 1). For systems consisting of a multitude of sector fields and quadrupole lenses one finds a quality factor

$$\begin{aligned} Q_m = R_m(2x_{00}2a_{00}) &= \left(\frac{i}{1+\eta}\right)\sum_{i=1}^{i}\left(\frac{A_{B\alpha i}}{2\rho_{B0i}} + \frac{\eta A_{E\alpha i}}{(1+2\eta)\rho_{E0i}}\right) \\ &\approx \sum_{i=1}^{i}\frac{A_{B\alpha i}}{2} \,. \end{aligned} \tag{1}$$

The approximate sign here corresponds to relativistically slow ions while $2\eta = 1/\sqrt{1-(v/c)^2}-1 \approx (K/m)$ 1931.5 with K in MeV and m in amu. Here $A_{B\alpha i}/\rho_{B0i}$ and $A_{E\alpha i}/\rho_{E0i}$ characterize for the i_{th} sector field the ratio between $A_{\alpha i}$, the area enclosed by the most inclined particle trajectories from the center of the particle source and ρ_{01} the corresponding radius of curvature of the optic axis, where the indices B and E characterize magnetic and electrostatic sector fields, respectively. Note here that the quadrupole lenses in the system modify the quality factor Q_m only in so far as the width of the ion beam increases or reduces the $A_{\alpha i}$ values in the i_{th} sector field. Note here further that with increasing size of such a system of fixed shape

- the Q_m–value increases linearly with the system size since the $A_{\alpha i}$ incrase quadratically and the ρ_{0i} linearly with this size;
- the electrode potential $\pm V_0$ in an electrostatic sector field is independent of the size of the system since with increasing ρ_{0i} the field strength E_0 decreases linearly, while the electrode separation $2G_0$ increases linearly;
- the required ampere turns in a magnetic sector field are independent of the system size since with increasing ρ_{0i} the flux density B_0 decreases linearly, while the magnet air gap $2G_0$ increases linearly. For this reason also the product of the used power for the magnet and the cross section of the magnet coil only increases linearly with the system size.

Usually thus the costs of a recoil separator increase only less than linearly with the size of the system while the performance, i.e. the mass resolving power times the transmission, improves quadratically. As example note that with increasing size the source area of a recoil separator increases quadratically while its mass resolving power stays constant.

2.1 Mass Separators Consisting of Magnetic and Electrostatic Fields

Analogously to the mass–to–charge Q_m–value one can define an energy–to–charge Q_K–value[26] or a charge Q_q–value[8] with:

$$\begin{aligned} Q_K = R_K(2x_{00}2a_{00}) &= \left(\frac{1}{1+\eta}\right)\sum_{i=1}^{j}\left[\frac{A_{B\alpha i}}{2\rho_{B0i}}(1+2\eta)+\frac{A_{E\alpha i}}{\rho_{E0i}}(1+\frac{2\eta^2}{1+2\eta})\right] \\ &\approx \sum_{i=1}^{j}\left[\frac{A_{B\alpha i}}{2\rho_{B0i}}+\frac{A_{E\alpha i}}{\rho_{E0i}}\right] \end{aligned} \tag{2a}$$

$$Q_q = R_q(2x_{00}2a_{00}) = \sum_{i=1}^{j}\left[\frac{A_{B\alpha i}}{\rho_{B0i}}+\frac{A_{E\alpha i}}{\rho_{E0i}}\right] \tag{2b}$$

where again the approximate sign corresponds to relativistically slow ions.

2.1.1 Energy or Charge Refocusing Recoil Separators

If the Q_K–value vanishes, the energy–to–charge resolving power vanishes, i.e. ions of different energy–to–charge ratios are focused to the same point as long as their mass–to–charge ratios are identical. In other words the double–focusing condition of Refs. 2–6 is fulfilled. For such a two–sector–field system (see Figure 2) thus the used areas in the two–sector–fields have ratios of 1 to 2 if both sector fields have the same radii of deflection ($\rho_{B0} = \rho_{E0}$). This factor of 2 is correct for relativistically slow ions and it decreases with increasing ion energy.

If the Q_q–value vanishes the charge resolving power vanishes, i.e. ions of different charges are

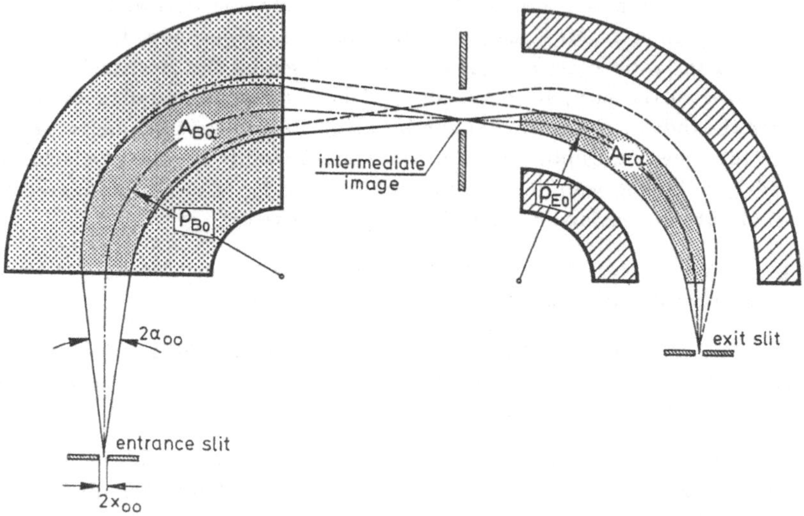

Figure 2: An angle– and energy–focusing mass spectrometer. Note that in this case the ratio of the areas in the two sector fields, divided by the corresponding radii of the optic axis, must have the value of 2.

focused to the same point as long as their masses and energies are identical. In other words the condition for an angle–focusing velocity filter of Refs. 7–10 is fulfilled. For such a two–sector–field system thus the used areas in the two–sector–fields are equal if both sector fields have the same radii of deflection ($\rho_{B0} = \rho_{E0}$).

A two–sector–field system, that uses additional quadrupole lenses, thus for different quadrupole settings can fulfill either one of the conditions (2a) or (2b) and thus can either act as an angle– and energy–refocusing (double–focusing mass spectrometer) or as an angle– and charge–refocusing (angle–focusing velocity filter) device.

2.1.2 Recoil Separators with Momentum– and Energy–Dispersions

Arranging magnetic and electrostatic sector fields perpendicularly to each other, the momentum–to–charge dispersion is perpendicular to the energy–to–charge dispersion and ions of equal mass–to–charge ratios are focused to a line where the particle energy determines the ion's position along this line[11–14]. In this case the two–sector–fields focus the ions in perpendicular directions like crossed cylinder lenses so that a stigmatic image can be obtained without any other lenses. In such systems ions of quite different energies can be analyzed since differently than in the cases of section 2.1.1, aberrations due to a large energy spread do not directly reduce the mass–resolving power since the corresponding dispersion is perpendicular to the mass dispersion.

2.1.3 Problems with High Electrostatic Fields

The mass separators as discussed in section 2.1 in principle could mass analyze all possibly formed recoil ions. However, the corresponding electrostatic sector fields require voltages which increase drastically with increasing ion energies. This is especially detrimental since technologically it is rather difficult to keep high voltages on sector electrodes without creating sparks between the electrodes or between one of the electrodes and the vacuum tank. The problem seems to be[27]

1. to avoid electron emission from some spot of the electrode surface, i.e. avoiding spots of locally low work function. This postulates a very clean vacuum where the vaccum should especially be free of even traces of organic material;

2. to avoid the creation of craters due to a once struck arc, i.e. using electrode and vacuum tank materials that have high melting points and —though this seems to be of lesser importance— good heat conductivity;

3. to use high voltage vacuum feedthrough that effectively avoid sparks along its insulating surfaces or use high voltage generators inside the vacuum housings[28].

Note here that the limiting quantity for high voltage performance in vacuum is the potential difference $2V_0$ between the two electrodes multiplied with the electrode separation $2G_0$. The best value reported so far[14] is to my knowledge $2V_0 2G_0 < 200\ 000\ Vm$ under operating conditions. This value was achieved with stainless steel electrodes and turbomolecular pumps.

2.2 Time–of–Flight Mass Separators

The advantage of recoil separators as described in Section 2.1 is that one can build recoil separators of low and medium Q_m–values in a cost saving manner by reducing the angles of deflection in the different sector field. Their disadvantage is that high electrostatic field are always technically difficult. An alternative is to use a laterally achromatic focusing system consisting of magnetic devices only but to design this system such that faster ions are sent on a detour which is dimensioned properly[15,17] to make the ion flight time independent of the velocity variation so that the system becomes an energy–isochromat. Such systems have a high transmission as all strong focusing devices and at the same time a good mass resolving power[18,19]. A certain disadvantage is that these systems cannot easily be built in low cost versions and that they cannot achieve a physical separation unless one uses very fast pulsed ion deflectors.

Energy isochronous time–of–flight mass separators can also advantageously be built as a ring (see Figure 3) so that the ions can be passed around the ring repeatedly and the total ion flight time is increased allowing even higher mass resolving powers at equal transmission[26,29].

In case there already exists a ring structure, for instance, a storage ring designed for a variety of experimental investigations one often can tune the quadrupole strengths in the ring such that the ring becomes energy isochronous[30], i.e. it is operated at the transition energy, where, however, here one additionally asks for a good lateral, achromatic focus at the detector position, in order that a small ion detector can be used which is essential for good timing properties.

3 Gas–Filled Magnetic Separators that make use of Charge-exchange Processes

In collisions with residual gas atomes fast ions usually loose all electrons with orbital velocities smaller than the ion velocity v. Thus the ion charge q becomes proportional to v/Z^a where Z is the proton number of the ion under investigation. The magnetic rigidity of such an ion thus becomes

$$B\rho = \frac{mv}{q} \propto \frac{m}{Z^a} \, .$$

Filling the vacuum chamber of a magnetic sector field analyzer with a residual gas of about 1 *mbar* one thus has a m/Z^a–separator. In such a system ions of larger $Z = Z_0(1 + \delta_Z)$ have a smaller magnetic rigidity and thus are found at the same position in the image plance as ions of smaller $m = m_0(1 - \delta_m)$ and proton number Z_0. Looking at a nuclear chart one consequently cannot expect an isobaric separation ($N + Z = const.$) but rather one ($N + bZ = const.$, with $b \ll 1$) that corresponds to lines between isobars and isotones[21]. Note here that gas–filled seprators have all only modest mass resolving powers, since the angle end energy straggling in the collision events cause the final image to be blurred. However, such systems can achive very low cross contaminations and thus seem well suited to separate ions of drastically different ion masses m and proton numbers Z as, for instance, is the case when heavy ion reaction products must be separated from the primary beam[22].

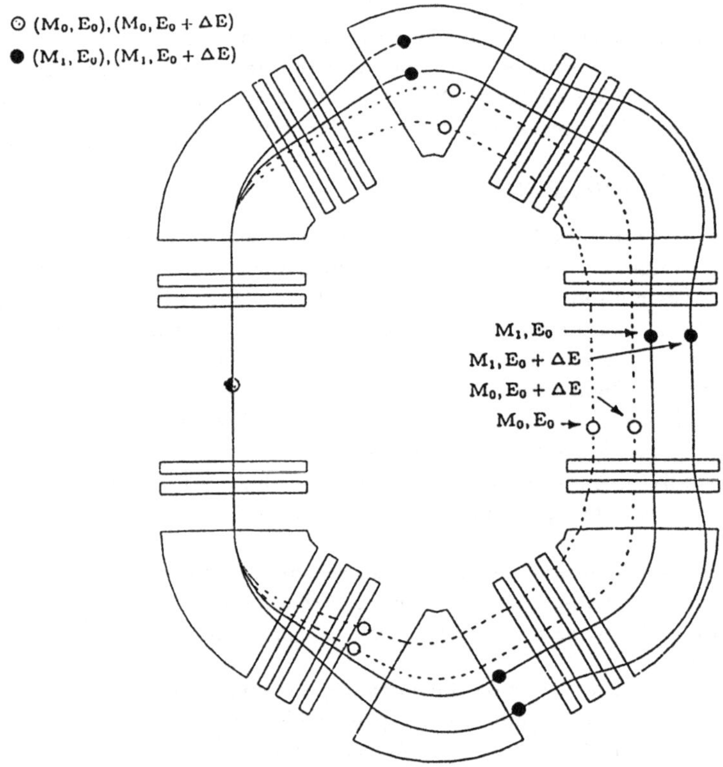

Figure 3: An energy–isochronous time–of–flight mass analyzer as achieved by retuning the quadrupole strengths in the ESR storage ring of the GSI in Darmstadt, W–Germany. Note that low and high energy ions of low (black) and higher (white) masses are indicated, illustrating the energy–isochronous behaviour of the time-of-flight mass analyzer system.

4 Recoil Separators that use Intermediate Energy Absorber Foils

High energy ions are fully stripped when passing through matter. Thus the mass–to–charge spectra of recoil separators described in section 2, which actually produce superimposed somewhat shifted mass spectra for the different ion charges, become much simpler. However, the increased energy makes it very difficult to use electrostatic fields. One idea to mass analyze such recoil ions and still refocus ions of slightly different energies to one point, is to perform a double momentum analysis by two magnetic dipole analyzers where the second stage compensates the momentum dispersion of the first stage yielding an overall achromatically focusing device. Placing now at the intermediate image an energy loss foil in which all ions loose a certain amount of their kinetic energy the reference ions can be focused to their original focus-point again, if the magnetic fields in the second stage are reduced accordingly. If the energy loss foil would be shaped like a wedge[23] the more energetic ions would loose more energy so that their relative energy deviations from a reference ion can be made to be the same before and after the energy loss foil. Thus, as in the original achromat also ions of slightly different initial energies are focused to the same point again. Since the energy loss depends on the proton number Z as well as on the ion mass $N+Z$ the system separates one isotope of a specific neutron N and proton Z number from all others and focuses them independent of their initial energies to the same spot.

Assuming that the ions to be investigated initially have all about the same velocity, the first magnetic stage allows only ions to pass for which $N/Z = constant$. The second stage in principle does the same, however, the energy loss in the absorber foil, that is mainly proportional to Z, rotates the corresponding line on the nuclear chart to one for which $N = const. + bZ$ with $b \ll 1$, so that it approximates an isotone $N = const.$ Optimally, the crossing point $N/Z = const.$ and $N = const. + bZ$ then selects one isotope only.

4.1 Recoil Separators with Double Magnetic and Single Electrostatic Sector Analyzers

If the ions initially have different velocities a recoil separator according to Refs. 23,24 focuses nuclei of more or less one isotone to one point. However, if somewhere in the system an electrostatic sector field could be included[25] a certain energy selection can be achieved. If the dispersion of this electrostatic field would be perpendicular to the momentum dispersion of the two magnetic stages different spectra of systems described in Refs. 23,24 would be formed besides each other where each spectrum would be characterized by a different initial ion energy. If the electrostatic sector field would be arranged behind the second magnetic stages the magnetic stages only need air gaps as in any system of Refs. 23,24 and the electrodes in the electrostatic analyzer only need to be separated enough to allow the line height of the magnetic mass spectra to pass.

4.2 Recoil Separators using Perpendicularly Arranged Magnetic Analyzers

An alternate arrangement —which to my knowledge has not yet been described— would be to arrange two magnetic stages such that their dispersions are perpendicular to each other and avoid the use of an electrostatic field analyzer. In this case a planar energy absorber foil arranged between the two systems would already purify one specific isotope since the dispersion in one plane would be proportional to $(N/Z)v_0 = const.$ and in the other direction to $(N - bZ)v_f = const.$ where v_0 characterizes the initial and v_f the final velocity of a specific ion. For ions of one initial velocity $v_0 = v_{00} + \Delta v$ one would find approximately $v_f = v_{f0} + \Delta v$ so that ions of the same N and Z but different Δv would be found on different points of a diagonal.

Determining in such a system the final lateral xy-position as well as the final ion velocity v_f one can within limits assign specific N and Z values to each ion also if the initial ion velocities v_0 differed considerably. However, this requires that the intensities behind such a recoil separator stay so low that individual ions can be recorded. Since the ions in such a recoil separator are very energetic their positions and velocities can very well be determined by two position sensitive detectors (for instance, parallel plate detectors). Since the total beam detector usually will be large it may also be advantageous to focus the ions behind a first position sensitive detector by quadrupole lenses to a smaller time detector, possibly followed by other detectors for decay products.

5 Image Aberrations of Recoil Separators

For all the recoil separators discussed above rather large aperture angles are desirable as well as rather wide initial energy spreads. Thus the calculations of image aberrations is of high importance. Such calculations advantageously are performed with the spectrometer design program GIOS[31,32] were it is important to note that because of the condition of symplecticity, i.e. the existence of the canonical equations of motion, there are certain relations between the different image aberrations, i.e. between the different elements of a transfer matrix[26,33].

For such calculations it is important do describe the expected field distributions as precisely as possible. Such field distributions are usually determined from systematic and if necessary by iterative finite element methods following assumptions about the geometry of pole faces or electrodes. Because of numerical inaccuracies such calculations produce maps of field values which all deviate slightly, for instance, by a few 10^{-4}, from the field distribution one would expect in a real experiment. For numerical ray tracing caluclations this has two detrimental effects

1. the field values involved violate Maxwell's equations slightly (a few 10^{-4});
2. the assumed pole faces or electrodes would produce slightly different fields (a few 10^{-4}).

The first of these effects is usually dominating since many effects on the ion trajectories noticeably depend on the existence of a fringing field but vary only slightly[26,34] when the geometry of the pole faces or electrodes is varied as is the extension and distribution of the fringing field. For all numerical trajectory calculations it thus is of high importance that the corresponding fields globally as well as locally fulfill Maxwell's equations as precisely as possible.

For the field distribution in main field regions in which they field distribution is calculated algebraically from a general theory Maxwells's equations are always easily fulfilled. In case of numerically calculated field distributions this postulate is not so simple. Having determined some table of approximate values for a potential distribution one thus should make certain that in case of a Cartesian grid the relation $\Delta V = \partial^2 V/\partial x^2 + \partial^2 V/\partial y^2 + \partial^2 V/\partial z^2 = 0$ is fulfilled as exactly as possible.

Apprixımating a smooth potential distribution locally by a polynomial one thus must postulate:

$$V(x,y,z) = \sum_{i,j,k=0}^{\infty} a_{ijk} \frac{x^i y^j z^k}{i!j!k!}$$

$$a_{i+2,j,k} + a_{i,j+2,k} + a_{i,j,k+2} = 0$$

which describes all $a_{i,j,k}$ except three for each order $\kappa = i + j + k$. Analogous expressions[13,26,37] should be used for interpolating in rotationally symmetric systems.

When calculating particle spectrometers and separators by modern ion optic programs[31,32] one usually can predict all properties of systems even of complex geometry quite satisfactorily[36]. Thus for most planned recoil separators a numerical optimization is most recommendable since the calculated properties actually can be achieved in reality. In most cases one even can avoid lengthy adjustment procedures as long as the system was built with sufficient precision.

6 References

1. F. W. Aston; Phil. Mag. **38** (1919) 709
2. W. Bartky and A. J. Dempster; Phys. Rev. **33** (1929) 1019
3. J. Mattauch and R. Herzog; Z. Phys. **89** (1934) 786
4. H. Ewald, E. Konecny, H. Opower and H. Rösler; Z. Naturforschg. **19a** (1964) 194,200
5. K. A. Arifov, H. D. Belayev and V. I. Kogan; Dokl. Akad. Nauk SSSR **204** (1972) 586
6. T. M. Cormier and P. M. Stwertka; Nucl. Instrum. and Meth. **184** (1981) 423
7. W. Wien; Wied. Ann. **65** (1898) 449; Ann. Phys. **8** (1902) 260
8. H. Wollnik; Nucl. Instrum. and Meth. **95** (1971) 453
9. H. Wollnik, G. Münzenberg and H. Ewald; Nucl. Instrum. and Meth. **111** (1973) 355
10. G. Münzenberg, W. Faust, S. Hofmann and P. Armbruster; Nucl. Instrum. and Meth. **161** (1979) 65
11. J. J. Thomson; Philos. Mag. **20** (1910) 752; **21** (1911) 225
12. S. Neumann and H. Ewald; Z. Phys. **169** (1962) 224
13. H. Wollnik; Thesis, TU Munich (1964) unpublished
14. E. Moll et al.; Proc. EMIS–7 (1970) 241, ed. H. Wagner and W. Walcher; Nucl. Instrum. and Meth. **123** (1975) 615
15. W. Poschenrieder; Int. J. Mass Spectr. Ion Phys. **6** (1971) 413
16. H. Wollnik and T. Matsuo; Int. J. Mass Spectr. Ion Phys. **37** (1981) 209
17. H. Wollnik; Nucl. Instrum. and Meth. **186** (1981) 441
18. J. M. Wouters et al.; Nucl. Instrum. and Meth. **A240** (1985) 77
19. H. Wollnik, J. M. Wouters and D. J. Vieira; Nucl. Instrum. and Meth. **A258** (1987) 331
20. C. B. Fulmer and B. L. Cohen; Phys. Rev. **109** (1958) 94
21. H. Lawin et al.; Proc. EMIS–7 (1970) 270, ed. H. Wagner and W. Walcher
22. H. Ghiorso; private communication (1984)
23. J. P. Dufour et al.; Nucl. Instrum. and Meth. **A248** (1986) 267
24. H. Geissel et al.; Proc. Acc. Rad. Beams Workshop, Vancouver (1985); Proc. 8th High Energy Heavy Ion Study, LBL–Rep–24580 (1987) 460; GSI–proposal (1987)
25. W. Mittig; private communication (1988)
26. H. Wollnik; *"Optics of Charged Particles"* (1987) Acad. Press, Orlando
27. G. F. Steib and E. Moll; J. Phys. D: Appl. Phys. **6** (1973) 243
28. H. Enge; private communication (1978)
29. H. Wollnik; Nucl. Instrum. and Meth. **A258** (1987) 289
30. H. Wollnik; Nucl. Instrum. and Meth. **B26** (1987) 267
31. H. Wollnik, J. Brezina and M. Berz; Nucl. Instrum. and Meth. **A258** (1987) 408
32. H. Wollnik, B. Hartmann and M. Berz; AIP Conf. Proc. (1988), in print
33. H. Wollnik and M. Berz; Nucl. Instrum. and Meth. **238** (1985) 127
34. H. Wollnik; Nucl. Instrum. and Meth. **38** (1965) 56
35. H. Wollnik, T. Matsuo and H. Matsuda; Nucl. Instrum. and Meth. **102** (1972) 13
36. C. Geisse; Thesis (1987) Universität Giessen, unpublished

GENERAL PROPERTIES OF RECOIL MASS SPECTROMETERS

J. D. Larson
10011 E. 35th Street Terrace
Independence, MO 64052, USA

Abstract:

The recoil mass spectrometer designed by Cormier at the University of Rochester has resulted in a new generation of instruments. Efforts to improve the design have increased the range of acceptance for solid angle, energy and mass but limits to effective resolving power seem to have been reached. The author offers opinions on realistic choices for building a good spectrometer.

1. Constructing a recoil mass spectrometer

A symmetric system is easiest to understand. Begin with any dipole, i.e. any device capable of uniformly bending a beam of moving charged particles. In a dipole particles are dispersed on the basis of (p/q) in magnetic fields, (E/q) in electric fields, (v/q) in crossed-field (ExB) Wien velocity selectors and (m/q) in double-crossed (Ex2B) mass selectors, where p, E, v, m and q represent particle momentum, energy, velocity, mass and charge, respectively. (The Ex2B example turns out to be less than the perfect choice for various technical reasons; consequently, it will be treated here more as a singularity than as a solution.) Next, let a mirror-symmetric pair of the same dipole types face each other so that their dispersion patterns fan out towards each other and cross at the symmetric center. Finally, join the fans into connected trajectories, to first order, using a lens centrally located at the plane of mirror symmetry, as shown schematically in Fig. 1a. This method probably was used first by Isaac Newton who dispersed white light into colors and then merged the color spectrum back to white using two glass prisms with a lens in between. Today such "achromatic" systems are used routinely for beam transport.

If we replace the center lens in Fig. 1a (all or in part) with a dipole element that continues to smoothly connect the various dispersion trajectories, then the system retains dispersion cancellation in the selected property (p, E or v). However, if the central dipole element (that replaces the lens) is dispersive in some other property, then particles of different mass (for a given p, E or v) will necessarily take different trajectories through the central dipole and become separated in proportion to m/q as shown in Fig. 1b.

Particles having different m/q values will appear to diverge in angle from a central pivot point. How this angular dispersion transforms into the final focal plane depends on the nature of the focal system downstream from the source of dispersion which, in turn, depends on how the beam of particles is delivered to the central region by upstream lenses, including focusing effects by the dipoles. If the incoming beam of particles is focused to a crossover point at the center, then the downstream focus from this crossover to the final focal plane will necessarily be point-to-point. As a consequence, all m/q values will reconverge to a point focus in the focal plane and spatial dispersion will cancel out; however, dispersion in angle in proportion to m/q will persist as illustrated in Fig. 1c. If instead the beam envelope is made very large and parallel at the m/q pivot, then the downstream focus for these rays will be parallel-to-point in which case the focus from the m/q pivot to the focal plane will be more nearly point-to-parallel. This produces substantial spatial dispersion in proportion to m/q as illustrated in Fig. 1d.

Large spatial dispersion is desirable in a spectrometer; consequently, the situation shown in Fig. 1c ordinarily would be rejected. However, the arrangement of dipoles is the same in both Figs. 1c and 1d so it is not really the dispersion which has changed but only its manifestation in angular rather than spatial coordinates. Traditional implementations start from a fixed object point and end with dispersed point images in the focal plane. To process angular dispersion it is appropriate to start with an initially parallel beam (or measure incoming particle angles) and then measure the final angles produced by dispersion. This offers opportunities for reducing aberrations because if both position and angle are measured, it becomes practical to correct for instrumental aberrations and for finite phase space volume occupied by the beam.

A large, parallel beam envelope truly is required in the central region of the traditional spectrometer. At first glance, the quadrupole lenses shown in Fig. 2a appear to provide an ideal solution; a large range of angular acceptance is compressed into a small parallel beam passing through the dipoles. In fact, this converging-diverging/diverging-converging (CD-DC) lens arangement couples almost entirely into angular dispersion (the focus is almost point-to-point from center to ends) and very little spatial dispersion is realized.

There appear to be two basic alternatives for obtaining substantial m/q dispersion. The first is to reverse the quadrupole lens polarities to DC-CD as shown in Fig. 2b. This provides the required highly parallel beam conditions in the center and good m/q dispersion in the focal plane. However, the range of angular acceptance in the dispersion plane becomes severely restricted if the beam is to clear moderate size gaps in electric dipoles. The beam envelope sketched in Fig. 2b is appropriate for weak focusing dipoles. Stronger dipole focusing will converge the beam toward the axis. One way to accomodate this is to split the central dipole and pull it apart, allowing the beam to cross over in between as shown in Fig. 2c. Polarities in the second half are reversed, producing the mirror-antisymmetric LARA configuration [1].

The second alternative is to expand the beam in the central dipole as shown in Fig. 2d. For a symmetric beam envelope (solid curves) the spatial m/q dispersion is variable ranging from zero for a central crossover to large values when the beam is greatly expanded. As indicated by dashed lines in Fig. 2d, the final quadrupole lenses may be omitted as it is obvious that a focus will occur without lenses. In this non-symmetric LNL mode [2] the position of the focus and the magnification depend on front quadrupole settings; spatial dispersion is fixed by the dipoles and not variable. In both Figs. 2c and 2d, dipole focusing helps to turn the focal plane more nearly perpendicular to oncoming particles.

In symmetric systems the intrinsic spatial resolution depends only on dispersion because the system magnification is unity. Very generally, however, resolution is only a function of upstream lenses and dipoles. To understand this, consider prticles leaving the last dipole. Their trajectories project either forwrd or backward to some intermediate real or virtual focal plane in which the relationship of dispersion to magnification is already established. Downstream lenses do not change this, they only serve to reproduce the intermediate image at the final focal plane with possibly different orientation and magnification as illustrated in Fig. 3.

2. Aberrations

Every spectrometer comes equipped with focal aberrations. The following second-order aberrations are the main concern:

Aberration	Mnemonic
(X/AA)	"crescent"
(X/BB)	"out-of-plane"
(X/DD)	"walk"
(X/AD)	"wobble"
(X/AG)/(X/G)	"depth-of-focus"
(Y/BD)	"vertical chromatic"

[The notation is from program GIOS and means that the X (or Y) position at the focus is affected by variables listed after the divisor slash. Acceptance angles A and B are measured in the dispersion and perpendicular planes; D and G are fractional deviations of energy and mass from central values.] Ordinarily, (X/AA) and (X/BB) are negligibly small before any second-order corrections are applied. Energy-dependent (X/DD) and (X/AD) are large initially. If energies are measured at the detector then (X/DD) can be corrected later by data reduction; therefore, (X/AD) is the most troublesome aberration and it becomes important to seek a first-order layout that minimizes the consequences of (X/AD). The depth-of-focus aberration (X/AG)/(X/G) is related to the tilt angle of the mass focal plane. If this is large then (for a detector mounted perpendicular to the axis) the focus will degrade for masses displaced from the central value. The vertical chromatic aberration (Y/BD) usually dominates the vertical spot size; smaller (Y/BD) allows a smaller detector to be used.

Sextupole corrections to reduce (X/AD) and (X/DD) must be applied within the dipole region where energy dispersion is available. Both (X/AA) and (X/BB) grow in magnitude as the energy-dependent aberrations are reduced. Often the best compromise occurs when (X/DD) is near zero, (X/AD) is small but not yet zero and (X/AA) dominantes. A cut in angle A then reduces the principal aberration quadratically while reducing solid angle acceptance only linearly.

Some designs permit (X/DD, (X/AD) and (X/AG) to be reduced nearly to zero. A strong temptation arises to dispose of the remaining (X/AA) and (X/BB) aberrations by applying sextupole fields in the vicinity of the entrance quadrupole lenses. Second-order computer codes show that this works wonderfully well; however, a more complete analysis generally reveals that sextupole fields at the entrance couple strongly into higher order aberrations and after important second-order aberrations are eliminated the final image is no better and often worse. Applying only the minimum of second-order correction necessary to reach a compromise between competing aberrations works better and if particle position, angle and energy can all be measured at the detector, it is preferable to perform all corrections through data analysis and not in the hardware.

3. Some recoil mass spectrometer designs

One can argue that an electric dipole provides the best discrimination for separating beam particles from recoiling ions because these two types are likely to differ more in energy than in momentum. It follows that the University of Rochester arrangement [3] of electric dipole, magnetic dipole, electric dipole (EME) represents one of the best choices for filtering out beam particles. If a simpler quadrupole doublet lens pair is substituted at the entrance to replace Rochester's triplet lens then (after futher comparative analysis) the following five configurations are survivors:

	Hardware configuration	Operating mode	Figure
(1)	Qc Qd EME	LNL	4a
(2)	Qc Qd EME Qd Qc	LNL mode	4b
(3)	Qd Qc EME Qc Qd	LARA mode	5a
(4)	Qd Qc EME Qc Qd Qc	Mixed mode	5b
(5)	Qd Qc EMME Qc Qd	LARA	5c

Qc and Qd refer to converging and diverging quadrupole lenses. The goals in each case are large ranges of acceptance and good resolution. Base modes that provide maximum solid angle acceptance are indicated by dashed curves in Figs. 4a and 5a. Dipoles (especially magnets) contribute substantial focusing; otherwise, for weak dipoles Fig. 2b is the only realistic choice. (Actual layouts and beam envelopes may differ significantly from Figs. 4 and 5.)

Case 1 (Fig. 4a) is the Legnaro (LNL) design [2]. The strengths of this design are simplicity, large acceptances in energy, mass and solid angle, moderately good resolution and relative ease of second-order correction. The principal weakness is poor control of the vertical focus.

Case 2 (Fig. 4b), the "FMA" [1], is the LNL design made symmetric (or nearly symmetric, as desired). It retains or slightly improves upon most of the good features of case 1 and significantly improves the vertical focus (by reducing both vertical magnification and chromatic aberration). Case 2 has some difficulty transmitting the full mass dispersion through the restricted apertures of the final quadrupole lenses. The depth of focus is good and correction probably is not needed for the focal plane tilt.

Case 3 (Fig. 5a) has the entrance lens polarities reversed and is operated in "LARA mode" with a nearly parallel beam entering the central magnetic dipole which overfocuses to a crossover preceding the final doublet lens. This is not good for the focal plane tilt and some other aberrations become worse. Case 3 shares with case 5 the advantages of relatively large solid angle acceptance and excellent vertical focus control; however, the angular acceptance may be skewed five times larger in the transverse plane compared to the dispersion plane.

Case 4 (Fig. 5b) becomes the choice when doublet and triplet lens combinations are mixed. The front doublet is not operated in the point-to-parallel "LARA mode"; instead some solid angle is sacrificed for more dispersion and the beam expands toward the central dipole mimicing the "LNL mode" (but note the reversed doublet lens polarity). Case 4 holds something of a middle position with average to good properties throughout.

Case 5 (Fig. 5c) is the Munich LARA [1] arrangement which requires an additional magnetic dipole and about 50% more length. Each half of LARA is a complete mass spectrometer with moderate to good properties. The combination of back-to-back spectrometers doubles the intrinsic resolution and brings nearly all properties into the good-to-excellent category. Unfortunately, the aberration-limited effective resolution is not doubled but instead remains comparable to the other cases (at the same solid angle acceptance). Overall, LARA is superior as a recoil spectrometer and it opens up new instrumental opportunities at the central crossover.

COMPARISON OF FIVE REALISTIC RECOIL MASS SPECTROMETERS

	(1) LNL CD-EME	(2) FMA CD-EME-DC	(3) DC-EME-CD	(4) DC-EME-CDC	(5) LARA DC-EMME-CD
Solid angle	moderate	moderate	high	moderate	high
Eff. resolution	good	good +	moderate	good +	good +
Intr. resolution	moderate	moderate	moderate	good	doubled
Dispersion	good	good +	good	good	good +
Aberrations	moderate	moderate	worse	moderate	moderate
Focal depth	very good	very good	poor	moderate	good
Mass range	good +	moderate	good	good	good
Vertical control	poor	moderate	excellent	moderate	excellent

References

[1] Descriptions of recoil mass spectrometers of the types discussed here will be found in other contributions to this volume.

[2] P. Spolaore et al., Nucl. Instr. and Meth. A238 (1985) 381.

[3] T. M. Cormier and P. M. Stwertka, Nucl. Instr. and Meth. 184 (1981) 423.

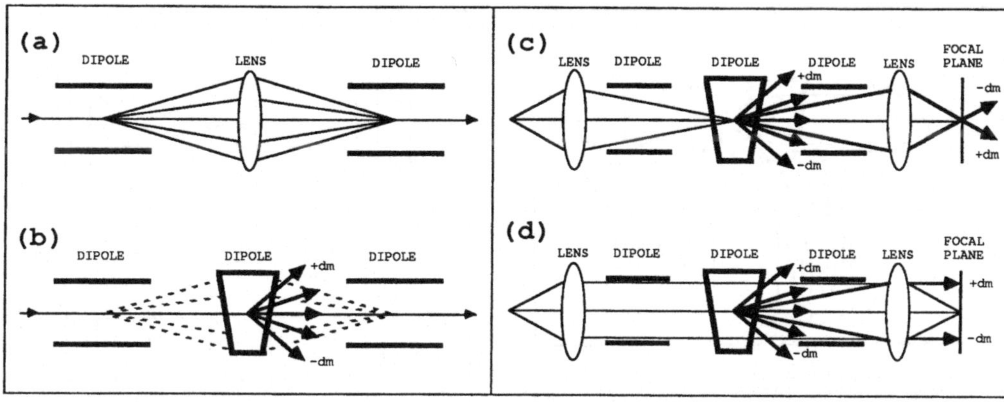

Fig. 1. Dispersion cancellation (a) using lens and (b) using mixed dipole types to obtain mass separation. Focus of masses from central pivot to focal plane (c) point-to-point and (d) point-to-parallel.

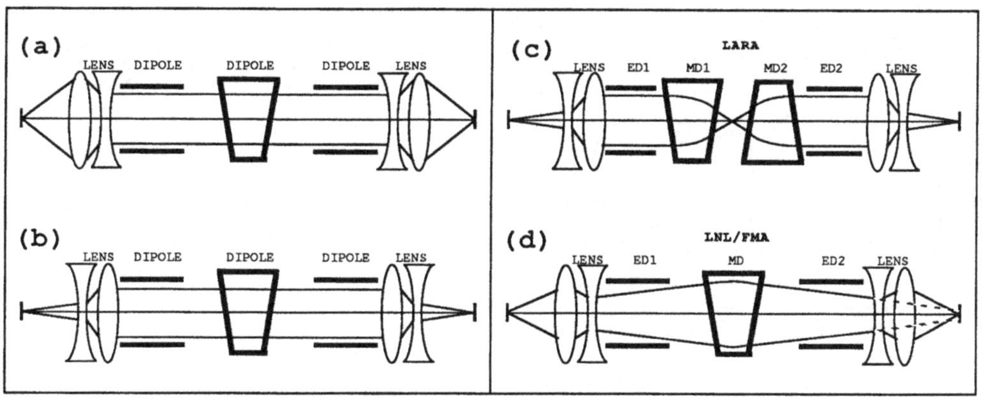

Fig. 2. Matching with quadrupole doublet lenses to weak dipoles (a) false solution with low spatial m/q dispersion and (b) good solution with high dispersion. Matching to strong dipoles (c) with split central dipole and (d) with expanded central envelope.

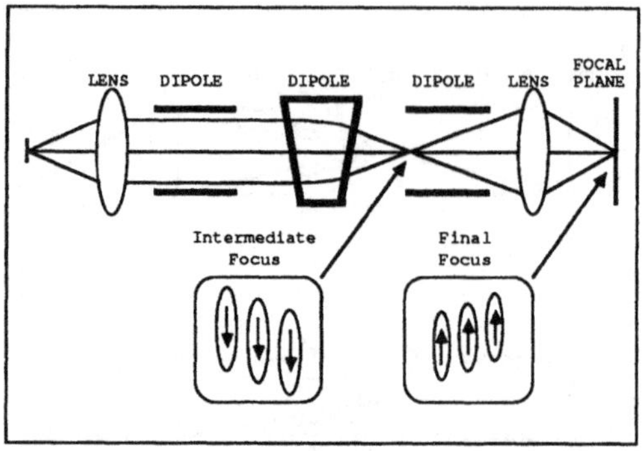

Fig. 3. Downstream lenses magnify image and dispersion equally; resolution fixed.

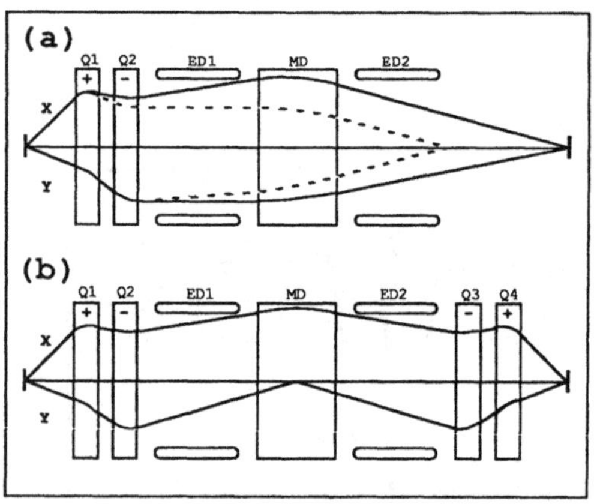

Fig. 4. Possible mass spectrometers with CD quadrupoles (a) LNL and (b) FMA.

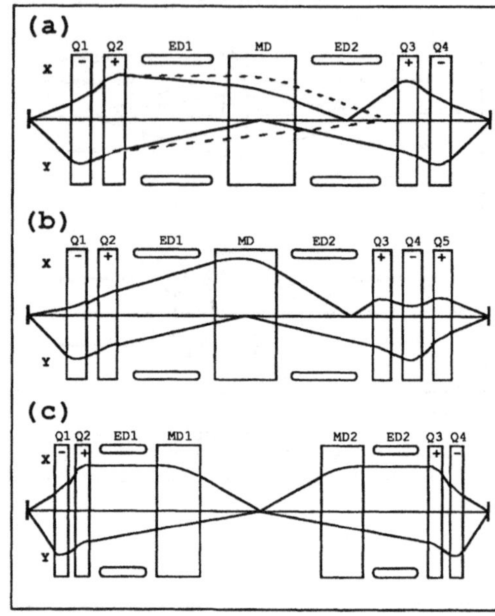

Fig. 5. Possible mass spectrometers with DC quadrupoles (a) final doublet (b) final triplet and (c) LARA.

The Recoil Separator System at GSI - Description, Experiments and Further Plans

F.P. Heßberger, G. Münzenberg, P. Armbruster, G. Berthes, W. Faust, S. Hofmann, W. Reisdorf, K.-H. Schmidt
Gesellschaft für Schwerionenforschung mbH, Postfach 110552, D-61 Darmstadt

H. Ewald, K. Güttner
II. Physikalisches Institut, Justus von Liebig Universität, Arndtstraße 2, D-63 Gießen

1. Introduction and Motivation

The velocity filter SHIP (**S**eparator for **H**eavy **I**on Reaction **P**roducts) at GSI was designed for the investigation of products from complete fusion reactions (evaporation residues, ER) using heavy projectiles (i.e. $A \geq 40$). Main points in the design of SHIP were to obtain both, a high transmission for the ER **and** an effective suppression of the projectile beam /1,2/. The basis of the method to separate ER from the projectile beam is due to their velocity difference, which is characterized by

$$v_{ER} \simeq v_{CN} = \frac{m_p}{m_{CN}} \times v_p \qquad [1]$$

v_p, v_{CN} and v_{ER} denote the velocities of the projectiles, compound nuclei (CN) and ER, respectively, while m_p and m_{CN} denote the masses of the projectiles and the CN. Since the intensity of the ER is peaked in beam direction, SHIP is aligned in direction of the primary beam. It accepts an angular width of $\pm 1.5°$ in axial and in radial direction, a relative velocity width of $\pm 5\%$ and a charge width of $\pm 10\%$.

In order to improve the experimental possibilities, SHIP recently was extended by the DQQ - spectrometer NASE (= **NA**ch**SE**parator), which is placed behind SHIP / 3 /. In combination with SHIP, NASE can be operated in two modes; the 'high resolution' mode allows an $\frac{m}{q}$ separation of the ER with a resolution of $\frac{m/q}{(\Delta(m/q))} = (130 \pm 12)$, while in the other operation mode ('m/q focussing' or 'charge focussing' mode) one renounces the good mass resolution for the benefit of a high transmission through NASE of up to 80%. This latter operation mode is of special interest as an additional stage of separation in cases where the overall background rejection of SHIP is less then 10^{10} (in general for $A_T \leq A_P$) as well as a possibility to collect the ER outside the 0° direction, if background radiation from the accelerator may disturb the measurements, e.g. for γ - decay studies of short living isomers, which decay during the 5ms - beam bursts ('macro pulses') of the UNILAC accelerator.

In a third operation mode we plan to use NASE as an independent gas-filled $B\rho$ - separator. Here we hope to profit from the larger acceptance angle of 10 msr and the large charge and velocity acceptance, which is necessary for very asymmetric projectile - target combinations ($\frac{A_T}{A_P} \leq 0.3$) where the transmission of SHIP is reduced due to the large angular spread of the ER.

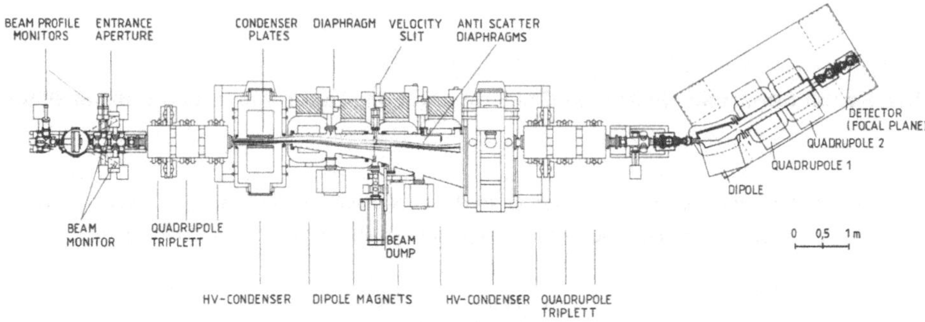

Figure 1: Top view of SHIP (partly opened to show the separation of projectiles and evaporation residues) and the DQQ-spectrometer NASE

2. Technical Description of SHIP and NASE

The whole set-up is illustrated in fig. 1. In the following we want to present a more detailed description.

2.1 SHIP

The technical layout of SHIP is characterized by the following requirements.

a) complete separation of the ER from the projectile beam

b) collection of all charge states of the ER

c) stigmatic focussing

d) high transmission

e) velocity dispersion at the exit

Taking these points into account, the machine was designed as a two stage velocity filter with the field arrangement:

quadrupole triplett - HV condenser - four dipole magnets - HV condenser - quadrupole triplett.

The electrical condenser, which could be built as a plate condenser due to the small deflection angle, was chosen as the first dispersive element. Since the projectiles always have a smaller radius of curvature than the ER, they are always less deflected. So even in cases of a high voltage break down, the primary beam will never hit the condenser plates in this arrangement. This is an important feature for experiments where beam currents of up to 1 pμA (6×10^{12} particles per second) are used, since not only severe damage of the plates could be caused if they would be hit by the primary beam, but also the detectors would be destroyed by the enormous background of projectiles scattered at the HV plates.

The plate condenser is followed by two dipole magnets. They are C-type magnets, the entrance and exit field bounderies are parallel. So it is possible to vary all deflection angles and also the dispersion from zero to the maximum value. For experiments needing deflection voltages exceeding the maximum attainable value, this is an important option. The separated ions appear to pass undeflected through the field arrangement as long as the deflection angles are small.

Since deflection fields are only weakly focussing a magnetic quadrupole tripett has to be added to each stage of the three dispersive fields. The quadrupoles introduce chromatic aberrations into the ion optical system. Since it is not possible to compensate these aberrations in quadrupole tripletts, we have minimized them by optimizing the distances between the quadrupoles as well as the object and image positions. The quadrupoles are short compared to their diameter, so they are expected to have large multipole components, which were corrected for by manufactoring the poles with appropriate step profiles. By this, the multipole components, including the hexadecapole a_{10}, were eliminated / 2 /.

The first filter stage is followed by a velocity slit, where the ER are separated from the primary beam. The components of the second stage of the filter are arranged symmetric to those of the first one with the plane of the velocity slit as the mirror plane.

2.2 NASE

The requirements to increase the experimental possibilities at SHIP as discussed in sect. 1, motivated us to add an additional separation stage to SHIP. Early considerations are discussed in ref. 4. The final concept, which was realized is described in ref. 3. Central point of the consideration was to compensate the velocity dispersion of SHIP by a dipole magnet and thus obtain an $\frac{m}{q}$ dispersion and to use a magnetic quadrupole dublett to achieve stigmatic focussing. The $\frac{m}{q}$ resolution of such a system is limited due to the small dispersion in the electrical fields of SHIP. In ref.3 a value of $\frac{m/q}{(\Delta(m/q))} = 136 \pm 23$ was obtained from ionoptical considerations. The layout of the system was essentially determined by the optional operation mode as a gas filled separator. Since a separation of the primary beam just in the beginning of the system was desired, a field arrangement dipole - quadrupole duplett (DQQ) had to be chosen. Thus horizontal focussing of the dipole magnet turned out to be necessary to limit the size of the quadrupole aperture. This was realized by an inclination of the pole edges of the dipole magnet. From calculations inclinations of 30° at the entrance and exit turned out to be the optimum values.

The technical data for the different components of SHIP and NASE are summarized in table 1.

	SHIP	***NASE***
axial and radial aperture	2.7 msr	10 msr
accepted charge width	± 10 %	
accepted velocity width	± 5 %	
Electric fields		
Plate distance	15 cm	
Plate dimension	$50x60cm^2$	
hv generators	± 330 kV	
$(F\rho)_{max}$	20 MV	
Dipole magnets		
Pole face	$40 \times 50cm^2$ (3 magnets)	$0.2m^2$
	$40 \times 65cm^2$ (1 magnet)	
Angle of Deflection		30°
Inclination of the field bounderies		30°
Gap	15 cm	9 cm
max. magnetic field	0.7 T	1.5 T
$(B\rho)_{max}$	1.2 Tm	2.1 Tm
Quadrupole magnets		
Aperture radius	7.5 cm	11 cm
Iron lenght	25 cm	50 cm
Max. gradient	9.5 T/m	9 T/m

Table 1. Technical data for SHIP and NASE

3. Some Ion Optical properties of SHIP and NASE

3.1 The operation modes of SHIP

Since SHIP is a two stage filter, the dispersive elements can be operated in a symmetric or an antisymmetric deflection mode. In the first mode, equivalent elements deflect into the same direction, in the second they deflect in the opposite one. Concerning the quadrupole elements two main modes of beam focussing are possible: parallel beam or intermediate focus in the plane of the velocity slit. Thus four principle operation modes of SHIP are possible:

1 symmetric, parallel beam, 2 symmetric, intermediate focus

3 antisymmetric, parallel beam, 4 antisymmetric, intermediate focus

For practical purpose the last mode is chosen, because it reflects some advantages.

1. optimum separation at the velocity slit due to the intermediate focussing and a maximum transmission
2. additional separation within the second stage since this mode exhibits velocity dispersion at the exit slit.

In fig. 2 we show the position distributions for the 4n and 5n evaporation residues produced in the $^{40}Ar+^{174}Yb$ complete fusion reaction at E/A = 4.62 MeV/u for the operation modes 3 and 4. The dashed lines represent the experimental results, the full lines the results from a Monte-Carlo calculation / 2 /. Since there is no velocity dispersion in mode 3, a narrow position distribution is obtained, the transmission, however, is reduced by a factor of 8 $\pm$ 1, which is fully in line with the calculations, where a reduction of a factor of seven is obtained. The calculations further show, that the main losses take place at the velocity slit.

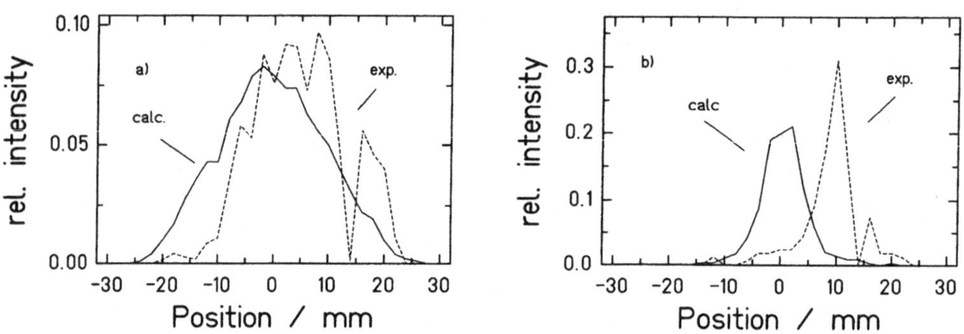

Figure 2: Position distribution of ER for the different focussing modes; a) mode 4 and b) mode 3

3.2 Velocity, charge and angular acceptances

Other important criteria for the operation of SHIP concerning the transmission are the velocity, charge and angular acceptances. In figs. 3 and 4 we compare measured and calculated distributions, where relative transmissions as functions of the SHIP setting for different velocity ranges and charge states are shown. The transmission was normalized to one for the maximum value. It should be noticed, that the curves represent the acceptance of SHIP folded with the velocity or charge distributions of the ER.

In fig. 3 we have plotted the velocity acceptances for ^{157}Hf, produced in the $^{103}Rh(^{58}Ni,p3n)^{157}Hf$ reaction at 5 MeV/u (dashed line) and $^{209,210}Ra$ produced in the $^{174}Yb(^{40}Ar,5n,4n)^{209,210}Ra$ reactions at 4.8 MeV/u (dotted curve). The acceptance is somewhat boader for ^{157}Hf than for $^{209,210}Ra$ Reasons for this behavior are a) the higher momentum transfer due to the evaporation of a proton, and b) a larger energy loss straggling due to a thicker target (^{103}Rh : 0.8 mg/cm^2, ^{174}Yb : 0.3 mg/cm^2), although the total momentum of the CN is higher in the ^{58}Ni + ^{103}Rh reaction. The experimental trend is fully established by the Monte-Carlo calculations (full line). In fig. 4 the charge acceptances for the p3n (^{157}Hf, dashed line) and the 2p3n (^{156}Lu , dashed-dotted line) from the ^{58}Ni + ^{103}Rh reaction at 5 MeV/u are compared with calculations (full line). The trend is also established. It should be noticed that the maximum transmission for ^{157}Hf and ^{156}Lu occurs at different SHIP settings, which indicates different average charge states of these ER.

The experimental results confirm that the required values of velocity and charge acceptance of SHIP are obtained in practice. Especially in reactions with light targets the critical limit in the transmission is given by the angular acceptance of SHIP. A simple improvement is achieved by a reduction of the distance between target position and entrance aperture, i.e. to increase the angular acceptance. The result of such a procedure is shown fig. 5 for the ER from the reaction ^{40}Ar + ^{174}Yb at 4.6 MeV/u. A disadvantage, however, is an increase of the background, which is undesired in experiments with low cross-sections. Here a additional separation using the 'm/q focussing' mode of NASE (see sect. 3.4) is expected to bring some advantage.

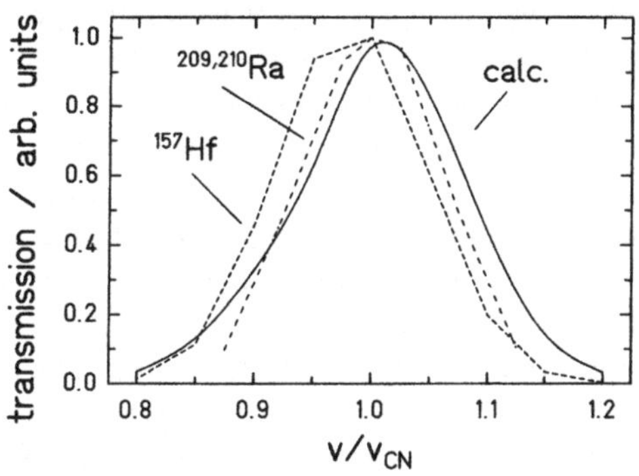

Figure 3: Velocity acceptance of SHIP. The explanations are given in the text.

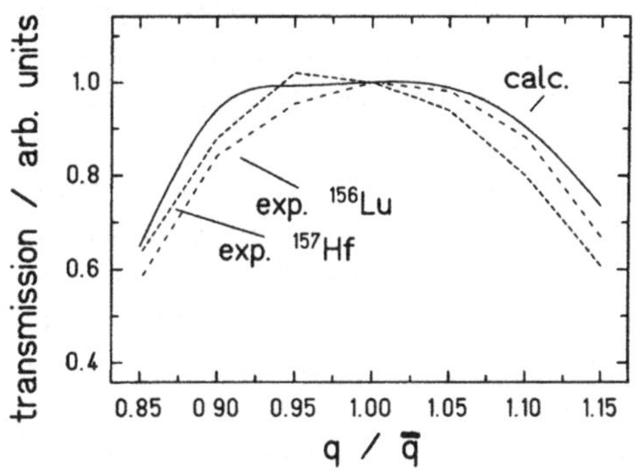

Figure 4: Charge acceptance of SHIP. The explanations are given in the text.

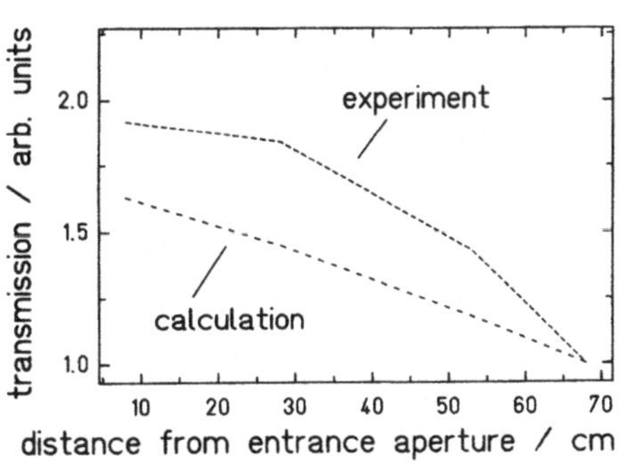

Figure 5: Transmission of SHIP as a function of the target position.

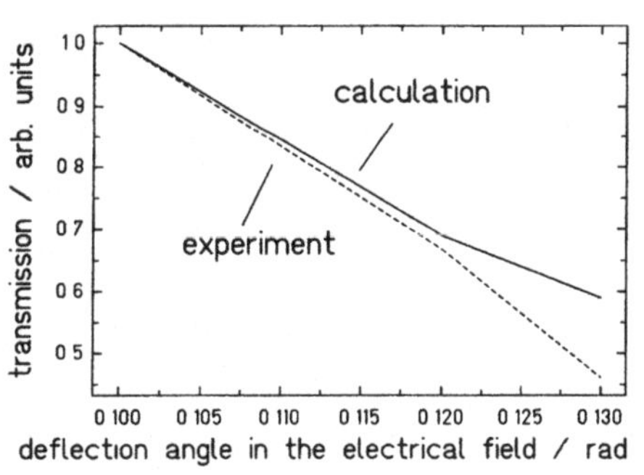

Figure 6: Transmission of SHIP as a function of the deflection in the electrical field

3.3 The background reduction

Beside the transmission, the background reduction is the most important feature in the operation of SHIP in practice. Generally we have to deal with three types of background:

1. scattered target-like nuclei, passing SHIP with v_{ER}
2. scattered projectile-like nuclei, passing SHIP with v_{ER}
3. projectiles passing SHIP with their full energy

Since SHIP is a velocity filter the quality of separation is expected to decrease with decreasing velocity difference between projectiles and ER, i.e. with increasing projectile mass, according to formula (2)

$$\frac{(v_p - v_{CN})}{v_{CN}} = \frac{m_t}{m_p} \qquad [2]$$

For small velocity differences between projectiles and ER the separation of both beams in the plane of the velocity slit is small. Because of chromatic aberrations of the quadrupoles the image in the plane of the velocity slits is poor for those primary particles having a magnetic beam rigity higher than that of the CN. Those defocussed and weakly deflected projectiles may pass the velocity filter and arrive at the exit of SHIP. These poorly focussed particles must have "exotic" charge values considerably lower than the average values.

The background of types 1 and 2 on the other hand is difficult to explain, since nuclear reaction kinematics for two body reactions exclude for elastic or inelastic scattering velocities equal to those of the CN. So 'high order' processes, like multiple scattering or eventually scattering at target frames etc. accompanied with high losses in kinetic energy have to be considered. The intensity of this background is strongly dependend on the beam focussing at the target position and on the quality of the targets. Due to many improvements in these directions since the beginning of UNILAC operation, especially for asymmetric systems with $Z_p \simeq 18-28$ a drastic reduction of background has been achieved. At present the background rates are below 100 Hz for beam intensities of up to 6×10^{12} particles per second. In some cases a considerable background reduction was obtained by increasing the dispersion, i.e. using higher field values, which is however accompanied by a reduction of the transmission, as shown in fig. 6. The reason for this is that the ER beam passes the region of the third dipole magnet very close to walls of the magnetic chamber, so it will hit at higher deflection fields.

3.4 The operation modes of NASE

As already mentioned, NASE can be operated in a 'high resolution' mode and in a 'm/q (charge) focussing'

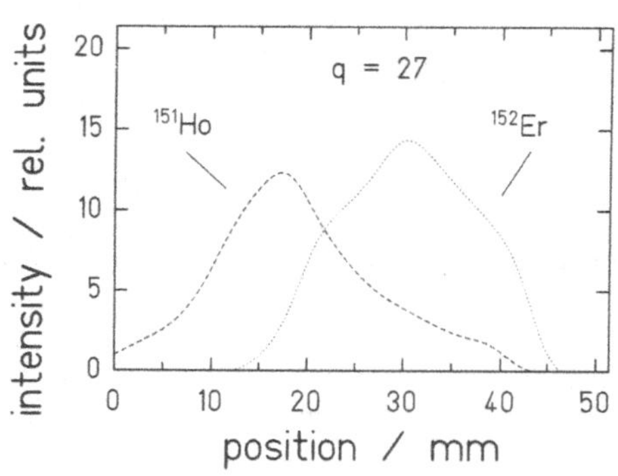

Figure 7: Position distribution of ER using the 'high resolution mode' of NASE

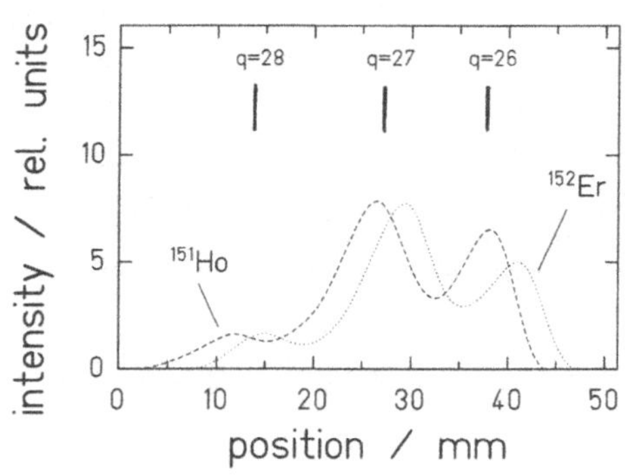

Figure 8: Position distribution of ER using the 'm/q focussing' mode of NASE

The data are taken from ref. 3. They have been smoothed for a better presentation.

mode. The m/q resolution of NASE was measured using the α decays of ^{151}Ho and ^{152}Er, which were produced via β decay chains from 2 and 3 nucleon deexcitation channels of the CN ^{154}Hf in bombardments of ^{96}Ru with ^{58}Ni. Fig. 7 displays the position distribution of the α decays on a $(45 \times 8)mm^2$ position sensitive surface barrier detector for a selected charge state of 27^+ and a selected mass number of A = 151.5. A mass resolution of $(\frac{m}{\Delta m})_{FWHM} = 130 \pm 12$ was obtained. In this operation mode about 20 % of the reaction products having the selected mass, that pass SHIP are detected behind NASE / 3 /.
A considerable higher transmission through NASE of up to 80 % can be achieved in the 'm/q focussing' mode. In this operation mode the SHIP focus is shifted into the dipole magnet of NASE with the help of the last quadrupole triplett of SHIP. The procedure enables an 'optimum minimizing' of the $\frac{m}{q}$-dispersion and thus focus several charge states to the detector position as shown in fig. 8, while background of projectile-like and target-like particles that pass SHIP is suppressed efficiently / 3 /.

4. The experimental program

4.1 Heavy element research

An important part of the experimental work during the last decade was concentrated on the synthesis of the heaviest elements on the borderline of nuclear stability. A detection technique, which allowed to follow the radioactive decay of an ***individual*** nucleus was developped. This enabled us to identify unambigouosly isotopes of the heaviest three elements with atomic numbers 107, 108, and 109 by observing only very few events / 5 /. For $^{266}109$, e.g. only ***one*** decay chain was observed in a first experiment performed in 1982. In a recent experiment this result was confirmed by observing two more decay chains.
In the next future our work will be concentrated on the synthesis of element 110 in the reaction ^{62}Ni + ^{208}Pb. Extrapolations of production cross sections for element 104 - 109 - isotopes in Pb-based systems lead to an expected formation cross section of $\simeq$ 1 pbarn for element 110. To succeed in this direction we have to increase the efficiency and sensitivity of our system. Modifications discussed presently are a) to position the target closer to the entrance aperture to increase the transmission, b) to use NASE in the 'm/q - focussing' mode to get out of the 0° direction and for additional background reduction, and c) to use a detector system with a higher detection effeciency.

4.2 Proton radioactivity

The search for nuclei decaying from their ground state by proton emission had been successful for the first time, when in an irradiation of ^{96}Ru with ^{58}Ni a proton line was observed at SHIP that could be attributed to ^{151}Lu / 6 /. In the following years two more proton emitters ^{150}Lu and ^{147m}Tm were identified at SHIP / 7 /, while proton decay of ^{147}Tm was measured by the on-line mass separator group at GSI / 8 /.
In bombardments of ^{58}Ni,^{54}Fe with ^{58}Ni a group from München observed two proton lines; based on reaction arguments, they assigned them to ^{113}Cs and ^{109}I / 9 /. In the first experiment performed with NASE these mass assignments were confirmed by a direct mass measurement. A search for proton radioactivity of ^{105}Sb turned out to be unsuccsessful, however, only a window for the Q_p value can be given to $320keV \leq Q_p \leq 460keV$ / 3 /.

4.3 Fusion of massive systems around the barrier

Another important project at SHIP is the investigation of fusion of massive systems at bombarding energies close to the barrier. I will not deal with it further, since it is part of the contribution of W.Reisdorf.

4.4 Investigation of highly excited nuclei

Another interesting aspect is the investigation of highly excited nuclei with $E^x >> 100$ MeV. In a first experiment at SHIP we investigated the reaction products having CN velocities from bombardments of ^{139}La,^{142}Nd and ^{144}Sm targets with ^{40}Ar projectiles at 15 MeV/u.

Applying α spectroscopy after implantation of the reaction products into surface barrier detectors, we could identfy ≃ 15 isotopes with N = 84,85 having A and Z differences to the compound system of e.g. ΔA = (27-35) and ΔZ = (8-15) in the ^{144}Sm induced reaction / 10 /. In an experiment, which is presently under preparation, we want to study the A-, Z-, and v - distributions of the reaction products in more detail. Further we want to try to attach the α decays to the corresponding implanted nuclei by α correlations for at least some of the isotopes. This we regard as a first step to measure the spectra of particles emitted during the decay of a highly excited nucleus for a well defined end product.

4.5 Laser - spectroscopy of ER

After the first successful test of NASE interest in applying Laser spectroscopy of shortlived isotopes behind NASE arose. The advantages of using this system (compared to on-line mass separators) are mainly for a) isotopes from refractory elements and b) isotopes or isomers with high primary production cross sections but lifetimes < 1 s. In a first experiment the applicability of this method has been demonstrated in an experiment with ^{160}Yb , which was produced in a $^{124}Sn(^{40}Ca, 4n)^{160}Yb$ reaction / 11 /. We plan to continue these experiments with Hf - isotopes, which will be produced using a ^{48}Ti beam instead of ^{40}Ca.

5. Outlook

The operation of SHIP during the last twelve years has demonstrated many advantages of a recoil separator for the investigation of heavy reactions as well as for heavy ion reaction products. The experimental possibilities at SHIP are by far not exausted, moreover they have been extended by NASE. Presently we are busy to increase the sensitivity and efficiency of the complete system and to get NASE running in the gas filled mode. So we are optimistic to make interesting experiments for at least twelve more years.

References

/1/ G.Münzenberg, W.Faust, S.Hofmann, P.Armbruster, K.Güttner, P.Armbruster: Nucl. Instr. Methods 161 (1979) 65

/2/ W.Faust: Thesis, Gießen 1978, GSI Report P-4-78

/3/ G.Berthes: Thesis, Mainz 1987, GSI Report 87-12

/4/ B.Thuma: Thesis, Gießen 1980, GSI Report 81-4

/5/ see e.g. G.Münzenberg: Rep. Prog. Phys. 1988,51,57-104

/6/ S.Hofmann, W.Reisdorf, G.Münzenberg, F.P.Heßberger, J.R.H.Schneider, P.Armbruster: Z.Phys. A 305 (1982) 111

/7/ S.Hofmann, Y.K.Agarwal, P.Armbruster, F.P.Heßberger, P.O.Larsson, G.Münzenberg, K.Poppensieker, W.Reisdorf, J.R.H.Schneider, H.J.Schött: Proc. of the 7th Int. Conf. Atomic Masses and Fund. Const. AMCO 7, Darmstadt-Seeheim 1984, 184

/8/ O.Klepper, T.Batsch, S.Hofmann, R.Kirchner, W.Kurcewicz, W.Reisdorf, E.Roeckl, D.Schardt, G.Nyman: Z.Phys. A 305 (1982) 125

/9/ A.Gillitzer, T.Faestermann, K.Hartel, P.Kienle, E.Nolte: Z.Phys. A 326 (1987) 107

/10/ F.P.Heßberger, G.Münzenberg, U.Gollerthan, E.Hanelt, W.Morawek: GSI Sci. Rep. 1986, GSI 87-1, 55

/11/ S.Dutta et al. Contribution to GSI Scientific Report 1987, in press

DEVELOPMENTS AT THE DARESBURY RECOIL SEPARATOR

A N James

Department of Physics, University of Liverpool

Liverpool, L69 3BX U.K.

Introduction

The Daresbury Recoil Separator has been taking heavy ion beams from the tandem accelerator for experiments since early 1986. One of the major functions of the Recoil Separator [1] is to provide channel selection for in beam γ-ray spectroscopy of compound nucleus residues following heavy ion reactions. The principal detectors used for this work are the standard focal plane detector [1] for the recoiling ions and the POLYTESSA array of twenty escape suppressed germanium detectors [2] for the prompt γ-rays produced in the reaction (Figure 1). The focal plane detector consists of a position sensitive channel plate readout from the secondary electron emission caused by the ion's passage through a thin carbon foil and a two parameter (ΔE and E) ion chamber. Examples of the use of this system can be found in references [3] and [4].

The beams presently available from the 20MV tandem at Daresbury allow the use of inverse reactions for residues with masses less than about 140 but higher masses can only be formed using low mass beams and consequently the emittance of the recoiling ions rarely matches the separator acceptance and the recoiling ions have low velocity. Low velocities ($v/c < 0.05$) imply that energy loss measurements (ΔE in the ion chamber) cannot be used to select atomic number (Z) and more complicated techniques (such as recoil-γ-γ coincidences) become essential for clean channel selection.

The Recoil Separator was conceived as a device which could be used in the future for recoiling ions formed near the Coulomb barrier using beams of very heavy nuclei. The energies of the heavy ion beams at Daresbury are soon to be boosted by the addition of superconducting radio frequency cavities and consequently by 1990 the Recoil Separator will have to operate at its full designed ratings. There is no problem with the magnets, but the electrostatic system is normally operated only up to 150kV per plate whereas 300kV per plate was designed and will be essential with the higher velocity very heavy recoiling ions.

Developments currently in hand are aimed at improving the recoil-γ-γ efficiency of the system, but improvements in the electrostatic system will be necessary by 1990.

Figure 1.

The POLYTESSA array of twenty escape suppressed germanium detectors drawn back to show the target chamber in front of the Daresbury Recoil Separator. The four rings of five detectors each are at angles of 143°, 117°, 101° and 79° to the incident beam.

Detector Developments

An example of recoil-γ-γ data is shown in figure 2 where it is compared with other spectroscopic data at the frontiers of γ-ray spectroscopy in heavy ion reactions. The advantages of the recoil-γ-γ technique are obvious, but also it is apparent that higher efficiency is very desirable. Two developments which are under way are described below. The first, a change in the germanium detector

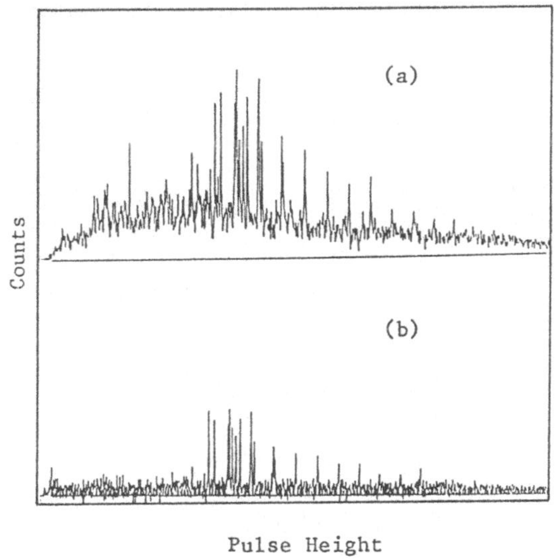

Figure 2.

Spectrum of counts from a germanium detector in coincidence with both a gamma-ray in the ^{152}Dy low deformation prolate rotational band and a) ^{152}Dy selection by multiplicity and total energy using the TESSA3 bismuth germanate ball, or b) the recoil separator signal for mass 152 charge state 21. The peak efficiency ratio (TESSA3 selection:Recoil Separator selection) is (10:1), but note the significant background reduction using the recoil separator.

array geometry, should improve the total E2-E2 γ-γ coincidence rate by a factor of two. The second, an attempt to use three charge states, will improve the efficiency by a factor of three.

Experience using the ion chamber ΔE signal to differentiate between recoils and low energy beam like particles and to differentiate between different Z values [4] has led to the splitting of the first anode to give two samples of the energy loss rate instead of one. Although a full experiment has not been analysed the high energy loss events in the ^{80}Y data [4] would probably have been identified simply by the extra energy loss datum, see figure 3.

New detectors have also been developed for experimental work using the mass separated beams at the output of the separator. Work on α and p decaying nuclei using the RIDS system of two dimension position sensitive silicon detectors [5] has started and because this system provides a measure of vertical position its output contains useful information about the separator performance. As an example the vertical transmission of the separator is compared in figure 4 with that

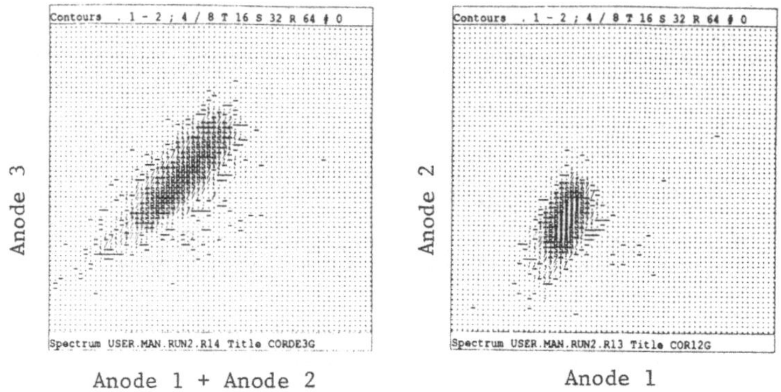

Figure 3.

Data taken from an experiment in which a ^{50}Cr target was bombarded with a 195MeV ^{56}Fe beam. Recoiling ions of ^{103}Cd were selected by a window on one γ-ray detected in the Germanium detectors. The right hand panel shows the correlation between the pulse heights in the two ΔE sections of the ion chamber while the left hand panel shows, for the same events, the correlation between the sum of the two ΔE sections and the remaining energy. In both panels a few events (about 1 in 500) fall to the lower right hand side of the main peak and these are the same events in both cases. They correspond to the high energy tail (~ 1 part in 500) observed in the data shown in reference [4] which causes background in the atomic number signal.

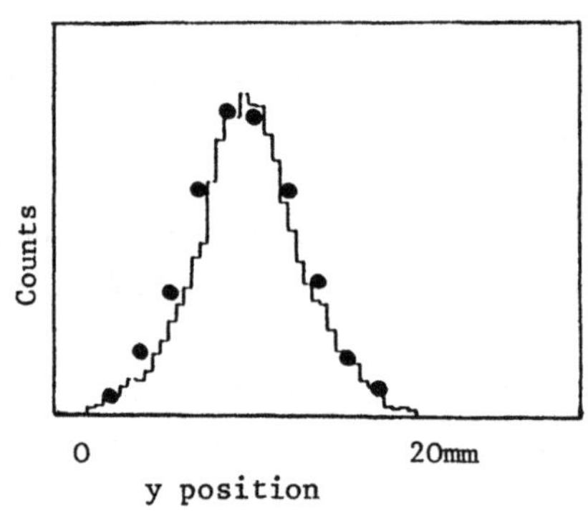

Figure 4.

The vertical distribution of recoiling ions detected in a two dimensional position sensitive silicon detector 20mm by 20mm. The dots show the results of a ray tracing computation using the programme TURTLE.

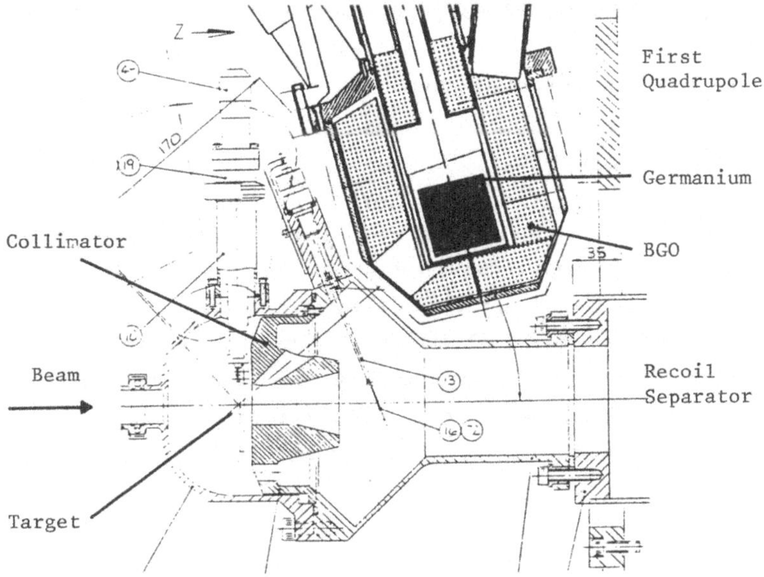

Figure 5.

The special BGO suppression shield and Germanium detector mounted at 40° on the first quadrupole of the Recoil Separator. The γ ray detection performance of the array is improved by the heavy metal collimator and this constrains the position at which the Recoil Separator Optic axis intersects the target.

predicted using the TURTLE [6] ray tracing programme. In searching for proton decays a useful technique is to search in a focal plane position corresponding to mass A for the occurrence of the α decay lines from mass A-1. In some experiments the sextupoles of the Recoil Separator have been adjusted to ensure that the second order tails from mass A extend more to low masses than to high masses, thus reducing the background of mass A-1 under the mass A peak.

Forward Angle γ-ray Detectors

The angular distributions expected in compound nucleus formation generally have a minimum near 90° so much of the POLYTESSA array (figure 1) is not as well positioned as it could be. A new special geometry bismuth germanate suppression shield has been designed [7], figure 5, which will give five detectors at 40° instead of the five detectors at 79°. A consequence of this change is that the optic axis of the separator must now be precisely positioned to intercept the centre of the POLYTESSA sphere because the new detectors have heavy metal collimators (necessary for clean γ-ray spectra) near the target spot. It is now known (using RIDS) that the optic axis is 4mm high and this will be adjusted by lowering the second quadrupole by 1mm.

Three Charge State Operation

Reactions in which light beams (e.g. Calcium) bombard heavy targets (A>100) have a large velocity difference between beam and recoils and an alternative mode of operation of the recoil separator becomes possible. By reducing the crossed field velocity filter field strengths it is possible to still separate beam particles from recoils at the intermediate focal plane while still accepting a ± 6% range in momentum/charge. The significance of this is that the mean charge state will exceed Q=20 and so three charge states from each mass will be transmitted.

Operating in this mode will make it impossible to refocus all velocities of one A/Q to a single line. Simulations using TURTLE [6] have shown that to retain a transmission of ±6% in magnetic rigidity (plus the velocity dispersion necessary to deflect beam particles) it is necessary to abandon all concept of a final focal plane. The whole acceptance can however be passed through a waist downstream of the last quadrupole where a detector measuring both position and angle of each ion may be placed.

The detector must have as little multiple scattering as possible since the ray angle is the most important factor determining the mass. It will consist of two 50mm x 50mm position sensitive carbon foil detectors separated by 250mm. A data acquisition system providing rapid calculations of the mass from position co-ordinates and time of flight will produce a mass signal with resolution estimated as 1 in 250 determined mainly by multiple scattering in the foils of the detector. An ion chamber with energy and energy loss sections will provide additional information to discriminate against beam like particles.

High Voltage Operation

The voltages required on the electrostatic system have only rarely exceeded 130kV per plate (2.6MV m^{-1}) during the past two years. Occasionally operation to 150kV has been required. Conditioning is performed automatically under computer control using a final ramp increment of 10 volts performed every 0.2 seconds, excessive load current rapidly trips and protects the supply. No discharge damage has ever been seen on the Titanium plates. An hour or so conditioning per week

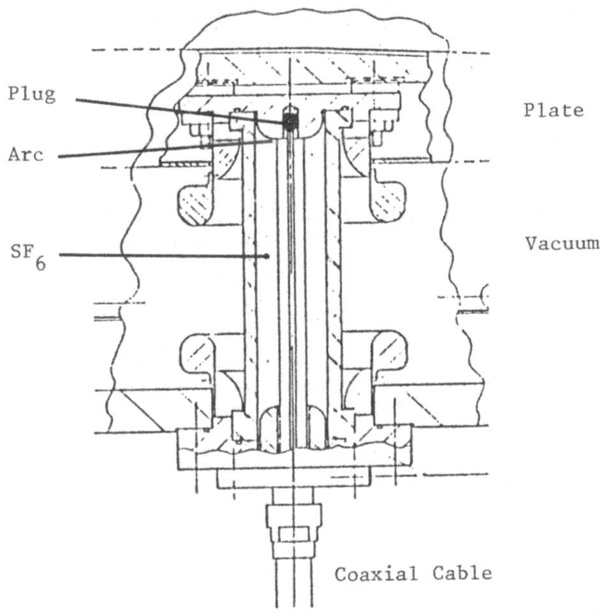

Figure 6.

The mounting of the electrostatic plates is done with ceramic insulating cylinders with an SF_6 gas filling. The mounting shown also acts as the voltage supply connection. The solid polyethylene insulator of the coaxial cable is carried through to the plug into the plate. The position where corrosive arcing has occurred is indicated.

maintains operation to 130kV but additional conditioning is required to reach higher voltages. Normally dust particles are rapidly cleaned away, but twice the interior condition of the high voltage chambers required cleaning before satisfactory high voltage performance was achieved (shreds of target debris carried by imperfect operation of vacuum systems?)

After one of the occasions where a large number of supply trips at high voltage had been experienced there was a progressively worsening failure of the supply feed through into the vacuum chamber. The failure was due to corrosive damage spreading through the SF_6 region (figure 6) from the metal to metal touching contact at the plate end. It appears that the rapid vacuum discharges cause voltage surges which discharge across small gaps before the surge can penetrate to the actual D.C. contact deep in a hole in the plate mounting block. A new contact system to avoid this problem will be necessary when longer conditioning runs become necessary to reach 300kV per plate.

One test to 200kV has been made, several hours were required to reach 200kV per plate, but the voltage was then held without problem for four hours.

References

[1] A.N. James, T.P. Morrison, K.L. Ying, K.A. Connell, H.G. Price and J. Simpson
Nuclear Instruments and Methods A267 (1988) 144-152

[2] P.J. Nolan, D.W. Gifford and P.J. Twin
Nuclear Instruments and Methods A236 (1985) 95-99

[3] K.L. Ying, P.J. Bishop, A.N. James, A.J. Kirwan, D.J.G. Love, T.P. Morrison, P.J. Nolan, D.C.B. Watson, K.A. Connell. A.H. Nelson and J. Simpson
J. Phys. G.; Nucl. Phys. 12 (1986) L211-L215

[4] C.J. Lister, M Campbell, A.A. Chishti, W. Gelletly, L. Goettig, R. Moscrop, B.J Varley, A.N. James, T.P. Morrison, H.G. Price, J. Simpson, K.A. Connell and O. Skeppstedt
Physical Review Letters 59 (1987) 1270-1273

[5] P.J. Woods, S.J. Bennett, B.R. Fulton, R.D. Page, K.A. Connell, J. Groves, J. Simpson, A.N. James and W.D.M. Rae
Nuclear Structure appendix to Daresbury Annual Report 1986/87 p127

[6] K.L. Brown and Ch. Iselin
CERN Report CERN 74-2

[7] J. Simpson, M. Bates, C. Brookes and P.J. Nolan
Nuclear Instruments and Methods, A269 (1988) 209-212

THE L.N.L. RECOIL MASS SPECTROMETER: FIRST TESTS.

P. SPOLAORE

INFN, Laboratori Nazionali di Legnaro, Padova, Italy.

Abstract.

After the installation of the basic hardware was completed, the L.N.L. recoil mass spectrometer underwent an initial series of tests, including the high voltage conditioning of the electrostatic dipoles and particles (α and heavy ions) analysis. The "raw" performances, with the still uncalibrated instrument, are described and compared with calculations.

1. Introduction.

With the beginning of 1988 the L.N.L. recoil mass spectrometer [1] was made available for some testing of the hardware and the optical performance. Fig. 1 shows the installation as in June 1988.

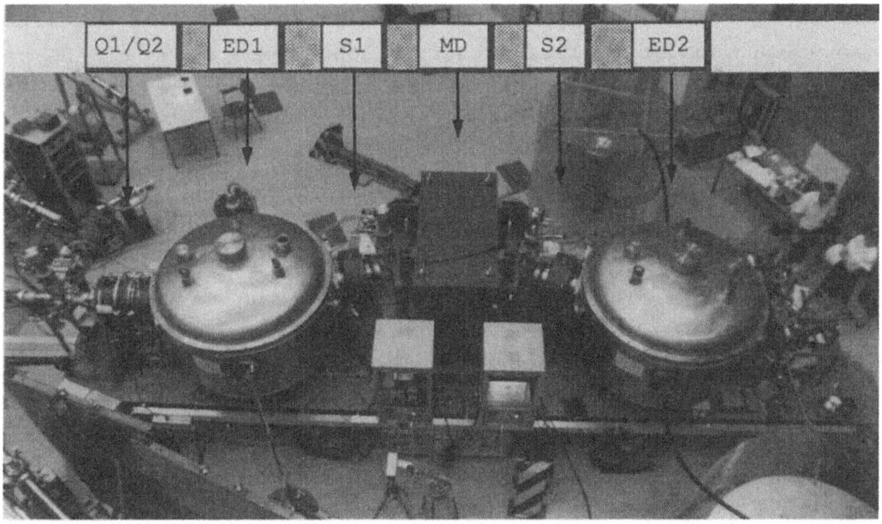

Fig. 1. Spectrometer installation as in June 1988. Optical components are: one quadrupole lens doublet (Q1/Q2), two electrostatic dipoles (ED1,ED2), one magnetic dipole (MD) and two sextupole lenses (S1,S2).

Table 1. Design specifications of the LNL recoil mass spectrometer.

m/q range	± 7 %
Energy range (ΔE)	± 20 %
Solid angle acceptance (Ω)	> 10 msr
Mass resolution at: Ω = 5 msr, ΔE = ±10 %	1/280
Ω = 10 msr, ΔE = ±20 %	~ 1/200
Ω = 2.5 msr, ΔE = ±5 %	~ 1/400
Magnification (dispersion plane)	-4.0 to -1.0
Mass dispersion at the focal plane	9 to 11 mm/%
Energy resolution via TOF	< 0.1 %

The design specifications of the instrument are reported in table 1 for reference. The principal features are the very wide acceptance in angle, energy and mass, the angular rotation of the instrument from 0° to more than 45° and the first order energy focusing both in position and angle (energy dispersion is cancelled after ED2, to the first order) that makes it possible to vary the focal plane position along the longitudinal axis. This last feature provides a large flexibility in the detector area, useful to accomodate different detector systems, and the choice of different focal-plane positions allows to make use of the corresponding sensible variations of some of the spectrometer parameters (see table 1) in order to optimize the set-up for different experimental requirements; the quadrupole doublet is used to adjust the space focus to the various detector positions. The good achromatism allows to perform precise energy measurements using the TOF technique, while a very crude energy and time measurement, provided even by very simple detectors, resolve the charge state ambiguity giving a pure mass measurement at the focal plane [1]. Large detector systems (e.g. 4π arrays) can be used around a withdrawn target position by giving up part of the solid angle acceptance and using the spectrometer at a fixed angle relative to the beam.

The principal tests performed up to now regarded the high voltage conditioning of the two large electrostatic dipoles (EDs) and a general check of the spectrometer "raw" performances, before the precise (and much time consuming) calibration is carried out.

2. High Voltage Conditioning of the Electrostatic Dipoles.

The electrostatic dipoles (fig. 2) have cylindrical-sector shaped electrodes with a radius of curvature of 4 m at the middle of the gap and

20° of angular deflection. The gap width is 150 mm.

Considering the capability of the LNL XTU-tandem accelerator and the research program proposed in connection with the spectrometer the upper limit of electric rigidity of the particles analyzed at the central trajectory has been initially set to 6 MeV/q, corresponding to an electric field of 30 kV/cm and to a total potential difference of 450 kV across the dipoles gap. In order to guarantee a reliable operation at 450 kV the design goal has been set to 500 kV; this voltage has already been reached with a model condenser, having the same 15 cm gap and titanium electrode surfaces, but smaller size [2].

Fig. 3 shows part of the spectrometer EDs first conditioning. The base vacuum in the two vessels is in the 10^{-8} mbar range when they are valved off and around $1 \cdot 10^{-7}$ mbar in operation. In the single-electrode conditioning (fig. 3a) one electrode is energized while the other one is grounded through its switched-off power supply; in the symmetric conditioning (fig. 3b) the anode and cathode are constantly kept at about the same voltage during conditioning. Fig. 3b shows that, as expected, the voltage level of 330 kV is still far from the saturation region; the few slope oscillations of the curve correspond to changes in the control of the conditioning process. The conditioning procedure (as well as other testing operations) could not be run in the desirable regular way because of the interference with the other activities going on in the same experimental room; this interference is essentially due to the intense

Fig. 2. Inside view of the electrostatic deflectors assembly. The materials are: titanium for the plate surfaces, aluminum for other field-exposed components of the supporting structure and stainless steel for the tank. The strip over the concave electrode is 1 m long.

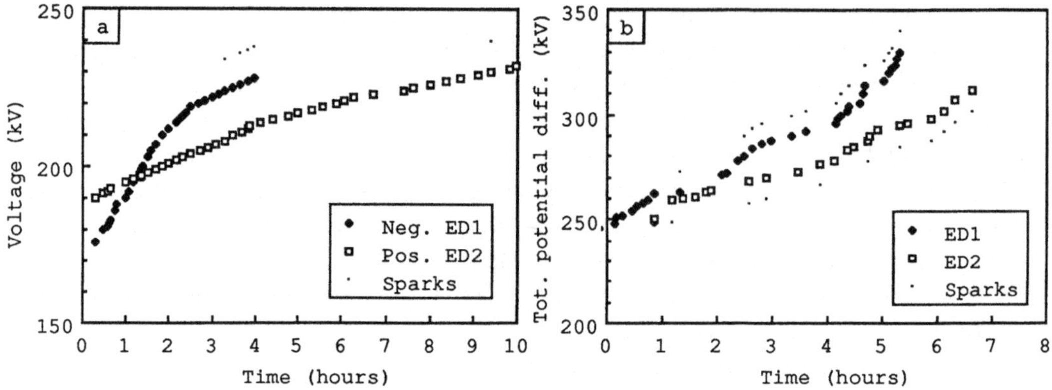

Fig. 3. Part of the first conditioning of the electrostatic dipoles of the RMS; a) single-electrode conditioning; b) symmetric conditioning.

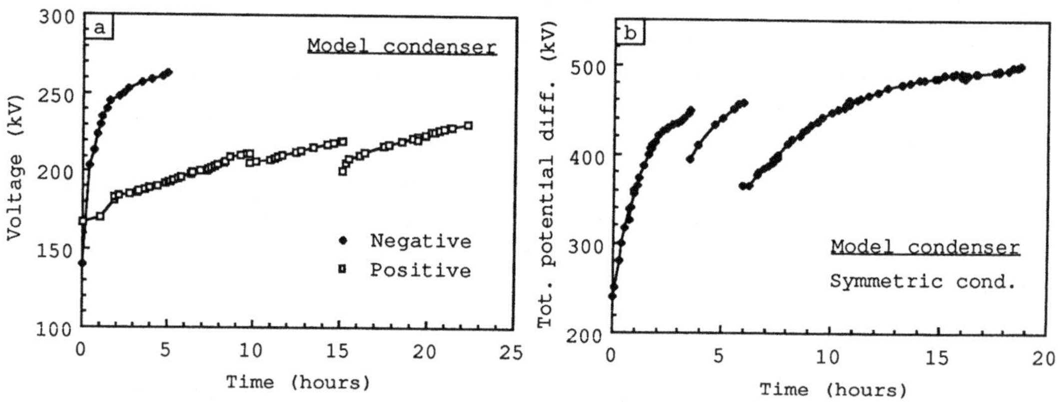

Fig. 4. First conditioning of the titanium plates of the model condenser [2]: a) single electrode conditioning; b) symmetric conditioning.

X-ray emission during conditioning, which is partially reduced by 1 cm thick lead shields only in a section of the area.

The maximum voltage reached up to now has been of 330 kV and 310 kV on ED1 and ED2 respectively, and 230 kV on negative and positive single electrodes. In all cases the conditioning was interrupted because of time limitations. Fig. 4 shows, for comparison, the corresponding conditioning curves obtained with the mentioned model condenser. Because of the basic similarities between the behaviour in the two cases it can be reasonably expected to reach the design voltage also with the spectrometer units.

3. α-Source and Beam Tests.

The large acceptance and flexibility of the instrument allows to "trade" between acceptance and mass resolution. To control this feature, besides the already mentioned variability in some of the optical parameters (see par. 1), there are presently two variable hardware constraints: a collimator at the entrance of the quadrupole doublet and a couple of horizontal slits at the middle of the magnetic dipole. The collimator can be used to cleanly cut down the solid angle, while the slits help in limiting the energy acceptance, even if at that location the interaction between energy and angle effects is large, so that a clean cut can be obtained only from a detector energy-measurement. These slits, however, are useful also to clean from multiscattered low-energy particles and to define, for calibration purposes, the central path through the spectrometer while using a suitable "pencil" beam.

The initial setting of the electric and magnetic dipole fields has been determined using a ^{32}S beam of 32 MeV, with charge state 11^+ after stripping on a carbon foil at the target location, detected by a segmented beam stopper at the focal plane. A collimator of 1 mm horizontal width was placed just in front of the carbon foil to define the optical-source dimension and position, the angular spread was reduced to 0.5 msr (symmetrically in the vertical and horizontal planes) by the quadrupole collimator and the central dipole slits were closed to ~4 cm. Yet, the final calibration of the dipoles setting will be determined while minimizing the residual energy dispersion at the focal plane (over the available longitudinal range of about 80 cm) in order to realize the energy double focusing (in position and angle) requirement: this condition is in fact influenced by the actual deflection angles, that depend from the effective field boundaries extent; for this purpose also an adjustement of the magnetic dipole shim angles, partially controlled by the field clamps, might be called for.

Alfa particles of 5.805 MeV from a ^{244}Cm source have been used to investigate the focal-plane-spot size and shape. The source is ~2 mm wide (horizontally) and was tilted to about 45° in order to obtain a smaller optical-source width: ~1.4 mm. The focal plane detector was positioned at a location corresponding to a theoretic horizontal magnification of -1.3 and mass dispersion of 11 mm/%, in order to obtain a good resolution, with a consequent limitation of the solid angle acceptance to about 5 msr. All of the hardware constraints were full open. A position sensitive Si detector, having a sensitive area of about $50 \cdot 8\ mm^2$, horizontally oriented, was used. The limited vertical width of the detector cuts down the total solid angle to ~3 msr, yet mantaining the principal optical aberrations (that develope in the horizontal plane) as if the full 5 msr were accepted. The detector resolution is close to 0.5 mm.

Fig. 5 shows a comparison between the experimental and calculated (using the computer code GIOS [3]) position spectra with ≈5 msr (figs 5a,

5b) and with 0.5 msr defined by the quadrupole collimator (figs 5c, 5d). No second order correction was applied except for the magnetic dipole curved boundaries. The calculated set-up reproduces the real instrument in all known details. The side shoulder of the peak in figs. 5a, 5b is essentially due to the second order aberration in angle, therefore it is reduced quadratically by closing down the angular acceptance; this aberration can be completely controlled by one sextupole lens, but the optimum setting of S1 and S2 will correspond to a compromise between the correction of three principal aberrations: (x/θ^2), $(x/\delta_E{}^2)$ and $(x/\theta\delta_E)$, i.e. the position change in proportion to the square of angle θ, square of fractional energy deviation from the central value δ_E and product $\theta\cdot\delta_E$ [1]. Two wires, 10 mm apart, have been placed just in front of the detector for calibration and show up in some of the measured spectra as small negative peaks. Also considering the experimental error, the FWHM is a little larger than expected in the case of 0.5 msr: most probably an

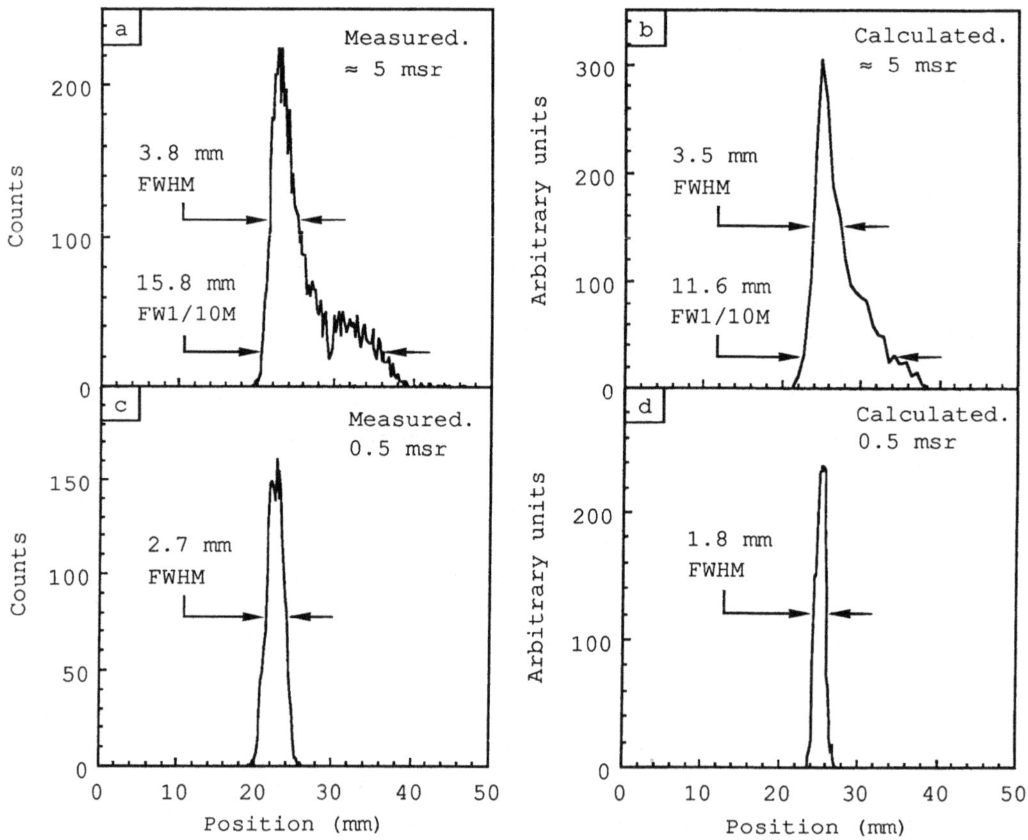

Fig. 5. Measured (a,c) and calculated (b,d) position spectra for α particles. The spectrometer setting is not optimized.

indication of a slight misalignement of the source position.

The evaporation residues with A = 86,87 from the known fusion reaction $^{32}S + ^{58}Ni$ [4] at a beam energy of 116 MeV in the laboratory have been used to measure the mass resolution in the present not optimized setting, at $\theta_{LAB} = 0°$. Due to a delay in the installation of the sliding-seal target chamber, the spectrometer is temporarily fixed at 0°. In order to obtain a well defined optical object the target (100 μg/cm²) consisted of a vertical strip of 1 mm horizontal width, evaporated on a 20 μg/cm² carbon foil. The detector set-up was the same used with the α-source.

In figs. 6a, 6b the measured and calculated position spectra of the evaporation residues are compared. With the sextupole lenses off, 5 msr of solid angle acceptance (defined by the quadrupole collimator), an estimated energy spread of about ±10 % and the MD halfway slits fully open the horizontal image spot size was observed to be of 3.3 mm FWHM and 10 mm FW1/10M (calculated values are respectively 2.5 mm and 9.0 mm), that

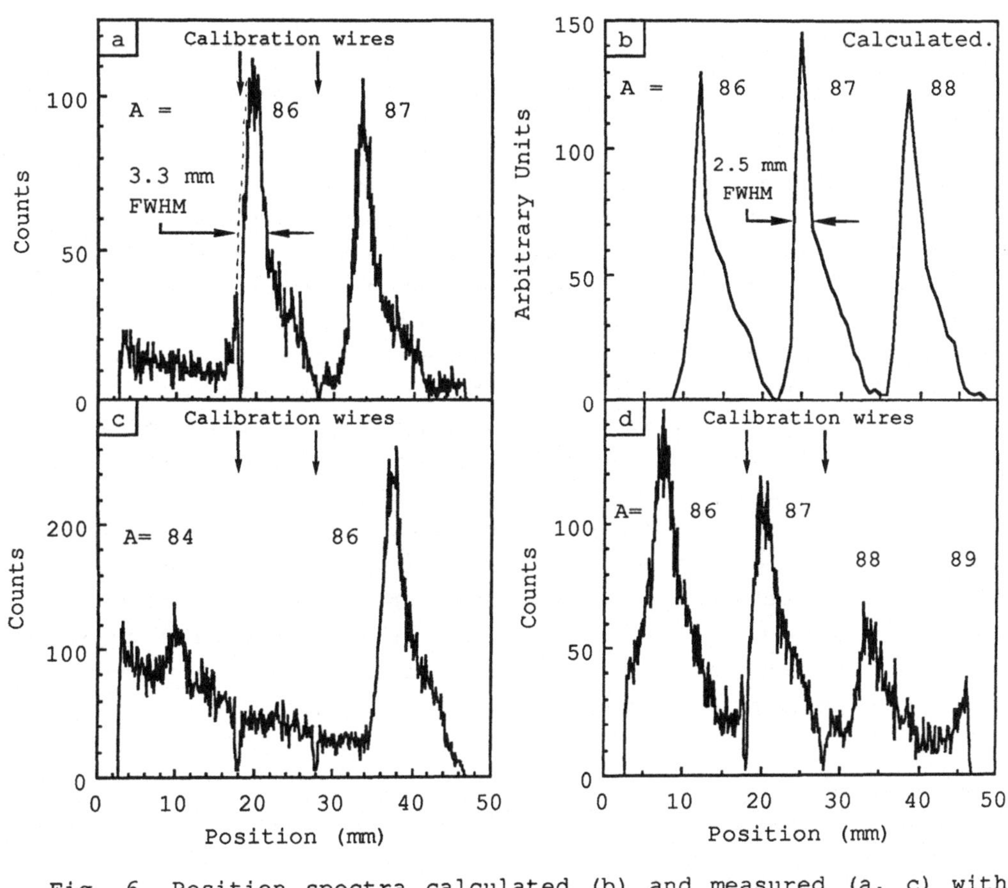

Fig. 6. Position spectra calculated (b) and measured (a, c) with the reaction $^{32}S + ^{58}Ni$; measured spectrum with $^{32}S + ^{nat.}Ni$ (d). $\Omega = 5$ msr.

correspond to a mass resolution of about 1/360 FWHM and 1/120 FW1/10M (calculated 1/450 and 1/125). The mass dispersion, at the choosen focal plane position, resulted to be of ~12 mm/% (calculated ~11). The double first order energy focus is assumed in the calculations but not yet verified in the real instrument.

For the mass identification also the theoretically next probable mass peak (A = 84, evaporation of 6 particles) was detected (fig. 6c) and a short run was made with a natural Ni target (fig. 6d), obtaining the expected double pattern produced by ^{58}Ni and ^{60}Ni, with the natural intensity ratio of about 2. Note that the Si detector horizontal width is only about 1/3 of the spectrometer available focal plane and the four peaks cannot be observed simultaneously, but a small dipole field variation shifts the pattern showing the full peak at mass 89.

The rather high background is largely attributed to the critical beam focusing on the target, as the beam was heating some unremouvable apertures in front of the target.

4. Conclusions.

The first stage of the installation of the LNL spectrometer has been succesfully concluded with a series of preliminary tests.

The first operation of the high voltage units is quite satisfactory; time limitations set the present operational-voltage to the level of about 300 kV, which allows the analysis of particles of electric rigidity up to 4 MeV/q (plus energy-range acceptance). Some improvement is needed for the beam-transport monitoring and focusing on the target, after which a first estimation of the background suppression factor will be possible.

The particle analysis shows a good general performance and a very reasonable agreement with the calculations. The next energy-focus calibration could sensibly improve the mass resolution.

The present set-up, with a mass resolution of 1/360 at about one half of the maximum acceptance of the instrument and with the addition of a focal plane gas detector (which will give the energy-loss information and a wider sensitive area) is already suited for some physics experiments. In fact the fine tuning of the instrument is planned in parallel with the first experimental work.

References.

[1] P. Spolaore et al., Nucl. Instr. and Meth. A238 (1985) 381.
[2] P. Spolaore et al., Nucl. Instr. and Meth. A268 (1988) 397.
[3] H. Wollnik et al., unpublished.
[4] A.M. Stefanini et al., Phys. Lett. B 162 (1985) 66.

THE ARGONNE FRAGMENT MASS ANALYZER

Cary N. Davids and J. D. Larson
Physics Division, Argonne National Laboratory, Argonne, IL 60439 U.S.A.

I. Introduction

The Fragment Mass Analyzer (FMA) is being constructed for use in heavy-ion experiments with the ATLAS superconducting linear accelerator system at Argonne National Laboratory. It is a triple-focussing recoil mass spectrometer, designed to separate nuclear reaction recoils from the primary heavy-ion beam and disperse them in M/Q (mass/unit charge) at the focal plane.

The main ion-optical elements are two electric dipoles and a magnetic dipole, with the electric dipoles (E) symmetrically placed before and after the magnet (M). The elements are spaced such that the position energy dispersion (x/δ_E) and the angle energy dispersion (θ/δ_E) of the system both vanish (energy focus). Here δ_E stands for $\Delta E/E$, the fractional energy deviation from the central energy. The energy-dispersionless EME configuration was first used by Cormier and Stwertka [1,2] for the Recoil Mass Spectrometer at Rochester, and was also chosen by Spolaore et al. [3] for the Recoil Mass Spectrometer (RMS) at the Laboratori Nazionali di Legnaro (LNL) in Padua. Although the energy dispersion is cancelled, the M/Q dispersion is not, and the device performs as a mass spectrometer. When used in the energy-dispersionless mode, the EME combination of elements offers some distinct advantages over designs containing Wien filters[4] (superimposed EM) or a single electrostatic element [5] (ME). These advantages include wider M/Q and energy acceptances and superior rejection of the primary beam ($\sim 10^{-12}$). The energy focus condition implies isochronous orbits for particles having the same M/Q and energy but with differing emission angles at the target, an important feature when used with pulsed beam accelerators like ATLAS.

Additional magnetic elements like quadrupole singlets (Q) and sextupoles (S) are commonly used in these devices to accomplish the geometrical focussing and to provide some second-order corrections. The Rochester device has the configuration QQQEMEQQQ, while the LNL RMS has the configuration QQESMSE. Other configurations that have been proposed include QQEMMEQQ by Wilhelm[6] for the LARA spectrometer at Munich, and QQEMEQQQ by Cole et al.[7] for a spectrometer at Oak Ridge.

II. Ion Optics

After a design study where a number of such configurations were compared, the symmetric configuration QQEMEQQ was chosen for the FMA. Figure 1 shows an outline of

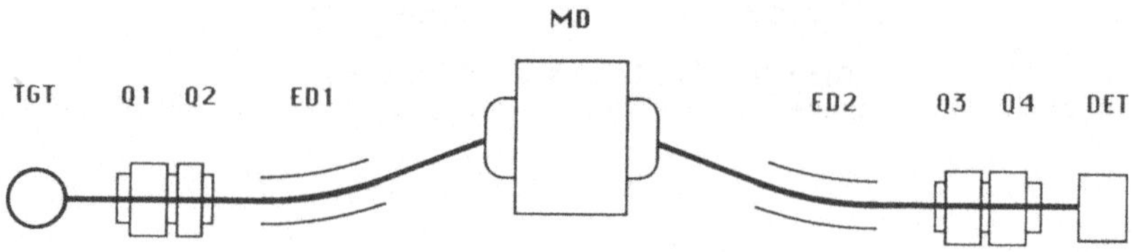

Fig. 1 Outline of the FMA. Beam is incident from the left.

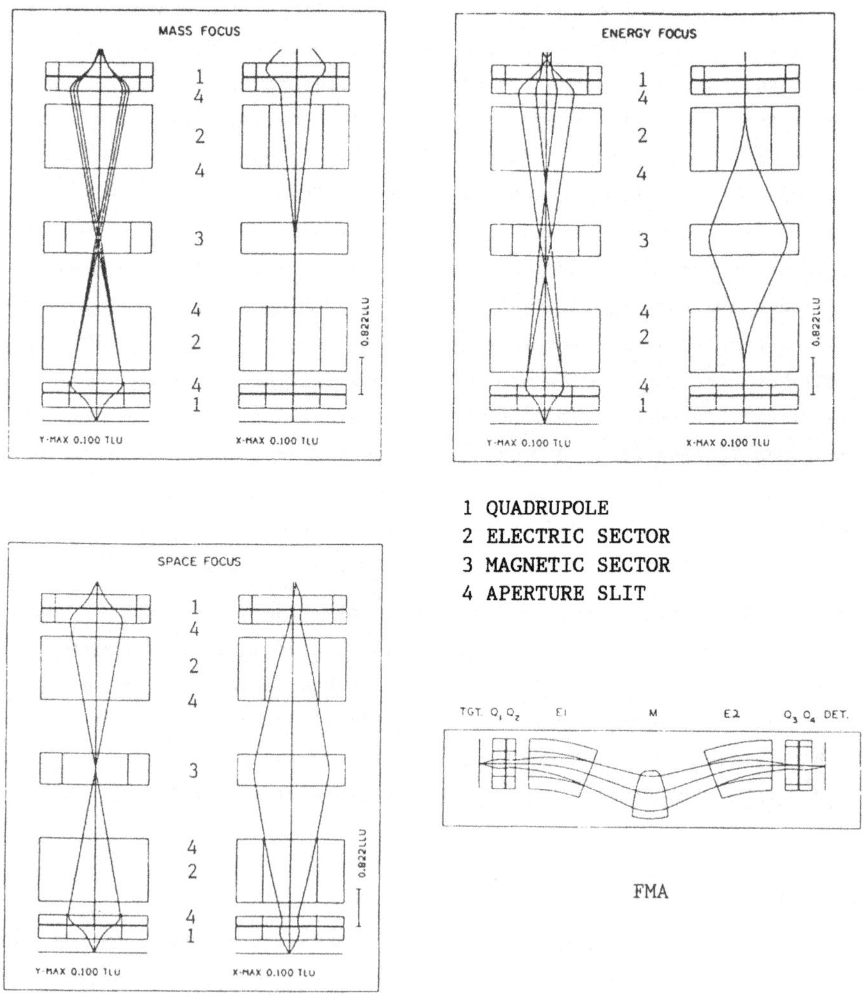

Fig. 2 Calculated ion optics for the FMA.

the device. The calculated ion optics are shown in Figure 2, and Figure 3 shows a calculated M/Q spectrum near mass 100. The matrix ion-optical code GIOS was used for these calculations. The dispersion used in Fig. 3 was 10 mm/%. This gives a spacing of 10 mm between adjacent masses at A = 100, which is easily within the capability of modern position-sensitive detectors.

There are a number of similarities and a number of differences between the ion optics of the FMA and the LNL RMS. The magnetic dipoles of both devices use a 7° pole edge rotation angle, and also curved pole edges to introduce a second-order correction. In the case of the RMS, the pole edge angle provides some vertical focussing in addition to that available from the quadrupole. In the FMA, a vertical crossover in the magnet means that all vertical focussing is done by the quadrupoles. For both devices, the divergence in the horizontal dispersion direction resulting from the 7° pole rotation angle serves to provide more space between the electric and magnetic dipoles than is mandated by the energy focus condition. This allows room for bellows, pumping ports, and diagnostic equipment.

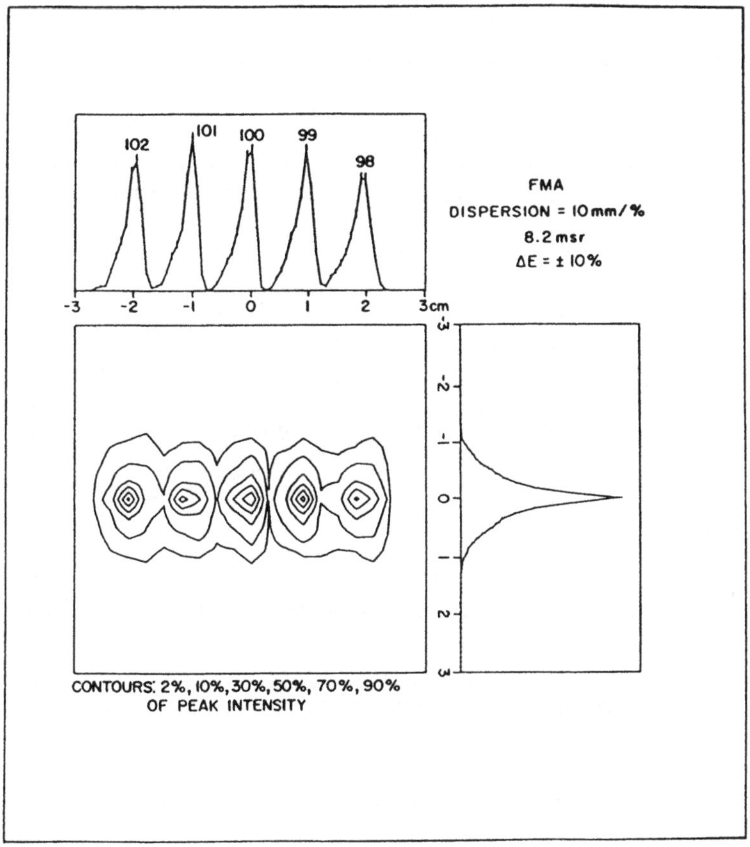

Fig. 3 Calculated M/Q spectrum at mass 100.

The extra quadrupole doublet in the FMA introduces an additional degree of freedom, allowing the M/Q dispersion to be made variable. The RMS has two sextupole magnets located in the spaces between the magnetic and electric dipoles. These are used to rotate the focal plane so that it is perpendicular to the optic axis of the device. In the FMA the depth of focus is sufficiently great that this rotation was felt to be unnecessary, and thus no sextupole lenses are used. Finally, the extra vertical control introduced by the second quadrupole doublet in the FMA provides a vertical beam spot size smaller than that of the RMS by about a factor of three. This is offset by a slightly smaller M/Q range at the focal plane, even though the quadrupole has a 15 cm bore.

There are three principal second-order aberrations in the FMA: (x/δ_E^2), $(x/\theta\delta_E)$, and (x/θ^2). The first has been set to zero by choosing the radius of curvature of the pole edges to be 2.8 m. Figure 4 shows that this action has reduced the second aberration considerably as well, but has raised the (x/θ^2) coefficient to a value which makes it the principal limitation on the M/Q resolution. This fact can be used as an advantage by the experimenters, since the M/Q resolution can be improved quadratically with decreasing horizontal acceptance angle, while the solid angle will only decrease linearly.

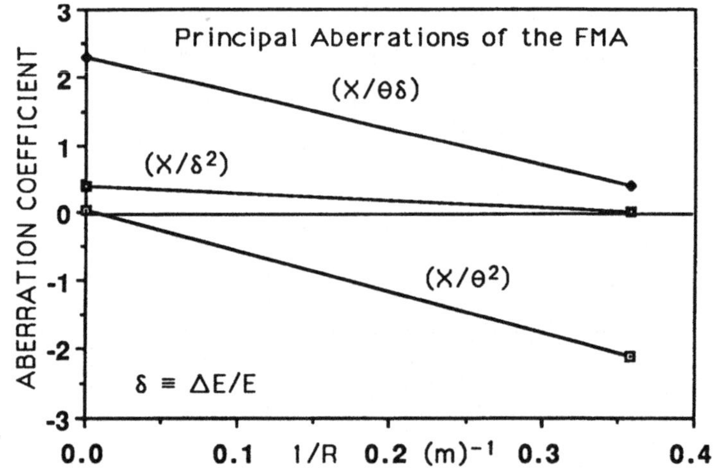

Fig. 4 Principal ion-optical aberrations of the FMA. The abscissa is 1/R, where R is the radius of curvature of the entrance and exit pole edges of the 40° magnet.

Table I gives some of the properties of the FMA. The solid angle, M/Q acceptance, and energy acceptance are slightly smaller than the corresponding values for the LNL RMS, almost entirely due to the fact that the electrode gap in the FMA electric dipoles has been set to 100 mm instead of 150 mm. All peak magnetic fields

Table I. FMA Dimensions and Capabilities

Overall length		8.2m	
Dipoles	MD	ED1, ED2	
Radius of curvature	1 m	4 m	
Deflection angle	40°	20°	
EFB* shim angles	7°	0°	
EFB radii	2.8 m	--	
Gap	12 cm	10 cm	
Maximum field	1.0 T	45 kV/cm	
Maximum rigidity (central trajectory)	1.0 T-m $\left(48.5\ \frac{\text{MeV - AMU}}{Q^2}\right)$	18 MV = (9 MeV/Q)	
Magnetic Lenses	Q1	Q2	Q_3, Q_4
Diameter (cm)	10	10	15
Effective length (cm)	30	20	30
Max. field at pole tip (T)	0.8	0.8	0.8
M/Q range		±6%	
Energy range (ΔE)		±15%	
Solid angle acceptance (Ω)		≈ 8 msr	
M/Q resolution (at Ω = 8 msr, ΔE = ±10%)		1/340	
Dispersion		variable, typically 10 mm/%	
Time dispersion		<0.06%	

*Effective Field Boundary.

have been chosen such that the upper bending limit on recoil ions is set by the voltages obtained with the electric dipoles.

III. Mechanical Design and Construction

The FMA will be mounted on a platform that will allow continuous positioning at angles between -5° and +45° to the beam direction. In addition, the entire device will be able to move in a radial direction over a total distance of approximately 0.5 meter. This will allow a variation in the distance between the target and the first quadrupole from about 10 to 60 cm, thus permitting the placement of large detector arrays around the target or larger solid angle for those experiments requiring it.

The problem of conducting high voltages into the electric dipole vacuum chambers has been solved in an unconventional fashion. Based on a similar installation for the MIT-Oak Ridge Velocity Filter,[8] the Cockroft-Walton multiplier stacks will be placed in SF_6-pressurized ceramic containers inside the vacuum chambers, making direct contact at the high-voltage end with the electrodes. By eliminating high-voltage bushings, damping resistors, and coaxial cabling from an external high-voltage supply, the amount of stored energy is greatly reduced, diminishing the potential for electrode damage during a spark. The multipliers are driven by a commercial RF high-voltage supply, and have proven to be extremely robust. In test runs, a total voltage of 480 kV (± 240 kV) over a gap of 94 mm between two aluminum test electrodes was achieved. In addition, a single electrode at 225 kV was bombarded by a 50-na 70-MeV $^{12}C^{5+}$ beam from ATLAS. No discernible voltage change at the level of a few parts in 10^5 was observed, while at the same time the supply current was observed to decrease by about 50 na, as expected.

To accommodate the expected high beam currents from the positive-ion upgrade to ATLAS, provision is being made for cooling the anode of the first electric dipole. It should be recalled that the primary beam will be dumped on this plate. Cooling will be accomplished by circulating a dielectric fluid through a suitable closed-circuit system in intimate contact with the rear surface of the electrode.

As has been mentioned previously, the energy focus condition manifests itself through the required separation distance between the electric and magnetic dipoles. All reasonable care will be taken to place these devices in the correct positions. This requires precise knowledge of the position of the effective field boundaries for each dipole. A limited adjustment capability will be provided by movable field clamps on the magnetic dipole. However, in order to most easily set up the energy focus, an additional adjustment is desirable. Using correction coils constructed of ordinary fiberglass circuit board, a variable quadrupole component will be superimposed on the dipole field of the 40° magnet. This adjustment is equivalent to a variation of the pole edge rotation angle, and also therefore of the focal length of the magnet. Tests will be done with an alpha source and heavy ion beams from the accelerator.

Finally, an FMA user group has been formed, with the intent of having some of the needed experimental facilities designed and constructed at user institutions. These include a neutron multiplicity detector for the target area, focal-plane detectors, a tape transport system, and a nuclear spectroscopy/nuclear moments facility to be located downstream from the focal-plane area.

IV. Status Report (June 1988)

A new 2700 ft^2 addition to the ATLAS target room is currently under construction. The FMA will be housed in this area, along with one or two other target lines. Bids for the 40° magnet, quadrupoles, electric dipoles, NMR, and magnet power supplies have been received from prospective vendors. It is expected that orders will be placed for these items during the summer, with delivery expected within one year. Construction of the support structure will begin as soon as specific details on supporting the major elements are received from their manufacturer. Work will continue on testing the high-voltage power supplies, with a full-wave prototype to be constructed later this year. Upon completion of the building in the spring of 1989, the support structure will be installed. Ion-optical elements will be tested and installed in late 1989, along with the vacuum system. It is hoped that preliminary tests of the whole system can begin in early 1990, and first experiments shortly thereafter.

This work was supported by the U.S. Department of Energy, Nuclear Physics Division, under contract W-31-109-ENG-38.

References

1. T. M. Cormier and P. M. Stwertka, Nucl. Instrum. Meth. 184, 423 (1981).
2. T. M. Cormier, M. G. Herman, B. S. Lin and P. M. Stwertka, Nucl. Instrum. Meth. 212, 185 (1983).
3. Paola Spolaore, James Daniel Larson, Cosimo Signorini, Silvio Beghini, Zhu Xi-Kai and Si Hou-Zhi, Nucl. Instrum. Meth. A238, 381 (1985).
4. A. N. James, T. P. Morrison, K. L. Ying, K. A. Connell, H. G. Price and J. Simpson, Nucl. Instrum. Meth. A267, 144 (1988).
5. S. Morinobu, I. Katayama and H. Nakabushi, Proc. 4th Intern. Conf. on Nuclei Far from Stability, Helsingor, June 1981, CERN 81-09, Eds. P. G. Hansen and O. B. Nielsen, p. 717.
6. W. Wilhelm, private communication (1987).
7. J. D. Cole, T. M. Cormier, J. H. Hamilton and A. V. Ramayya, Proc. 5th Intern. Conf. on Nuclei Far from Stability, Rosseau Lake, September 1987, AIP Conf. Proceedings 164 (1988), Ed. I. S. Towner, p. 835.
8. J. L. Blankenship, private communication (1986).

DESIGN OF THE LARGE ACCEPTANCE RECOIL-NUCLEUS MASS ANALYZER LARA

W. Wilhelm, T. Faestermann, H.-J. Körner
Physik-Department, Technische Universität München
James-Franck-Straße, D-8046 Garching

A Large-Acceptance Recoil-nucleus mass Analyzer (LARA) for the Munich tandem accelerator has been designed. The facility is planned in conjunction with the postaccelerator Tritron producing neutron-rich nuclei by spallation and incomplete fusion as well as neutron-deficient nuclei in fusion reactions. It can also be tested and used with the postaccelerator SchweIN. The device separates nuclear reaction products from the beam and disperses them in A/Q (mass/charge) at the focal plane. The special symmetric arrangement with an intermediate focus will give an excellent mass resolution, a fine one-to-one focus and high beam suppression also for inverse reactions (Fig. 1). It furthermore allows an identification of the nuclear charge Z via the energy loss in a degrader at the intermediate focus by two time-of-flight measurements. A large solid-angle acceptance of 16 msr, (10 msr for good mass resolution), energy acceptance of ± 10 percent and mass range of ± 5 percent at a primary beam suppression of 10^{13} makes it an ideal tool to investigate rare reaction channels and to produce short-lived exotic nuclei far from stability. The design goal for the electric rigidity is 20 MV to permit reactions with inverse kinematics for easier Z- identification of the recoils with beam energies presently available, and in the future the production of neutron-rich reaction products with Tritron beams.

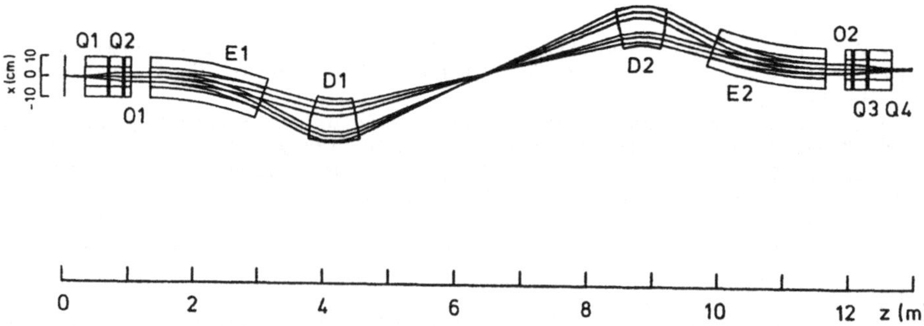

Fig. 1: LARA-geometry. The drawn rays deviate from the central energy by ± 10 percent. Q1 to Q4 are magnetic quadrupole lenses, E1 and E2 electric dipoles, D1 and D2 magnetic dipoles, O1 and O2 magnetic octupoles.

Fig. 1 shows a layout of the proposed instrument. Table 1 contains the relevant data of the ion-optical elements. The LARA consists of two, essentially complete mass spectrometers which operate back-to-back. The first half consists of a quadrupole doublet lens, a 20 degree electric dipole, and a -40 degree magnetic dipole element. About 2 meters behind the magnet, i.e. about 7 meters behind the target a focus is produced. Particles of a fixed A/q value are dispersed in the horizontal (x) direction, and are focussed spatially for different energies and emission angles (both in x- and y-direction) from the target. Thus the first half of LARA is a complete mass spectrometer with properties rather similar to the Legnaro design except for the loss in beam filtering power that a second electric dipole provides.

The second half of the LARA design is a mirror antisymmetric image of the first half resulting in an overall S shape. Following the intermediate focal plane is another two meter drift length to a second +40 degree magnet followed by the second -20 degree electric dipole and the corresponding second mirror image quadrupole doublet, also the second and third order corrections are symmetric. Because of the mirror symmetry of the whole instrument the spatial magnifications in horizontal and vertical directions are both unity, and the whole system is energy-dispersionless.

Table 1: Data of the ionoptical elements

	D1, D2	E1, E2
Radius of curvature	1 m	5 m
Deflection angle	40 °	20 °
Shim angles	8.4 °	-
Edge radii	D1: 1.67 m, - D2: -, 1.67 m	-
Gap	120 mm	100 mm
Maximum field	1 T	40 kV/cm
Maximum rigidity	$B\rho$ = 1 Tm ($E_{kin}A/q^2$ = 48.5 MeV)	$E\rho$ = 20 MV (E_{kin}/q = 10 MeV)

	Q1, Q4	Q2, Q3	O1, O2
Aperture	100 mm	100 mm	100 mm
Effective length	350 mm	200 mm	100 mm
Field at poles	-0.65 T	0.61 T	0.024 T

The whole spectrometer will be mounted on a rigid support. We provide the possibility to rotate the entire assembly around the target through angles between -5 degrees to

+40 degrees. Furthermore we foresee the possibility to enlarge the distance from the target to the first quadrupole from 30 cm to about 60 cm, at the cost of cutting the solid angle down to 4 msr. This arrangement would provide more space in the vicinity of the target which might be necessary to install larger equipment for coincidence experiments, like gamma ray ball assemblies or 4 π light particle ball structures, or both. The location of the detector system of the entire instrument, i. e. behind the last quadrupole doublet is determined by the setting in that doublet. In the standard version indicated in fig. 1 we use the mirror-symmetric arrangement, with a detector location 30 cm downstream from the effective boundary of the last quadrupole.

Originally all ion-optical calculations were performed with the program GIOS developed by H. Wollnik and his collaborators at Giessen. This program allows to calculate and optimize ion optical systems using a matrix formalism which includes all terms up to third order for deflections in the horizontal (mass dispersive) plane, and up to second order in the vertical direction. Later we performed extensive comparative calculations using the MIT program RAYTRACE. All results quoted in this paper have been subject to such a comparison.

In the following we discuss the standard mode of operation of the instrument, with a 30 cm distance between the target and the first quadrupole, and also between the last quadrupole and the focal plane detector. The first half of LARA, with the quadrupole doublet, the 20 degree electric dipole and the -40 degree magnetic dipole produces a focal plane, about 2.10 meters downstream from the exit boundary of the magnet. First order focussing begins with the quadrupole doublet. The first lens element is defocussing horizontally. Thus the angular acceptance is small in the horizontal and large in the vertical direction. This mode of operation is opposite to the choice made at Legnaro. Ion optical calculations have shown that with this mode the trajectories can enter the electric dipole in the dispersive horizontal plane as parallel rays, whereas in the mode chosen at Legnaro (and also in Argonne) the beam has to be slightly horizontally diverging behind the quadrupole lense for good ionoptical characteristics. But the parallel beam is essential for using a long electrostatic deflector with a large bending radius and a small gap size. Consequently the trajectories in the mass dispersive horizontal plane enter the magnet as parallel rays. The 40 degree magnet provides the mass dispersion, the energy focus and most of the spatial focussing power in the horizontal plane. For the full solid angle particles of one A/q value and one energy are focussed to a point in the focal plane. The relevant trajectories are shown in Fig. 2a. Particles of one A/q but different energies (plus/minus 10 %) are focussed to a vertical line (Fig. 2b). The image of a beam spot at the target position is enlarged by a factor (Y,Y) = 15 in the vertical and by a factor (X,X) = 1.34 in the horizontal direction.

The second half of the LARA is a mirror antisymmetric image of the first half. Because of this symmetry the spatial magnifications in horizontal and vertical directions are both

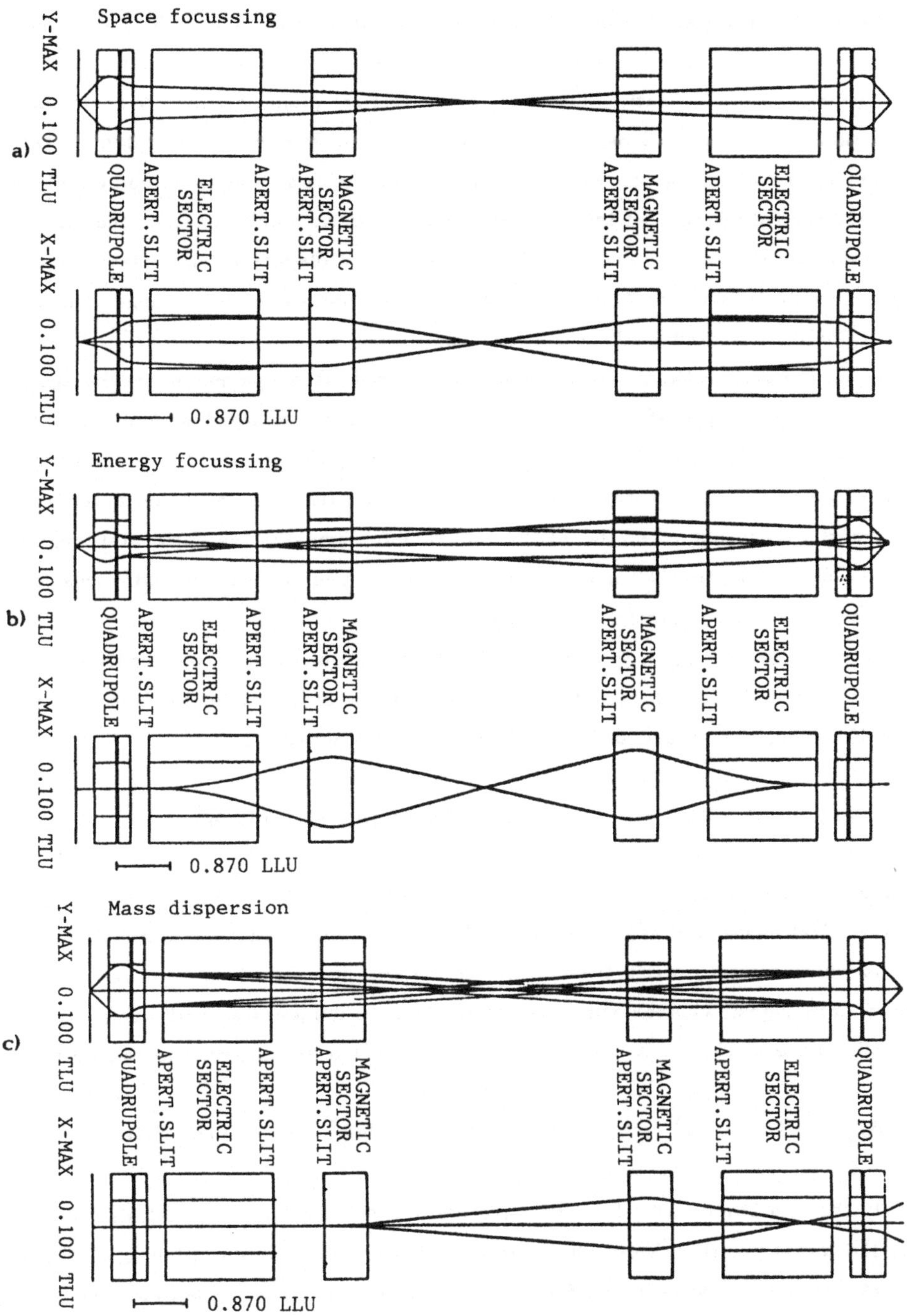

Fig. 2: The ionoptics of the LARA, longitudinal and transversal length units LLU and TLU are in meters. a) Spatial focussing vertical and horizontal. b) Energy focussing: The drawn rays deviate from the central energy by ± 10 percent. Note the strong vertical chromatic aberration at the intermediate focus and its compensation at the final focus. c) Mass dispersion: The rays have the central mass and ± 5 percent.

unity, and the total system is nearly energy dispersionless. Thus we achieve the small focal point which we aimed at for a number of experimental applications. A consequence of the mirror design is that chromatic aberration in the vertical plane is partially compensated for a useful range of energies. This can be seen in fig. 2b. The mass dispersion of the LARA instrument is shown in Fig. 2c. Mass is dispersed in the first magnetic dipole and spatially separated in the following drift length. The second magnetic dipole focusses the masses again to a crossover in the second electric deflector. The mass dispersion, in GIOS-notation (X,G), of the first half is 8.3 mm/% and -12.4 mm/% after the whole instrument. The resolving power (X,G)/(X,X) is doubled by the second half.

The magnification (X,X) = 1 is achieved for complete symmetry of the instrument. However, the drift length after the last quadrupole can be enlarged without disturbing the energy focus by adjusting the excitation of the lens. In this way the magnification and dispersion can be adjusted to the requirements of a specific experiment and detector system. A drift length of 1 m instead of the 30 cm, for example, increases the horizontal magnification to 1.55, the dispersion to -19.2 mm/%, and the vertical magnification to 2.87. This flexibility may be an important feature for some experiments.

As can be seen in fig. 1, in each half of the spectrometer a second order correction is made with a curved magnet boundary and also a third order correction with an octupole. In GIOS-Notation, the second order aberrations may be written as

$$(\Delta x)_2 = (X,AD)*A*D + (X,DD)*D^2 + (X,AA)*A^2 + (X,BB)*B^2 .$$

Here, A means the starting angle from the central trajectory in the horizontal bending plane, B is the vertical angle, and D is the energy deviation. $(\Delta x)_2$ is the horizontal deviation from the focus where a particle with these initial conditions hits. The first term, abbreviated AD, may have both signs, whereas the other terms have a unique sign and therefore cause an asymmetric shape of the mass lines. Similarly, the third order aberrations can be written as

$$(\Delta x)_3 = (X,AAA)*A^3 + (X,ABB)*A*B^2 + (X,DDD)*D^3 + (X,DYB)*D*Y*B.$$

Here only the important terms for LARA are shown. Y is half the vertical size of the beam spot at the target position. Terms linear in G, which is the mass deviation, are also omitted here; such terms may cause a broadening of the mass lines for increasing deviations from the central mass (depth-of-focus aberration). The main term of this type is the tilt angle ϑ of the focal plane,

$$\tan \vartheta = (X,AG) / ((A,A) * (X,G)).$$

The complete LARA in the symmetric case has a magnification of one and also an angle magnification (A,A) of unity, so the focal plane tilt in this case is simply (X,AG)/(X,G).

Calculations with GIOS and RAYTRACE have shown, that especially the AB^2 Term deteriorates the mass spectra, if large values of A and B are allowed, i.e. for large solid angle acceptance. But it is possible and recommended to make this spheric aberration to vanish with the help of an octupole shortly behind the quadrupole lenses. The effect of this octupole, as calculated with RAYTRACE, is shown in fig. 3.

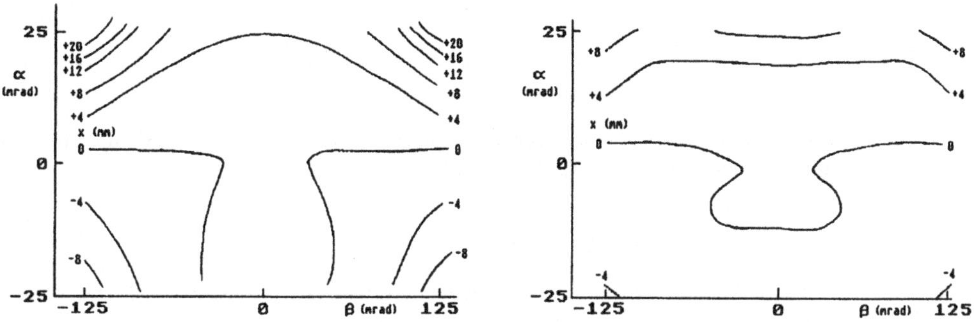

Fig. 3: Contour lines of equal deviation x from the focus, plotted versus the horizontal and vertical starting angles α and β, as calculated with the program RAYTRACE for particles with central mass and energy. Left side without octupoles, right side with octupoles adjusted to make (X,ABB) to zero.

If one looks only at the quality of the mass spectra at the final focus, one octupole would suffice to compensate the AB^2-Term. But for the quality of the intermediate focus it is better to have two octupoles symmetrically. The main second order aberration is the AD-Term. It can be made smaller with increasing curvature of the magnetic dipole boundary and could even be made zero with a curvature radius of 1.2 m at entrance <u>and</u> exit of the dipole. The focal plane tilt also decreases with increasing curvature and would vanish at a curvature radius of about 1 m at the entrance only. We have chosen a value of 1.67m at the entrance of the first dipole and no curvature at the exit (and symmetrically at the second dipole). With this sextupole correction, the D^2-Term and also the D^3-Term are zero. It is not recommendable to use a stronger correction, because all other terms, especially the A^2 and DYB increase with smaller radii. Some calculations with separate sextupole lenses didn't show any profit.

Fig. 4 shows the acceptance characteristics of the LARA in energy and mass-to-charge ratio. As mentioned above the acceptance in the vertical direction is large. Thus with the help of suitable slits or position sensing devices the angular distributions of the reaction products could be measured up to angles of ± 9.5 degrees. The solid angle can be as large as 16 msr, but good resolution is obtained at somewhat smaller values. The energy acceptance is ± 12 %, and the accepted mass to charge ratio (m/q) is ± 5 %.

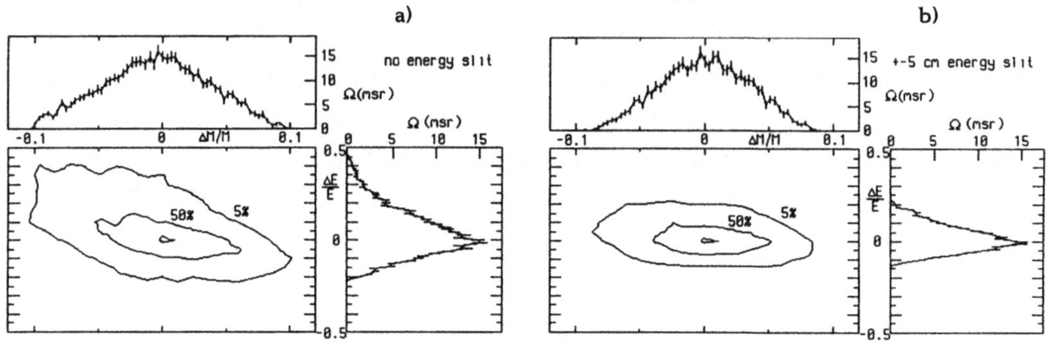

Fig. 4: Acceptance of LARA in energy and A/Q, a) without an energy-limiting slit, b) with a slit of ±5 cm at the entrance of the first magnetic dipole.

Calculated mass spectra for the intermediate and final focal plane are shown in Fig. 5. The calculations were performed for seven discrete mass values, A = 97 to 103 and charge states q = 20. The particles were assumed to originate from a 1 mm diameter target spot, with an energy spread of plus/minus 10 %. The assumed angular acceptance was ± 20 mrad horizontally and ± 125 mrad vertically, i.e. a solid angle of 10 msr. The mass resolution for this mode of operation is (m/Δm) = 340 at the intermediate focus and 450 in the final focal plane. The magnification at the final focal plane is unity, and the dispersion is 12.4 mm/%.

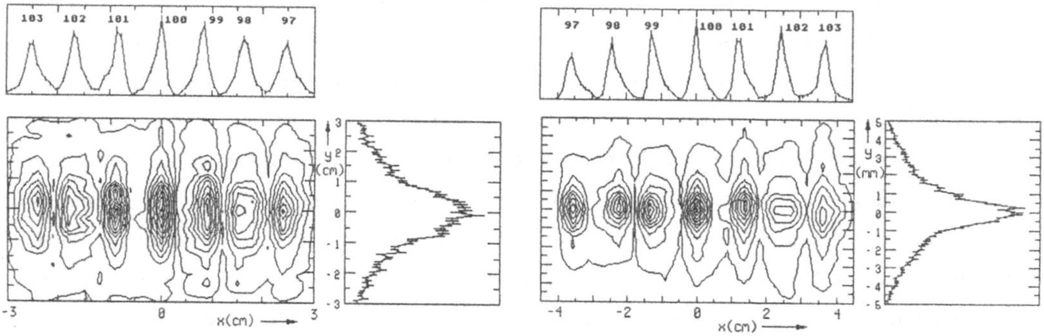

Fig. 5: Calculated mass spectra in the intermediate (left side) and the final focal plane (right side). For 7 discrete masses A = 97 to A = 103 rays were calculated coming from a spot at the target with 1 mm diameter within an energy spread of ±10 % and a solid angle of 10 msr (horizontal -25 to +15 mr, vertical ±125 mr).

Finally we want to point out some features, which are gained by the specific, highly symmetric LARA design:

The ionoptics with the first defocussing quadrupole in the dispersive plane provides the parallel beam in the electrostatic dipole and therefore the large acceptance and high electric rigidity.

The vertical focus and chromatic aberrations are extremely well controlled, one-to-one imaging from target to focal plane is achieved.

The effective mass resolution for particles with ± 10% energy spread and 10 msr solid angle is increased from 340 at the intermediate focus to 450 at the final, the dispersion from 8 mm/% to 12.4 mm/%, the intrinsic resolution i.e. dispersion divided by magnification is doubled.

By adjusting the last quadrupole doublet and the last drift length, the dispersion and magnification can be adapted to the experiment.

By inverting the polarities of the quadrupoles, especially of the final doublet lens, a "SHIP mode" is possible. Mass dispersion is lost with this mode at the final focus, but not at the intermediate. So it is possible to select the mass and three charge states at the intermediate focus with the help of a 3-slit device, and then collect all of these evaporation residues with large acceptance together to a small spot at the focal plane.

Increasing the first drift length to provide space for γ-detectors is possible, for example a 60 cm drift length with a solid angle of 3.8 msr. The use of inverse reactions is desired for such experiments because the recoils are collected in forward direction and the Z-identification is easier. The high limit of the electric rigidity of LARA is especially adapted to such experiments.

By using a slit in the intermediate focal plane the background of scattered beam particles should be extremely low ($< 10^{-13}$). This is again very important for the use of inverse kinematics: Tom Cormier has pointed out to us that inverse kinematics cannot be used in Rochester, as the beam particles are not stopped in the first electric dipole and cause tremendous background at the focal plane. But LARA with its nice intermediate focus should be able to handle this mode.

With an energy degrader and a time-measuring device at the intermediate focus it should be possible to derive information on the nuclear charge. By two time-of-flight measurements the energy loss in the degrader can be measured with high precision. If the degrader is very homogeneous, which could be achieved with a gas degrader, this energy loss has to be compared only with the energy straggling. Estimations have shown that for recoils of 1 MeV/nucleon degraded to about a third of the energy a Z-separation should be possible for $Z \leq 60$.

We thank Prof. H. Wollnik and his group in Giessen for very fruitful hints and discussions and the wonderful GIOS-program, and also J.D. Larson for intensive discussions, lessions how to use RAYTRACE, the calculations which led to the use of the octupoles, and his improvements at the GIOS- and RAYTRACE-programs. We thank S. Skorka, K.E.G. Löbner, K. Rudolph and E. Nolte for discussions. Financial support by the B.M.F.T. is acknowledged.

LIST OF PARTICIPANTS

Andrés M.V., Department of Physics, University of Sevilla, Spain
Anthon L., DANFYSIK A/S, Denmark
Bacelar J., K.V.I., University of Groningen, The Netherlands
Bedau J., BRUKER, F.R. Germany
Beghini S., Department of Physics, University of Padova, Italy
Bellwied R., Institut für Kernchemie, University of Mainz, F.R. Germany
Betts R., Argonne National Laboratory, Argonne, Illinois, U.S.A.
Bilwes B., Centre de Recherches Nucléaires, Strasbourg, France
Bizzeti P.G., Department of Physics, University of Firenze, Italy
Blocki J., Institute for Nuclear Studies, Swierk, Poland
Bortignon P.F., Istituto di Ingegneria Nucleare - CESNEF, Milano, Italy
Broglia R.A., University of Milano, Italy and Niels Bohr Institute, Copenhagen, Denmark
Bromberger J., HVEC, Burlington, Massachusetts, U.S.A.
Cavallaro Sal., Department of Physics, University of Catania, Italy
Corradi L., Department of Physics, University of Munich, F.R. Germany
Dasso C.H., Niels Bohr Institute, Copenhagen, Denmark
Davids C., Argonne National Laboratory, Argonne, Illinois, U.S.A.
Dermois O.C., K.V.I., University of Groningen, The Netherlands
Diaz J, IFIC, Faculty of Physics, Valencia, Spain
Ecker R.-D. Department of Physics, Technical University of Munich, F.R. Germany
S. Feminò, Department of Physics, University of Messina, Italy
Fernandez Niello J., Department of Physics, CNEA, Buenos Aires, Argentina
Ferrer Roca C., IFIC, Faculty of Physics, Valencia, Spain
Ferrero J.L., IFIC, Faculty of Physics, Valencia, Spain
Fick D., University of Marburg, F.R. Germany
Fröbrich P., Hahn Meitner Institute, Berlin, Germany
Funke F., Institut für Kernchemie, University of Mainz, F.R. Germany
Gillitzer A., Department of Physics, Technical University of Munich, F.R. Germany
Gomes P.R.S., Department of Physics, Federal University, Rio de Janeiro, Brasil
Gorla S., INFN, Laboratori Nazionali di Legnaro, Italy
Hessberger F.P., G.S.I. Darmstadt, F.G. Germany
Iachello F,. Yale University. New Haven, Connecticut, U.S.A.
James A., Oliver Lodge Laboratory, Liverpool, United Kingdom
Jänker P., Department of Physics, University of Munich, F.G. Germany
Kim H.J., Oak Ridge National Laboratory, Oak Ridge, Tennessee, U.S.A.
Landowne S., Argonne National Laboratory, Argonne, Illinois,. U.S.A.
Lanza E.G., INFN, Catania, Italy
Larson J.D., Independence, Missouri, U.S.A.
Lee L., State University of New York, Stony Brook, New York, U.S.A.
Lenzi S., Department of Physics, University of Padova, Italy
Li Zhichang, INFN, Laboratori Nazionali di Legnaro, Italy
Lilley J., Daresbury Laboratory, Daresbury, United Kingdom
Löbner G., Department of Physics, University of Munich, F.R. Germany
Lotti P., Department of Physics, University of Padova, Italy
Lunardi S., Department of Physics, University of Padova, Italy
Million B., Niels Bohr Institute, Copenhagen, Denmark
Montagnoli G., INFN, Laboratori Nazionali di Legnaro, Italy
Mueller P., Department of Physics, Technical University of Munich, F.R. Germany
Nagarajan M.A., Daresbury Laboratory, Daresbury, United Kingdom

Nagashima Y., Tandem Accelerator Centre, University of Tsukuba, Japan
Napoli D.R., INFN, Laboratori Nazionali di Legnaro, Italy
Pass C.N., Department of Nuclear Physics, Oxford University, United Kingdom
Pirrone S., INFN, Catania, Italy
Pollarolo G., Department of Physics, University of Torino, Italy
Potenza R., Department of Physics, University of Catania, Italy
Porto F., INFN, Laboratorio Nazionale del Sud, Catania, Italy
Predicatori F., INFN, Laboratori Nazionali di Legnaro, Italy
Rapisarda A., INFN, Catania, Italy
Rechatin J.L., IRELEC, Meylan, France
Rehm E., Argonne National Laboratory,Argonne, Illinois, U.S.A.
Ripouteau F., GANIL, Caen, France
Reisdorf W., G.S.I., Darmstadt, F.R. Germany
Ricci R.A., Department of Physics, University of Padova and INFN, Laboratori Nazionali di Legnaro, Italy
Romanski J., Department of Physics, University of Catania, Italy
Rossner H., Hahn Meitner Institute, Berlin, Germany
Rudolph K., Department of Physics, University of Munich, F.R. Germany
Ruitz J., IFIC, Faculty of Physics, Valencia, Spain
Sambataro S., Department of Physics, University of Catania, Italy
Sanchez F., IFIC, Faculty of Physics, Valencia, Spain
Scarlassara F., Department of Physics, University of Padova, Italy
Signorini C, Department of Physics, University of Salerno and INFN, Padova, Italy
Skorka S., Department of Physics, University of Munich, F.R. Germany
Smith W.J., DANFYSIK A/S, Jyllinge, Denmark
Smithson M., Daresbury Laboratory, Daresbury, United Kingdom
Soramel F., Department of Physics, University of Padova, Italy
Spolaore P., INFN, Laboratori Nazionali di Legnaro, Italy
Stefanini A.M., INFN, Laboratori Nazionali di Legnaro, Italy
Udagawa T., University of Texas, Austin, Texas, U.S.A.
Van Lummel W., K.V.I., University of Groningen, F.R. Germany
Vandenbosch R., University of Washington, Seattle, Washington, U.S.A.
Vigezzi E., Department of Physics, University of Milano, Italy
Vitturi A., Department of Physics, University of Trento, Italy
Vlastou-Zannis R., Department of Physics, University of Athens, Greece
Wilhelm W., Department of Physics, Technical University of Munich, F.R. Germany
Wirth G., G.S.I., Darmstadt, F.R. Germany
Wollnik H., Institute for Theoretical Physics, Giessen, F.R. Germany
Wolter H.H., Department of Physics, University of Munich, F.R. Germany
Zhang Huanqiao, Institute of Atomic Energy, Peking, People's Rep. of China
Zardi F., INFN, Padova, Italy

GPSR Compliance

The European Union's (EU) General Product Safety Regulation (GPSR) is a set of rules that requires consumer products to be safe and our obligations to ensure this.

If you have any concerns about our products, you can contact us on ProductSafety@springernature.com

In case Publisher is established outside the EU, the EU authorized representative is:

Springer Nature Customer Service Center GmbH
Europaplatz 3
69115 Heidelberg, Germany

Batch number: 10640288

Printed by Printforce, the Netherlands